ROADSIDE GEOLOGY OF UTAH

Felicie Williams,
Lucy Chronic, and
Halka Chronic

2014
Mountain Press Publishing Company
Missoula, Montana

Geologic road maps and many of the illustrations revised by Mountain Press
Publishing Company based on original drafts by the authors.

Roadside Geology is a registered trademark
of Mountain Press Publishing Company.

Library of Congress Cataloging-in-Publication Data

Williams, Felicie, 1953-2015 author.
  Roadside geology of Utah. — [Second edition] / Felicie Williams, Lucy
Chronic, and Halka Chronic.
      pages cm
  Includes bibliographical references and index.
  ISBN 978-0-87842-618-8 (pbk. : alk. paper)
  1. Geology—Utah—Guidebooks. 2. Utah—Guidebooks. I. Chronic, Lucy
M., author. II. Chronic, Halka, author. III. Title.
  QE169.C48 2014
  557.92—dc23
                        2014007728

Printed in Hong Kong by Mantec Production Company

MP Mountain Press
PUBLISHING COMPANY
P.O. Box 2399 • Missoula, MT 59806 • 406-728-1900
800-234-5308 • info@mtnpress.com
www.mountain-press.com

From Felicie to Mike, Amber, and Wes,
the best thing to ever happen to me.

From Lucy to Halka, who traipsed these roads before us,
and to my family, Chris, Betsy, and Haley,
who are perpetually wonderful.

This is not Highway Eighty-Nine

And it will not take you home,
But it can lead you away,

Through hundreds of miles of rangeland,
Full of dust and sage and wind,

Through jagged stabbing peaks,
Spattered with snow and goats, and gold mines,

Through sandy red cliffs
That come alive only in the chill of the night.

Highway eighty-nine could take you home,
But this is not highway eighty-nine.

—AMBER WILLIAMS

USING THIS BOOK

It is best to start by reading the first chapter, which is a minicourse in geology. Then review the second chapter, Great Events in Utah, an illustrated timeline showing key geologic moments in Utah's past. The geology of each region, whether the Colorado Plateau, High Country, or Great Basin, is summarized in each chapter introduction. When following a highway, refer back to these sections as needed. To better understand the geology of the parks in the final chapter, Something Special, it would be helpful to read the appropriate regional introduction.

The logs in this book follow main highways and a few less-frequented but geologically interesting routes. Geologic maps, photos, and figures help explain what can be seen.

ACKNOWLEDGMENTS

We would like to thank many people for their assistance with this book. Foremost is Halka Chronic, who wrote and illustrated the original edition, blazing the path for us. Then, a great tribute goes to all the geologists who came before us, walking the dusty hills and combining their knowledge to make sense of the state's amazing scenery. We give special thanks to those with whom we discussed particular areas of Utah and who helped review the manuscript: Carol Dehler, Jim Kirkland, Jim Davis, Adolph Yonkee, Marjorie Chan, C. G. "Jack" Oviatt, Joe Fandrich, and Don Baars. Judy Foster saved the day by helping with both artwork and photography. We have unbounded praise for the people who are behind the Utah Geological Survey website. It is incredibly useful. And, we would like to thank the staff at Mountain Press and especially our editor, James Lainsbury.

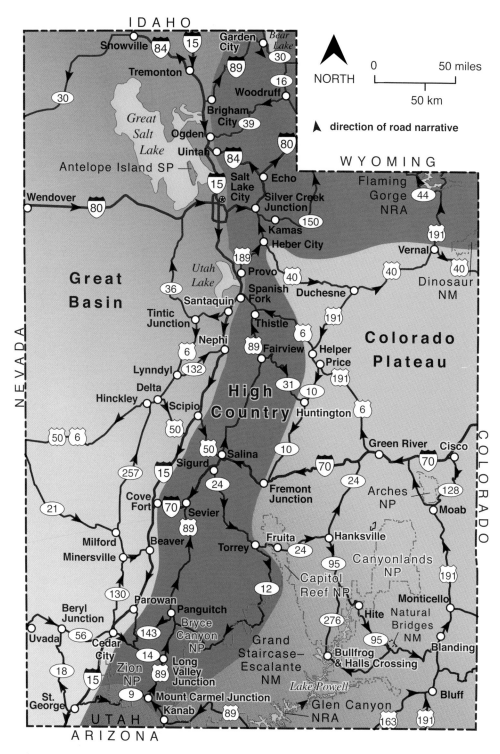

In each chapter, interstates are covered first, followed by US highways and then state highways. Arrows show the direction of travel for which each road guide is written. Individual highways are usually described in the same direction along their entire length, with neighboring highways described in the opposite direction. This way several highways can be followed in a loop from most locations. The larger parks are described in the final chapter, Something Special.

CONTENTS

ROCK KEY FOR MAPS

sedimentary and metamorphic rocks

Cenozoic

Holocene

- alluvium and lake deposits
- wind (eolian) deposits
- landslides
- marsh

−0.01*

Pleistocene

- Lake Bonneville deposits
- gravel
- glacial deposits
- Quaternary-Tertiary sediment

−2.6

Tertiary: Paleogene and Neogene

- Sevier River Fm., Castle Valley Conglomerate
- Browns Park Fm., Bishop Conglomerate, Salt Lake Fm.
- Uinta and Duchesne Fms.
- Green River Fm.
- Flagstaff, Colton, Claron, and Wasatch Fms.
- Cretaceous-Tertiary North Horn and Evanston Fms.

−66

Mesozoic

Cretaceous: most common formations

- Kaiparowits Fm., Mesaverde Gp.
- Mancos Shale, Tropic Shale, Wahweap Fm., Straight Cliffs Fm.
- Cedar Mountain Fm., Dakota Sandstone

−145

Jurassic: most common formations

- Morrison Fm.
- San Rafael Gp.: Page Sandstone, Carmel Fm., Entrada Sandstone, Curtis Fm., Summerville Fm., Arapien Formation, Twin Creek Limestone

Triassic-Jurassic

−201 Glen Canyon Gp.: Wingate Sandstone, Moenave Fm., Kayenta Fm., Navajo Sandstone, Nugget Sandstone

Triassic: most common formations

- Chinle and Ankareh Fms.
- Triassic undivided and Moenkopi Fm.

−252

*All ages in millions of years ago
Fm.: Formation
Gp.: Group

Paleozoic

Permian

- upper
- lower

Pennsylvanian-Permian

−299 undivided and Oquirrh Gp.

Pennsylvanian

- evaporites

−323

Mississippian

- upper
- middle
- lower

−359

Devonian

−419

Silurian

−444

Ordovician

−485

Cambrian

- upper
- middle
- lower

−541

Precambrian

- Neoproterozoic: Uinta Mountain Gp., Big Cottonwood Fm., and others

−1,000

- Paleoproterozoic and Mesoproterozoic undivided

−2,500

- Archean

Igneous rocks

- Quaternary basalt
- Tertiary volcanic rocks undivided
- Miocene volcanic rocks
- Oligocene volcanic rocks
- Tertiary basalt
- intrusive rocks porphyritic

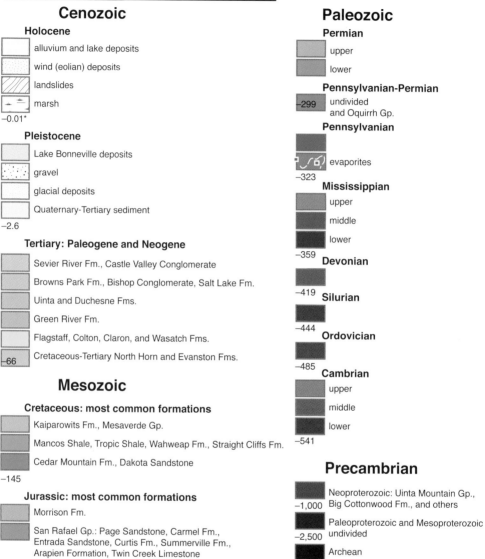

symbols used in stratigraphic diagrams

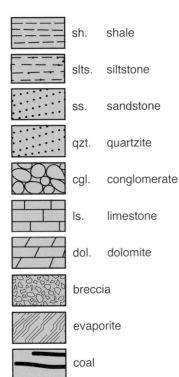

sh.	shale
slts.	siltstone
ss.	sandstone
qzt.	quartzite
cgl.	conglomerate
ls.	limestone
dol.	dolomite
	breccia
	evaporite
	coal

ABBREVIATIONS

Fm.	Formation
Gp.	Group
Ma	millions of years ago
Ba	billions of years ago

symbols used on maps

All faults and fold axes are dashed where uncertain and dotted where buried.

FAULTS

thrust fault
  teeth on overthrust side

normal fault
  upthrown side
  down-dropped side

lateral fault (or other fault in cross sections)
  arrows show relative direction of movement along fault

FOLD AXES

arrows point in the direction beds dip down

anticline

syncline

monocline

 cinder cone or other volcanic vent

 depression: volcanic caldera, crater, open-pit mine

 mine

 dinosaur bone quarry

dinosaur footprint site

hiking trail

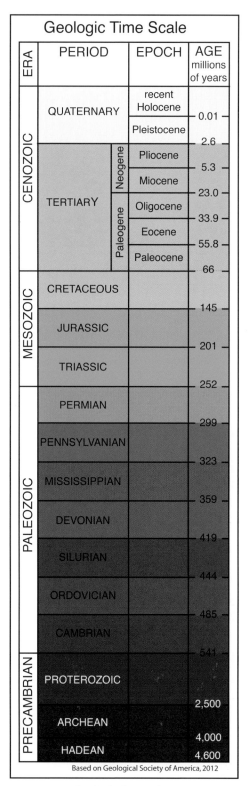

Geologic timescale.

GETTING STARTED

THE FACE OF THE EARTH

Utah's magnificent scenery is founded on its geology. Its mountains, mesas, rivers, lakes, and desert ranges—in as great a variety as you'll find anywhere—owe their locations, colors, and contours to the rocks. Geology even determines the sites of towns and cities and the paths of highways.

Geology, the study of Earth, began as a way to predict the locations of coal seams and mineral deposits. Through their work, geoscientists have revealed our world's inner workings, its immense age, the story of even its oldest rocks, and the evolution of its life. In the last century we have come to recognize that Earth's continents and poles are ever wandering, and its lands, seas, lowland plains, and mountains are constantly changing. Geologists add something new to geology's body of knowledge every year—indeed, every day—on their quest to understand more completely the planet we live on.

GEOLOGIC TIME

When geologists talk about Earth's history, they speak in terms of time: When did this rock form? How does it relate to rocks that formed earlier or later? The scale of geologic time can be hard to grasp, mostly because Earth and its rocks are so much older than humanity.

Early geologists had no way to tell exactly how old rocks were, but they were able to figure out the relative ages of rock layers. They understood that sedimentary layers accumulated over time; thus, the oldest rocks were at the bottom of a stack. Fossils also helped with age determination, because species evolve, changing over time, and many are found only in rock of certain ages. Geologists around the world have compiled information related to the age of rocks, and from this body of knowledge a geologic timeline, or timescale, was developed—and continues to evolve with new information.

On the geologic timescale, the largest intervals, called eras, are subdivided into periods, and periods into epochs. Eras were named according to the evolution of life: Paleozoic (ancient life), Mesozoic (middle life), and Cenozoic (recent life). The oldest rocks, which contain the least evidence of life, are called Precambrian because they precede the Cambrian, the earliest period of the Paleozoic era.

Most periods were named for the location in which their rocks were first studied or where they were especially abundant: Cambrian for Cambria, the Roman name for Wales; Jurassic for the Jura region of France; and Permian for

1

the province of Perm in Russia. There are some exceptions; for example, Creta-ceous derives from the Latin word for chalk—*creta*—and was named after the chalky white cliffs near Dover, England.

Era and period names are used fairly consistently worldwide, with only minor differences between continents. Epoch names, on the other hand, vary from one continent to another. In this book we use eras and periods; the only epoch names we use are those for the Cenozoic.

Since the discovery of radioactivity, geologists have learned to use radioac-tive elements and their gradual but steady decay to figure out the absolute ages of some rocks. When igneous rocks cool and harden, the radioactive elements of certain minerals begin breaking down, or decaying, into other elements. By comparing the amount of decay product with the amount of the original ele-ment left in these minerals, a reasonably precise date for when the rock cooled can be obtained. Radioactive dates from igneous rocks interbedded with sedi-mentary layers, combined with the fossil record, have given us a remarkably complete and accurate timescale.

LOOKING DEEPER

The current theory explaining the birth of our universe posits that 13 to 14 bil-lion years ago, a singularity—a speck of infinite density—suddenly expanded, spreading matter and energy. Tiny variations in the amount of matter from place to place caused variations in gravity, and gravity concentrated matter into larger and larger amounts, eventually building planets, stars, and galaxies.

The oldest mineral grains found on Earth so far are zircon; found in Austra-lia, the grains are estimated to be 4.36 billion years old. The oldest known rocks formed about 4.28 billion years ago. By then, Earth had changed from a ball of gas and dust into molten rock, and then, after heavier material had settled toward the center, into a relatively rigid mass with a thin but solid crust and an atmosphere.

Geologists have learned a great deal about Earth by studying the way seismic waves, generated by earthquakes, travel through it. Earth's core is composed of iron and nickel with a radius of 2,160 miles (3,480 km). Outside the core is the mantle, a seething layer of rock just under 1,800 miles (2,900 km) thick and composed mostly of magnesium, silicon, iron, and oxygen. The lower part of the mantle is thought to be solid. Its upper part deforms plastically, the way red-hot iron can be bent and shaped on a blacksmith's anvil, and it is semi-molten in areas, such as under midocean ridges, where submarine volcanic eruptions create new oceanic crust.

Heat and pressure are the signatures of the environment inside Earth: stu-pendous pressure is exerted by the weight of Earth's layers, and heat is left over from Earth's original gravitational compression and early meteorite impacts, as well as being generated by the ongoing decay of radioactive elements.

Earth's solid crust floats on the plastic portion of the mantle. The crust com-bined with the upper 40 to 60 miles (65 to 100 km) of the uppermost mantle (a brittle, chilled portion) is called the lithosphere. Relative to the size of Earth, the lithosphere is a very thin skin.

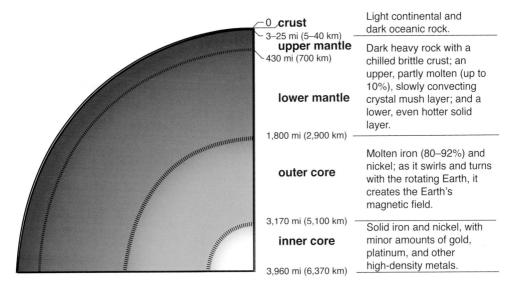

0 **crust**	Light continental and dark oceanic rock.
3–25 mi (5–40 km) **upper mantle** 430 mi (700 km) **lower mantle**	Dark heavy rock with a chilled brittle crust; an upper, partly molten (up to 10%), slowly convecting crystal mush layer; and a lower, even hotter solid layer.
1,800 mi (2,900 km) **outer core**	Molten iron (80–92%) and nickel; as it swirls and turns with the rotating Earth, it creates the Earth's magnetic field.
3,170 mi (5,100 km) **inner core** 3,960 mi (6,370 km)	Solid iron and nickel, with minor amounts of gold, platinum, and other high-density metals.

Cutaway of Earth.

The portion of the mantle just below the lithosphere is called the asthenosphere. It extends down about 430 miles (700 km) below the lithosphere and acts like boiling soup, rising, rolling over, and plunging downward again—but on a huge scale and extremely slowly by human standards. These convection currents transfer heat from the core to the surface; they move, at most, several inches (tens of cm) a year.

As the convection currents move horizontally at the top of their roll, they carry the film of the lithosphere across Earth's surface. The currents have gradually pulled the lithosphere apart, creating a dozen large, fairly rigid plates and a number of smaller ones, which the currents shove here and there—jostling them, pushing them upward, pulling them downward, bending and breaking their rocky layers, and constantly reshaping the face of the planet.

There are two types of crust: continental and oceanic. Most plates include both types. Under the oceans the crust is dark, dense basalt. The continental crust, on the other hand, includes lighter-colored, lower-density rocks that are richer in silica, such as granite. Continental crust is thicker than oceanic crust and, being less dense, floats higher on the asthenosphere.

Plates are edged by oceanic ridges, rifts, deep oceanic trenches, and lateral faults. Oceanic ridges form where the upward-convecting mantle pushes adjacent plates apart, allowing molten rock (magma) to erupt onto the seafloor, where it cools into bands of new crust that are broken and pushed aside by more upwelling magma. Rifts form where upwelling magma splits apart a continent. Oceanic trenches develop where plates converge and one plate is sinking beneath the other along what is called a subduction zone. Where continental crust and oceanic crust converge, the denser, thinner oceanic crust sinks into the mantle.

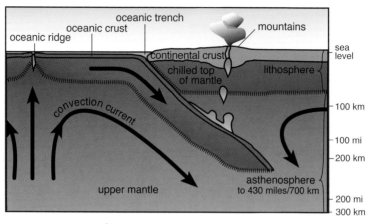

Cross section of a subduction zone.

Utah is part of the North American Plate. The Mid-Atlantic Ridge, in the middle of the Atlantic Ocean, defines the plate's eastern boundary. Here, new rock is steadily being added to the plate as it moves west, away from Europe and Africa. Propelled away from the oceanic ridge, the opposite edge of the plate meets the Pacific Plate, the Juan de Fuca Plate, and the Cocos Plate. This has resulted in the development of a subduction zone along the west coast of North America, except where the Pacific Plate and the North American Plate are sliding by one another along the San Andreas Fault.

As with all collisions, a certain amount of fender bending occurs when plates converge. Mountains are pushed up, parts of the continent are drawn down (or subducted) into the mantle, and islands are tacked onto continents. Earthquakes are frequent. When an oceanic plate is being subducted beneath a continental plate (such as where the Juan de Fuca and North American Plates are converging along the northwestern coast of the United States), water from the oceanic plate triggers the partial melting of the asthenosphere above it at depths of 60 to 90 miles (100 to 150 km). Part of the continental crust often melts as well. Both sources generate molten material, or magma, of different types. The magma rises upward to build arcing chains of tall volcanoes along the continent's edge. Variations in the types of volcanoes and resulting volcanic rock occur due to different rates of convergence, different angles of subduction, and variations in the magma sources.

Despite Utah's current inland location, away from the activity of plate margins, the state has seen a great deal of mountain building. Its mountains have been forming due to compression, extension, and igneous events, all of which are results of the voyage of the North American Plate across Earth's surface. Two orogenies, or mountain building events, were particularly important in Utah. During the Sevier Orogeny, layers of rock were thrust eastward great distances to build the Sevier Thrust Belt, a broad area that extended over all of Utah except the Colorado Plateau. And during the Laramide Orogeny, thrust faulting and uplifts bowed up rock layers of the Colorado Plateau, forming the Rocky Mountains and numerous broad uplifts in Utah.

ROCKS AND MINERALS

Earth's crust is made of rocks, such as sandstone, shale, granite, and basalt. Rocks, in turn, are made of minerals, such as quartz, feldspar, mica, garnet, and gypsum.

Minerals have definite chemical formulas and can be identified by their characteristic color, hardness, and way of crystallizing—for instance, as six-sided rods or small cubes or nondescript glassy masses. Some localities yield particularly beautiful minerals, such as azurite and malachite, two ores of copper found at

quartz	Glassy or milky white, hard enough to scratch glass and harder than a knife. Breaks like glass, with a curved surface. Forms grains in igneous and metamorphic rock and in sedimentary rock derived from them. Some varieties, such as rose quartz and amethyst, are tinted with minute amounts of metallic oxides. Quartz crystals are hexagonal, usually with end facets that form a point.
feldspar	A family of translucent light-colored minerals that can be scratched with a knife. Breaks along flat cleavage faces that then reflect sunlight. Abundant in igneous rocks, where it often forms phenocrysts, and common in metamorphic rocks. Because feldspar readily breaks down into clay, it is not usually found in sedimentary rocks.
mica	A family of minerals that separate into shiny, paper-thin, flexible flakes. Easily scratched with a knife, even with a fingernail. Biotite (black) and muscovite (white) occur in granite and other intrusive rocks as well as in metamorphic rocks such as schist and gneiss.
clay	A group of very fine-grained platelike minerals that absorb water between their plates so the plates slide easily. When wet, clay feels sticky. Many clays swell noticeably when wet.
hornblende	Black, needlelike crystals; very common in igneous and metamorphic rocks.
calcite	White or light-gray calcium carbonate, the chief component of limestone. Can be scratched with a knife but not with a fingernail. When dilute acid is dropped on calcite, it fizzes; if the calcite is powdered, lemon juice will work.
dolomite	Similar to calcite, but with both magnesium and calcium; fizzes only slightly when exposed to dilute acid.
halite or salt	Crystallizes from salt-saturated seawater or lakes. Found not only around the Great Salt Lake but also in much older rocks throughout eastern Utah.
hematite	Dark-red iron oxide; common as tiny grains that tint sandstone and shale red beds. A valuable iron ore when found in concentrated form.
limonite	Rusty yellow iron oxide that, in small amounts, tints rocks yellow or tan.
gypsum	A translucent white or gray evaporite precipitated from salt-saturated water. Soft and easily carved with a knife. Many of southeastern Utah's red rocks have veins of white selenite, a variety of gypsum.
pyrite	Iron sulfide, brassy and metallic; often forms small cubes in igneous rocks.
chalcopyrite	Copper iron sulfide, brassy and metallic with a bluish tint. A common ore of copper.

Common minerals, listed from most common to least common.

Utah's Bingham Canyon Mine. Utah's remarkable minerals can be seen in museums around the state. Many are available in rock and mineral shops.

Rocks are classed by their origin and texture. By origin, they fall into three categories: igneous, sedimentary, and metamorphic.

Igneous rocks originate in molten material that rises from deep within the Earth. While the molten material is below the surface, it is called magma; if it cools underground, magma becomes intrusive igneous rock. Because magma cools slowly underground, the mineral crystals that are forming the igneous rock have time to grow relatively large; thus, intrusive igneous rock is often coarse grained.

When magma erupts on Earth's surface, it is known as lava. Cooled lava becomes extrusive igneous rock, also known as volcanic rock. Because it cools relatively quickly, lava generally forms fine-grained igneous rock.

Both intrusive and extrusive rocks are further subdivided according to their mineral composition, from dark colored to light colored. Since they solidify from a wide variety of melted rocks, igneous rocks are parts of a continuum.

Sedimentary rocks form from broken-up or dissolved older rocks or animal shells. Sediments and dissolved material are carried by water, wind, or ice and are normally deposited in horizontal layers called strata. Over time the sediments lithify due to the pressure of overlying sediments or because the grains are cemented together by a naturally occurring cement, or both. Most

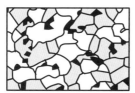

In intrusive rocks that cooled slowly, such as granite, the crystals are all about the same size and each is big enough to be visible to the naked eye.

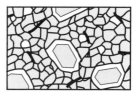

Porphyry contains large crystals scattered through a finer matrix. Porphyries develop from a period of slow cooling (large crystals) followed by fast cooling (small crystals, the matrix).

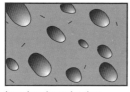

In volcanic rocks that cooled quickly, gas bubble holes, or vesicles, are common. The crystals are often microscopic in rocks with this texture.

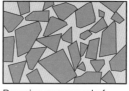

Breccias composed of volcanic rock are common because rocks are fragmented both during eruptions and later when they slide down the slopes of steep volcanoes.

Textures of igneous rock.

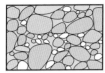 **Conglomerate**: Composed of gravel, it contains both large and small rock fragments rounded by having been bounced and ground against one another in running water. The matrix, the material between the pieces of gravel, is usually sand. The large particles in conglomerate were deposited in high-energy rivers, usually near a mountain range.

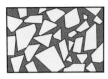 **Breccia**: Composed of angular rock fragments or those that are only slightly rounded. The matrix may be mud or minerals that grew later around the fragments. Breccias form in events such as landslides, in which the original rock was broken and deposited without much time spent in water. Breccia composed of predominantly volcanic material is considered a volcanic rock.

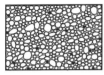 **Sandstone**: Composed of small but visible particles 0.0025 to 0.08 inch (0.0635 to 2 mm) in diameter. The grains are usually rounded, having been rubbed against each other during transport by wind or water. Rivers, the ocean, or wind sorted the sand, a process indicative of a very high-energy environment. Crossbeds in sandstone are big and dramatic when the sand was deposited by wind and flatter and smaller when deposited by water.

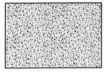 **Siltstone**: Composed of particles between 0.0025 and 0.00015 inch (0.0635 and 0.004 mm) in diameter. Though individual grains are hard to see, the particles still feel a little gritty. Clay, a finer sediment, feels very smooth. Silt and clay were deposited in low-energy environments such as swamps, deltas, or lakes.

Textures of sedimentary rock.

Dune sandstone, common in Utah, can be recognized by its long, sloping crossbedding as well as by its tiny, evenly sized, well-rounded grains—all characteristics of wind deposition. This photo was taken at Dinosaur National Monument. —Felicie Williams photo

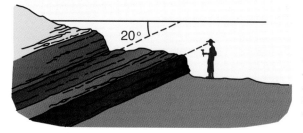

When sedimentary rocks are tilted, we speak of their dip, the angle between horizontal and their maximum downward slope.

sedimentary rocks were deposited in lakes, oceans, and river floodplains, though some, such as the coarse conglomerate of alluvial fans or sandstone formed of ancient dunes, accumulated on dry land. Individual layers of sedimentary rock are called beds, and the bedding plane between them signifies a break in deposition, or an unconformity.

Metamorphic rocks are sedimentary and igneous rocks that have been changed from their original forms by heat, pressure, and chemical-laden water. They may be only slightly altered, or they may be so altered that it is nearly impossible to tell what they once were. Rocks are greatly altered when they are buried deeply by thrust faulting or by other sediments deposited in a down-warped basin. If a metamorphic rock's sedimentary or volcanic origin is still clear, the general word *metasediment* or *metavolcanic* is often used to describe it.

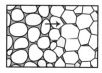

Quartzite and marble: Form when grains in sandstone and limestone grow until they are interlocked. These metamorphic rocks break through the grains instead of around them.

Slate: Lightly metamorphosed, fine-grained rock. The clay changes to very fine mica. The mica grows in the direction of least pressure, giving the rock foliation (dashed lines), which is a tendency to split along parallel planes. Foliation often cuts across original sedimentary bedding.

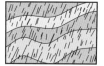

Schist: Fine-grained sedimentary rock that experiences greater pressure and temperature. Mica grains grow and merge until individual grains grow large enough to become visible.

Gneiss or migmatite: Coarse-grained rock that has been pressed and heated close to its melting point. The mineral grains are visible and have separated into light and dark bands. Parallel mica and hornblende form foliation. The rock is often intensely folded.

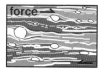

Mylonite: Rock formed during intense shearing, such as what occurs in a fault zone or in a fold. The rock has recrystallized with fine, dense, flowing texture even though it was never molten. Isolated large, rounded grains, called augen, look like eyes.

Textures of metamorphic rock.

igneous intrusive	**gabbro**	Very dark, coarse-grained rock with visible individual grains of dark minerals.
	diorite, monzonite	Coarse-grained rock that is intermediate in color between light and dark and has visible crystals of feldspar, black mica, and hornblende. Mineral composition is intermediate between gabbro and granite.
	granite	Common, light-colored, coarse-grained rock with visible crystals of quartz and feldspar and small amounts of mica or hornblende.
igneous extrusive, volcanic	**basalt**	Very fine-grained, gray or black rock with a high content of iron and magnesium; forms widespread flows or cinders.
	andesite, dacite	Dark-colored, fine-grained rock with a mineral composition that is intermediate between basalt and rhyolite.
	rhyolite	Extrusive equivalent of granite; light-colored and often pinkish. Contains a lot of silica and small amounts of iron and magnesium. Occurs as small, thick flows and beds of tuff or ash.
sedimentary from pieces of other rock	**shale, siltstone**	Very fine-grained rock; made of silt and clay cemented together. Usually breaks in flat slabs.
	sandstone	Composed of sand grains cemented together.
	arkose	Sandstone with at least 25% of the grains composed of feldspar, meaning it was deposited close to a rapidly eroding granitic or metamorphic source.
	conglomerate	Sand and pebbles deposited as gravel and then cemented together. May include cobbles and boulders.
	breccia	Like conglomerate but with broken, angular rock fragments. When volcanic, mapped as a volcanic rock.
	tillite, diamictite	Rock of completely unsorted material deposited by ice.
sedimentary biologic	**limestone**	Gray or white rock made primarily of calcite (calcium carbonate)—often from shells and skeletons of marine organisms. Fossils are often visible.
	coal	Black carbon-rich layers derived from plant material.
sedimentary chemical	**evaporite**	Formed by the evaporation of mineral-rich water. Includes gypsum, halite, potash, and phosphate from seas and lakes, and travertine in limestone caves and hot springs.
metamorphic	**marble**	Recrystallized limestone, often with visible calcite crystals. Often white or gray in color.
	quartzite	Recrystallized sandstone that breaks through the grains instead of around them.
	schist	Streaky rock with visible grains and abundant parallel mica and/or hornblende. Tends to split along the mica grains.
	gneiss	Coarse-grained rock with alternating bands of flaky and granular minerals. More intensely metamorphosed than schist.

Common rocks of Utah.

A Word about Formations

One recognizable kind of rock will often extend across a large distance. Where it is extensive enough to be mappable, it is called a formation. Several formations that are always found together are mapped as a group. Formation and group names are often taken from where the rocks were first recognized and described. Thus, the Bluff Sandstone is named for the town of Bluff, and the Green River Formation for the Green River. Some names used in Utah come from other states, for example, the Dakota Sandstone and the Mesaverde Group. When referring to rocks in a stratigraphic context, geologists use the divisions of lower, middle, and upper, as in the upper Cambrian sandstone, which lies above the middle Cambrian shale. The terms early, middle, and late are used when referring to occurences in a time context, as in the late Permian extinction.

Joints, Faults, and Folds

Sets of parallel cracks, called joints, are common in rock; often there are several intersecting sets. Joints form in many ways: fine clayey sediment shrinks and cracks as it dries; molten rock shrinks and cracks as it crystallizes; or when overlying layers erode or are removed due to faulting, underlying rocks are uncovered and the pressure on them is reduced, so they expand. The stresses of crustal movement due to tectonic activity cause many joints to form. Joints aid in the disintegration, or weathering and erosion, of rocks because they provide pathways for water and air.

Faults are breaks in rock along which a body of rock—a fault block—has moved, whether only a few feet or a great distance. Faults are classified by the angle of the plane of fault movement (fault plane) and the relative direction of movement. Normal faults and detachment faults generally form where tension pulls the Earth's crust apart. In contrast, reverse and thrust faults form where there is compression, such as where plates are colliding. Along lateral faults, one body of rock is sliding horizontally past the other. When faults have an inclined fault plane, the body of rock above the plane is the hanging wall, whereas that below the plane is the footwall.

Faults are most obvious in sedimentary rocks, where layers are visibly offset, and in regions of varied rock types, where one type butts up against another. Faults are harder to pinpoint in large masses of igneous or metamorphic rock, because the rocks on the opposite sides of the fault may look the same. Most faults are found through the careful mapping of rock types in a region.

Large faults with considerable displacement are rarely simple. Usually the rock near the fault has broken along many small, more or less parallel faults. A region with complex displacement is called a fault zone. In the fault zone, the rock may be completely shattered and ground into a flourlike material called fault gouge; or it may have flowed slowly due to great pressure, becoming a rock called mylonite, which has a distinctive, flowing, fine-grained texture and occasional large, rounded mineral grains.

Faults formed by tension during times of crustal extension or stretching

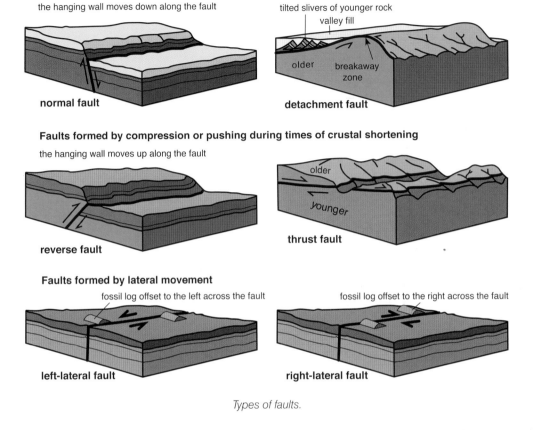

the hanging wall moves down along the fault

tilted slivers of younger rock

valley fill

older   breakaway
zone

normal fault

detachment fault

Faults formed by compression or pushing during times of crustal shortening

the hanging wall moves up along the fault

older

younger

thrust fault

reverse fault

Faults formed by lateral movement

fossil log offset to the left across the fault

fossil log offset to the right across the fault

left-lateral fault

right-lateral fault

Types of faults.

Faults are numerous in Utah and were critical to the development of many of its geologic and geographic features. For instance, tectonic compression during the Sevier Orogeny caused huge thrust faults to develop; they extended east to near the edge of the Colorado Plateau. East of this belt of thrust faults, or thrust belt, the Colorado Plateau also experienced compression-related faulting, though it was not as extreme.

Major normal faults—younger than the thrust faults and caused by tectonic uplift and tension—edge most of the mountain ranges in western Utah. Steeply to moderately inclined, some of these faults level out at depth and join gently sloping detachment faults. Others are planar. Utah's earthquakes show that many of its faults are active.

Folds are just what they sound like: folds or bends in rock. Given enough time and stress, even rocks can bend! Some folds are very slight—nothing more than a gentle warping of rock layers. Others are like accordion pleats, having been folded so tightly that the rock layers on either side of the fold are parallel. Most folds fall between these extremes.

Folds are classified as anticlines (upward folds), synclines (downward folds), or monoclines (steplike folds in which one side of the fold is higher than the other). Like faults, folds are most easily detected in layered sedimentary rocks. Most folds are the result of horizontal compression of Earth's crust; however, monoclines form when near-surface sedimentary rocks drape over deeply buried normal or reverse faults.

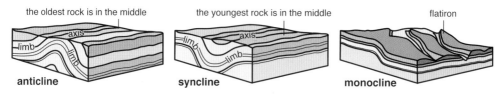

Types of folds.

GREAT EVENTS IN UTAH

The following pages contain a timeline for Utah's geology. For each era or period, the figures tell the state's position on the globe, the paleoenvironment of the time, and the timing of important tectonic and evolutionary events.

Precambrian: Hadean and Archean

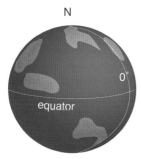

During Hadean and Archean time, the early continents develop from small areas of continental crust that thicken and collide. The whereabouts of rocks that will become Utah is unknown.

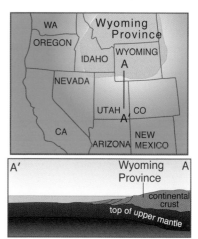

4,600 Ma	4,000 Ma	2,500 Ma
Hadean	Archean	

4.6 Ba: Earth forms from a disk of gas and dust orbiting the Sun.

4.5 Ba: The moon forms, probably from debris related to the impact of Earth with a Mars-sized body.

4.4 Ba: Zircon of this age from the Jack Hills of Western Australia is the oldest dated material that originated on Earth.

4,150–3,850 Ma: Meteorites pummel Earth for 300 million years, making it unlikely life could have started this early.

Few rocks on Earth are Hadean in age; Earth was mostly molten, its rocks reformed and reshaped continuously. The atmosphere was a toxic mix of carbon dioxide and nitrogen.

The world's oldest fossils, found in Western Australia, are possibly 3.5 billion years old. These primitive cells flourish without oxygen, using sulfur instead.

By the Archean eon, Earth has cooled enough for a few small continents to form. One is called the Wyoming Province for its location today. Sediments accumulate along its shores.

Utah's oldest rocks, formed between 2,650 and 2,500 Ma, are in the Raft River and Grouse Creek Mountains, in the northwest corner of the state. They are metamorphosed sedimentary rock and dark igneous rock (some is gabbro, with a probable origin in early oceanic crust) intruded by light-colored granite. By the end of the Archean, simple, single-celled life-forms evolve to where they begin to produce oxygen through photosynthesis, increasing the oxygen content of Earth's water and atmosphere.

Ma: millions of years ago
Ba: billions of years ago

13

Precambrian: Proterozoic

N

Australia

180° R o d i n i a

Antarctica

equator

rifting

UTAH

ice cap

Continents continue to collide, gradually forming the supercontinent Rodinia, which includes today's North America, Australia, and Antarctica. Stress across Rodinia forms a great number of northwest and northeast lineaments, weak and faulted lines in the crust. These are reactivated during later tectonic events. Rodinia then breaks into smaller continents.

Extensive glaciation may have been global in extent.

Australia-Antarctica continent

WYOMING

A

North

midocean ridge

Utah Hingeline

UTAH CO A'
ARIZONA NEW
MEXICO

Laurentia

Utah's Proterozoic rocks are varied, from heavily metamorphosed rocks caught up in mountain building to the sedimentary rocks of the Uinta Mountains.

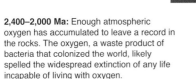

A
midocean ridge
stretched continental crust
Utah Hingeline
A'
Australia-Antarctica
Laurentia

2,400–2,000 Ma: Enough atmospheric oxygen has accumulated to leave a record in the rocks. The oxygen, a waste product of bacteria that colonized the world, likely spelled the widespread extinction of any life incapable of living with oxygen.

1,800–1,650 Ma: Two early continental provinces, the Mojave and Yavapai, accrete to the Wyoming Province. The line of suture, called the Wyoming Suture Zone, runs east-west just north of Salt Lake City. It is a weak zone that guides later tectonic events.

800–650 Ma: At least two extensive glacial episodes cover most of Earth with ice, spelling the end for much of the algal life that dominated up to this point. This probably frees up habitat for an explosion of life-forms. The Mineral Fork Formation tillite is a glacial deposit from this time frame.

770–670 and 600–550 Ma: Rodinia stretches and Australia and Antarctica split from the supercontinent, leaving Laurentia, which includes North America, Greenland, and parts of Europe. Western Utah and much of Nevada are pulled thin by the rifting. The boundary between the thinned crust to the west and thicker, unthinned crust to the east, today called the Utah Hingeline, has greatly influenced Utah's geology.

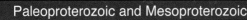

2,500 Ma

1,000 Ma   541 Ma

Paleoproterozoic and Mesoproterozoic   Neoproterozoic

1,800–1,650 Ma: Farmington Canyon Complex rocks, sediments, and volcanics of the Mojave Province are intensely metamorphosed during accretion. This resets their radiometric dates. Now gneiss and schist, none of these rocks contain fossils due to the metamorphism.

950 Ma: Utah's oldest known fossils are spherical microalgae found near the base of the Uinta Mountain Group.

770–740 Ma: In Utah an extensional basin opens along the Wyoming Suture Zone, where the Uinta Mountains are today. Vast quantities of mostly sandy sediment—the Big Cottonwood Formation and Uinta Mountain Group—fill the basin.

541 Ma: Plate collisions and rupturing have slowed to a crawl. Most continental surfaces, including mountains, have eroded down to sea level. The erosional surface, called the Great Unconformity, shows up on every continent.

541 Ma: The earliest shelled animals appear in abundance. They include simple, small arthropods, brachiopods, mollusks, and echinoderms. Their presence defines the division between Precambrian and Cambrian time.

1"/2.5 cm

Mats of microalgae formed stromatolites, mound-shaped laminated fossils common in Precambrian rocks. —Virtual Fossil Museum, www.fossilmuseum.net

14

Cambrian

N

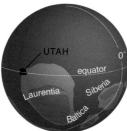

Laurentia, the continent that includes North America, lies rotated about 90 degrees from today's orientation, with what is today's west coast at about the equator. Laurentia drifts slowly northeastward.

deep-water deposition

WA

Because Laurentia was fairly flat, the sea transgressed and regressed great distances.

OREGON

IDAHO

NEVADA UTAH WYOMING

CO

A *Shallow sea with abundant fauna, including trilobites.* A'

NEW MEXICO

CA

ARIZONA ◀ North

Utah's Cambrian rock section is one of the most complete in the world, with exceptionally fossiliferous beds. Its fossils and sediments accumulated in shallow continental seas.

Ordovician

N

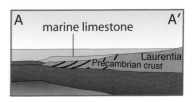

A marine limestone A'

Laurentia
Precambrian crust

Laurentia continues its slow northeastward drift. Utah's edge of the continent is tectonically stable, with no significant mountain building. Iapetus, the ocean between Laurentia and Baltica, starts to close, resulting in the earliest episode in the building of the Appalachians.

Utah's Ordovician rock section is also thick and fairly complete. The Ordovician was characterized by fluctuations between warm, wet intervals and ice ages, with corresponding changes in sea level. Glaciation was widespread by the end of the period.

541 Ma				485 Ma	444 Ma
early	Cambrian		late	Ordovician	

541 Ma: The Cambrian Explosion occurs, which is a dramatic increase in shelly fossils and in the diversity of life.

530 Ma: Jawless fish, the oldest vertebrates, appear; they become more common during Ordovician time.

Fossils of incredible quality form after slumps bury soft-bodied animals in western Utah. The fossils show astonishing diversity: phyla exist that have no known descendants after the Cambrian.

Middle Cambrian: Uncounted numbers of trilobites are preserved in the Wheeler Formation shale.

Elrathia kingii

510 Ma: Spores and cells of the earliest plants signal their colonization of land. They spread over the damp areas of Earth.

490 Ma: Widespread anoxia decimates trilobites and other marine animals in a global extinction. The cause for the drop in oxygen may be glaciation or global ocean turnover, which brings deep anoxic ocean water to the surface—where the animals live.

Ordovician: Trilobites, brachiopods, corals, echinoderms, bryozoans, gastropods, bivalves, ammonoids, conodonts, and graptolites peak in genera-level diversity.

440 Ma: The second greatest extinction in Paleozoic time is tied to the last Ordovician glaciation. Sixty percent of all marine animal genera disappear. Plankton species are particularly hard hit.

This is a partial fossil of Hallucigenia, *a small wormlike animal with spines down its back and spiky legs.* —John Adamek photo

This fossil has not been assigned to any life group; it is simply classed as enigmatic.
—Peter Watson photo

—Photos courtesy of the Virtual Fossil Museum, www.fossilmuseum.net

15

Silurian

N

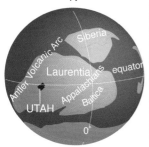

Baltica collides with Laurentia along what is now the eastern side of North America, building the Appalachian Mountains. The Antler Volcanic Arc approaches Laurentia from the west. The sea continues to transgress and regress on the shallow continental shelf of Laurentia and Baltica.

Early land plants colonize land.

Near the equator during Silurian time, Utah was covered with a shallow sea populated with corals, crinoids, and brachiopods. The rock deposited in this sea, the Laketown Dolomite, contains abundant fossils.

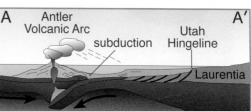

Devonian

N

Euramerica, the continent formed by the combination of Baltica and Laurentia, drifts south toward the southern continent Gondwana. The Antler Volcanic Arc collides with Euramerica during the Antler Orogeny.

A warm climate enhances the growth of extensive reefs and the first significant colonization of land by both plants and animals.

Early Devonian time was characterized by a shallow sea with abundant marine animals, resulting in limy sediment that turned to dolomite. During the later part of the period, the start of the Antler Orogeny on the coast caused increasing amounts of sand and clay to be deposited in the sediment. Devonian sediments thicken greatly west of the Utah Hingeline; the rock section there is up to 5,000 feet (1,500 m) thick.

444 Ma	419 Ma		359 Ma
Silurian	**Devonian**		

Climbing temperatures melt the glaciers of Ordovician time, raising sea level. Marine life recovers from the extinction that occurred at the end of the Ordovician.

425 Ma: Primitive plants move onto land. Otherwise, the land is mostly barren, with the exception of algae in wet areas and lichens.

Several groups of plants evolve and start to colonize land. Most of them are initially no more than 1 to 2 feet (0.3 to 0.6 m) tall.

Fish diversify and start to dominate the oceans. Bony fishes similar to modern fish and cartilaginous fishes like sharks both appear during the Devonian.

380 Ma: The Antler Orogeny starts in Nevada. In north-central Utah a small uplift erodes, forming local conglomerate.

Massive reefs are built by spaghetti-like stromatoporoids in Utah. These are most likely related to sponges.

The first full-blown forests have trees up to 33 feet (10 m) tall. *Eospermatopteris*, a tree fern, can be one of the tallest.

Much Silurian sediment erodes away, leaving only one Silurian formation, the Laketown Dolomite, in western Utah.

Eurypterids, or sea scorpions, became the dominant ocean predators. Some were more than 6 feet (2 m) long. They looked like scorpions and grabbed prey with their pincers.
—Virtual Fossil Museum, www.fossilmuseum.net

Two or more extinctions devastate the reef-building stromatoporoids and corals, many other shallow-water marine animals, and marine vertebrates, opening up habitats for new creatures to evolve during Mississippian time. More than 75 percent of all species disappear.

16

Carboniferous: Mississippian and Pennsylvanian

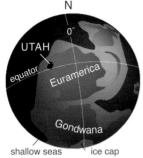

N

0°

UTAH

equator

Euramerica

Gondwana

shallow seas   ice cap

Mississippian

Euramerica continues to drift northward across the warm, humid equator. Gondwana also moves northward, and the ocean between the two closes. Shallow warm seas flood the continents, providing habitat for abundant ocean life, including brachiopods, bryozoans, forams, and fish.

WA   MT   **Pangaea**
North   Euramerica
OREGON   WY   Ancestral Rocky Mts
NV   IDAHO
Oquirrh Basin   CO   Uncompahgria
A   Paradox Basin   A'
UT
equator
CA   ARIZONA   NEW MEXICO

In northwest Utah the Oquirrh Basin slowly subsided, and more than 7,000 feet (2,130 m) of sediment accumulated in it. Ash and sediment from the Antler Orogeny to the west added to the fill.

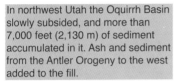

A   Antler Orogeny mountains   Utah Hingeline   Ancestral Rocky Mountains   A'
Oquirrh Basin   Paradox Basin
Pangaea

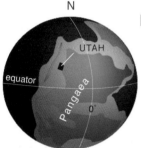

N

UTAH

equator

Pangaea

0°

Pennsylvanian

By Pennsylvanian time, Euramerica is colliding with Gondwana to form the single supercontinent Pangaea. Pressure from the collision pushes up the Ancestral Rocky Mountains along ancient faults.

The Oquirrh Basin continued to subside gradually and filled with more than 17,000 feet (5,200 m) of sediment and ocean fossils. The Paradox Basin, southwest of rising Uncompahgria, also subsided, filling with alternating layers of evaporites and black mud (deposited in restricted, oxygen-starved marine environments) and arkose (washed in from Uncompahgria).

359 Ma	347 Ma	331 Ma	323 Ma	299 Ma
Early Mississippian	Middle Miss.	Late Miss.	Pennsylvanian	

The extinctions of some reef builders and other species at the end of the Devonian opens up shallow marine habitat for the evolution of many new marine species.

Repeated glaciations throughout the Mississippian and Pennsylvanian cause sea level fluctuations and corresponding changes in sedimentation.

In what is now the eastern United States, vegetation in widespread swamps includes tree ferns, giant horsetails, and giant lycophytes. Over time the accumulation of organic material forms thick coal deposits.

The Mississippian-Pennsylvanian boundary, made obvious by changes in rocks elsewhere, does not stand out in Utah

320 Ma: Reptiles first appear and diversify rapidly. Amphibians are abundant and getting bigger—up to 20 feet (6 m).

300 Ma: The ancestors of birds and dinosaurs appear and rapidly evolve into several groups.

300 Ma: Lush vegetation has removed carbon dioxide from the air, leading to an oxygen surplus–35 percent compared to 21 percent today. This may explain the huge size of insects: 6-foot-long (2 m) centipedes, 3-foot-long (1 m) scorpions, 2.5-foot-long (0.8 m) dragonflies.

The amniotic egg (an egg covered in a hard shell), an evolutionary adaptation, allows four-legged animals to move inland away from water.

5 cm

Lycophytes included giant club mosses and scale trees much larger than any on Earth today. This bark section belonged to a scale tree that grew up to 130 feet (40 m) tall.
—Virtual Fossil Museum
www. fossilmuseum.net

0.5 cm

The netlike Fenestella lived attached to the seafloor. Colonies of bryozoans like this were externally similar to many modern corals, with many small animals living attached to each other. —Wikimedia Commons

Platyhystrix, a 2.5-foot-long (0.8 m) amphibian with a dorsal sail, lived near where Glen Canyon is today.

17

Permian

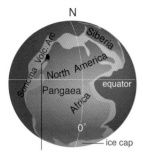

N

Utah is about 10 degrees north of the equator.

ice cap

Pangaea is complete. The Ancestral Rocky Mountain uplifts slow and then stop rising.

Collision of a volcanic island arc with North America, close to the northwest edge of the continent, starts the Sonoman Orogeny.

The climate in the interior of Pangaea is dry; the huge swamps of Carboniferous time are gone. Glacial episodes continue to affect worldwide sea level, but not as frequently as during the Carboniferous.

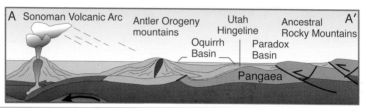

In Utah the Oquirrh and Paradox Basins continued to subside, the Oquirrh filling with as much as 19,000 more feet (5,800 m) of sediment. The Paradox, often restricted, filled with salt from saturated seawater. Uncompahgria continued to shed sediment. Sea level rose and fell frequently, so limestones (shallow-water) interfingered with shales (deeper water) and sandstones (shoreline).

Brachiopods, sponges, bryozoans, and fusulinids are common in some of the limestones of this time.

Just after mid-Permian time, the polar ice caps melted and sea level rose. Then the ice caps re-formed, lowering sea level again. The subsequent period of erosion extended into the Triassic Period.

299 Ma		272 Ma	260 Ma	252 Ma
early Permian			middle Perm.	late Perm.

The Permian starts with the low sea level and wet climate associated with an ice age.

Fossils of the amphibious *Eryops* of early Permian age have been found in the Cedar Mesa Sandstone near Canyonlands National Park.

The fantastic whorl of teeth of Helicoprion, *a sharklike fish from 290 million years ago, found in northwest Utah.* —Greg McDonald photo, Wikimedia Commons

A broad expanse of shallow sea transgresses from the west. The resulting Kaibab and Park City limestones are full of sea animal shells: bryozoans, brachiopods, gastropods, crinoids, and bivalves.

Conifers become the most common trees. In the drier climate, they and other seed-bearing plants have a great advantage over the spore-bearing fernlike trees that were dominant during the Carboniferous.

New varieties of land animals become increasingly adapted to life far from the ocean. The ancestral groups that will give rise to mammals, turtles, and dinosaurs all evolve.

Marine diversity is the greatest it will ever be.

The Permian Period ends with the greatest extinction of life after the Precambrian. It is the end of existence for 96 percent of marine species and 70 percent of terrestrial species, among them trilobites and two kinds of corals that had been very important reef builders. Of the species that survive, few will thrive again.

Several worldwide disasters, perhaps in combination, are suspected of causing the extinction: violent and large-scale volcanism in Siberia, the coming together of Pangaea and a subsequent shift in weather patterns, lower sea level, the venting of gas from the ocean floor, and a meteorite impact.

Triassic

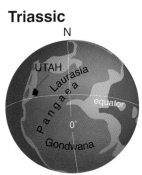

N

At the start of the Triassic, Pangaea straddles the equator and extends from pole to pole. It gradually splits into two continents: Laurasia to the north and Gondwana to the south. The climate of Pangaea is generally dry, and hot near the equator. After it splits in two, ocean currents change and the climate moderates and becomes wetter.

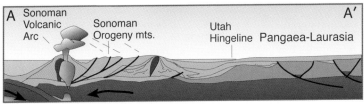

During the Early Triassic, Utah was on the broad coastal plain of Pangaea. Often below sea level, the sediment that was deposited formed limestone to the north (Thaynes and Dinwoody Formations) and dark-red shale and sandstone interbedded with thin limestone, gypsum, and conglomerate in the south (Moenkopi Formation). Marine deposits of this age contain fossils of animals such as conodonts, brachiopods, pelecypods, and ammonites. Continental deposits contain abundant swimming and walking tracks of reptiles and amphibians, and amphibian fossils.

By the Late Triassic, after a time of erosion, most of Utah was above sea level. The Ancestral Rockies had eroded down. With the breakup of Pangaea and a wetter climate, forests grew. On the coast, volcanoes developed. Their ash was carried to Utah to become colorful beds in the Ankareh Fm. in northern Utah and the Chinle Formation in the rest of the state. Both form fantastic badlands.

252 Ma	237 Ma		201 Ma
Early and Middle Triassic		Late Triassic	

Survivors of the Permian extinction slowly start to recolonize Earth.

The Moenkopi Formation contains the oldest Triassic reptile trackways known in North America; they were made when the land was above sea level.

Although distant, the active Sonoman Orogeny in Nevada sheds sediment into the lowlands in Utah. The Ancestral Rocky Mountains, now inactive, also continue to erode.

Utah sits in an embayment, often blocked from the open ocean by the Sonoman Orogeny mountains. As a result, the seawater evaporates, depositing salt and gypsum in the Moenkopi Formation.

230 Ma: Pterosaurs appear, the first flying vertebrates.

Small theropods leave trackways in the Chinle Formation in southern Utah.

The first sauropods appear. These will evolve into huge plant-eating dinosaurs such as *Apatosaurus*.

Conifers grow abundantly in Utah. They will be fossilized in the Chinle Formation's vibrant-colored ash beds. Silica in the ash dissolves easily and replaces the woody material of thousands of conifer logs.

210 Ma: Small mammals appear.

Another major extinction (or extinctions) decimates many creatures, including conodonts and many species of ammonites, reptiles, and amphibians.

—Lucy Chronic photo

Chirotherium tracks have a thumb, but it is on the outside, unlike human hands.
—Wikimedia Commons

230 Ma: The first theropods appear. These eventually evolve into meat-eating dinosaurs, such as *Tyrannosaurus rex*, and birds.

Crocodile-like phytosaurs (this is the skull of one) haunted the rivers in Utah.
—Wikimedia Commons

Jurassic

N

North America

UTAH

equator   Gondwana

South
America

0°

The Atlantic Ocean opens, and North America moves northwest, overriding the Farallon Plate. Several archipelagoes on the Farallon Plate accrete to the coast, enlarging the continent. In Middle and Late Jurassic time, rapid subduction causes the Nevadan Orogeny, resulting in a chain of volcanoes and, beneath them, magma chambers that will cool into the light-colored granite bodies of the Sierra Nevada.

WA

MONTANA
WYOMING

OREGON   IDAHO

CA   NEVADA   *shallow sea*   sand

A   *tidal flat sand*   *dune sand*   A'

accreting archipelagoes   UTAH   CO

North

Middle Jurassic   ARIZONA   NEW MEXICO

A   Sierra   Sevier Orogeny   A'
Nevada   mts.

Pacific Plate   midocean ridge   *alluvial fans*   *dunes*   old faults   Precambrian continental crust

Farallon Plate   North   American   Plate

metamorphism

The Early Jurassic in Utah was a time of eolian activity. Its rocks, the Glen Canyon Group, are the wind-deposited Wingate, river-deposited Kayenta, and wind-deposited Navajo. Today their massive layers above Triassic-age slopes provide dramatic scenery for much of Utah's plateau country.

During Middle Jurassic time, more than 6,000 feet (1,830 m) of evaporites and mud—the Arapien Formation—accumulated in a basin in central Utah. West of it marine limestone was deposited to become the Twin Creek Limestone. To the east, shoreline deposits became the San Rafael Group. Many erosional episodes interrupted layers of deposition, forming unconformities.

Late in Jurassic time, the Morrison Formation accumulated on a widespread plain that extended from south-central Utah across Colorado and north well into Canada. Occasional inundations of ash blew in from volcanoes to the west. Unconformities indicate periods of erosion.

201 Ma	174 Ma	164 Ma	145 Ma
Early Jurassic	Middle Jurassic	Late Jurassic	

The extinction at the end of the Triassic eliminates many large nondinosaur reptiles and amphibians, opening up habitat into which dinosaurs diversify and expand in numbers. Plesiosaurs and pterosaurs become abundant.

Both walking and swimming dinosaur tracks in the Moenave Formation near St. George are made along the ege of a freshwater lake called Lake Dixie.

The continent flexes downward due to the weight and pressure of mountain building in Nevada, so Utah is inundated by a shallow interior seaway.

Ocean-dwelling invertebrates found in the Twin Creek Limestone are limited in diversity; possibly the seaway was restricted and brackish. Oysters, crinoids, and brachiopods are the most common.

Tectonic pressure from the west starts to affect western Utah. Granitic magma intrudes the region deep beneath the surface, and deeply buried sedimentary rocks are metamorphosed.

Dinosaurs are abundant in eastern Utah, their bones preserved in the Morrison Formation. They include *Stegosaurus*, *Allosaurus*, and *Brachiosaurus*. Both Dinosaur National Monument and the Cleveland-Lloyd Dinosaur Quarry are in the Morrison Formation. Burrows and nests of many kinds of insects have also been found in the Morrison.

Dinosaur tracks like this Eubrontes *have their own names because we can't be certain which dinosaur made them.* —Mark A. Wilson photo, Wikimedia Commons

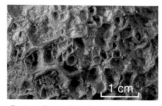

Oysters formed banks, great masses of them growing on one another and on old shells. —Mark A. Wilson photo, Wikimedia Commons

Allosaurus is Utah's State Fossil. —Wikimedia Commons

20

Cretaceous

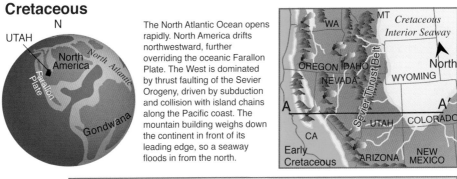

UTAH

North America

Farallon Plate

North Atlantic

Gondwana

The North Atlantic Ocean opens rapidly. North America drifts northwestward, further overriding the oceanic Farallon Plate. The West is dominated by thrust faulting of the Sevier Orogeny, driven by subduction and collision with island chains along the Pacific coast. The mountain building weighs down the continent in front of its leading edge, so a seaway floods in from the north.

Cretaceous Interior Seaway

WA
MT
OREGON IDAHO
NEVADA
WYOMING
North
A
A'
Sevier Thrust Belt
UTAH
COLORADO
CA
Early Cretaceous
ARIZONA
NEW MEXICO

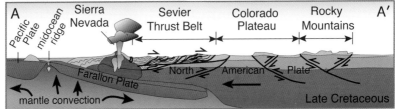

A — Pacific Plate — midocean ridge — Sierra Nevada — Sevier Thrust Belt — Colorado Plateau — Rocky Mountains — A'

Farallon Plate
North American Plate
mantle convection
Late Cretaceous

As pressure from the coast moved eastward, the Sevier Orogeny began. Thick layers of rock were thrust eastward to build mountains in western and central Utah—the Sevier Thrust Belt. Thrust faulting came in pulses, with one thrust active and then another. The crust across Utah shortened by at least 137 miles (220 km). Each uplifted range was soon flanked by a thick wedge of conglomerate.

In the warm climate near the Cretaceous Interior Seaway, swamps developed; these became coal. Dinosaurs wandered through forests and on plains between the mountains and the sea; Utah has one of the best Late Cretaceous dinosaur records in the world. Late in the Cretaceous the Laramide Orogeny began, and the seaway retreated northward.

Nevadan Orogeny   Sevier Orogeny   Laramide Orogeny

145 Ma   (Rocky Mts.) 66 Ma

Cretaceous

130–125 Ma: Early flowering plants appear.

During the Cretaceous, insects, which had diversified greatly during the Carboniferous, experience evolution that is closely tied to the evolution of flowering plants.

118 Ma: Rising sea level and the conclusion of the breakup of Pangaea isolate North America. Evolution on the continent begins to follow its own path.

100–90 Ma: The interior seaway expands southward; its shoreline sands will become the Dakota Sandstone, and mud deposited in the sea will become the Mancos, Mowry, and Tropic Shales.

Gryposaurus monumentensis, *a hook-beaked hadrosaur about 30 feet (9 m) long, lived in Grand Staircase–Escalante National Monument 75 million years ago.*
—Berkeley T. Compton photo

Dinosaurs roaming the edges of the receding seaway include hadrosaurs, alamosaurs, torosaurs, tyrannosaurs, and troodons. A lot of feathered dinosaurs and ankylosaurs have been found in the Cedar Mountain Formation recently.

The extinction at the Cretaceous-Tertiary boundary spells the end of 60 to 70 percent of marine species and 15 percent of those on land. Larger creatures, including the dinosaurs, are more likely to become extinct. Smaller birds and reptiles survive.

Hypotheses for the cause of the extinction include the Chicxulub meteorite impact, extensive volcanism in India, climate change associated with the breakup of Pangaea, and/or lowering sea levels.

Early and Middle Tertiary: Paleogene

N

UTAH

As North America drifts west, subduction of the Farallon Plate continues to exert eastward and upward pressure. This causes the Laramide Orogeny by flexing the continent into uplifts and basins, forming the Rocky Mountains.

Between 40 and 30 Ma, the North American Plate begins to override the midocean ridge between the Farallon Plate and the northward-moving Pacific Plate. This alters the direction of stress on the continent to a northward shear along the coast, starts the San Andreas Fault, and helps stretch the crust of the Southwest, making it thinner.

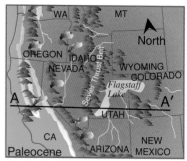

Paleocene

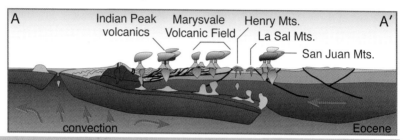

Eocene

During Paleocene time, erosion of the Laramide Orogeny mountains and the now tectonically quiet Sevier Thrust Belt glutted basins with sediment.

Sedimentation from the still-rising Laramide mountain ranges continued during Eocene time. Immense volumes of granitic magma fed intrusions and explosive eruptions that covered the surrounding region with tuff and ash.

Granitic intrusions and explosive volcanism continued into Oligocene time. Huge piles of volcanic tuff and breccia accumulated around active volcanoes.

Paleogene

66 Ma	56 Ma		34 Ma	23 Ma
Paleocene		Eocene		Oligocene

Late Cretaceous to early Paleocene: The Uinta Mountains and the San Rafael, Kaibab, Monument, and Circle Cliffs uplifts continue to rise. After the Cretaceous-Tertiary extinction, flowering plants and small mammals start filling niches that were opened up by the extinction.

62 Ma: In landlocked basins between uplifts, freshwater Flagstaff Lake reaches from the Uinta Basin (south of today's Uinta Mountains) southwest almost to Arizona. The lake's fossils include root casts, turtles, and other vertebrates.

50 Ma: With the end of the Sevier Orogeny thrust faulting, thrust sheets slide back westward along their original fault planes and along new normal faults.

40–20 Ma: A wave of volcanism builds towering volcanoes as the Farallon Plate sinks downward and a wedge of mantle opens above it.

Another lake, Lake Uinta, fills the Uinta Basin and part of older Flagstaff Lake. Its sediment, the Green River Formation, is famous for its superbly preserved fish and other vertebrates.

The Laramide Orogeny ends in Utah. Highlands continue to shed sediment. When the Laramide Orogeny ends and the tectonic pressure is released, the sedimentary rock overlying a series of granite intrusions in western Utah slides off the granite bodies. As a result, the intrusions rise, along with surrounding metamorphic rock, forming mountain ranges known as metamorphic core complexes.

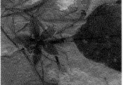

—Virtual Fossil Museum
www.fossilmuseum.net

This fossil fish (Knightia eocaena) and fossil flower (left) show the superb preservation of fossils in the Green River Formation.
—Wikimedia Commons

Most mammals and birds we are familiar with appeared during Paleogene time, along with some, like this entelodont, that did not survive. —Charles R. Knight artwork, Wikimedia Commons

22

Late Tertiary and Quaternary

For reasons not yet certain, but probably tied to the demise of the Farallon Plate, the Southwest continues to rise. Uplift combined with northward shear along the San Andreas Fault creates tremendous tension across western Utah and Nevada, starting Basin and Range–style extensional faulting. The land's height helps give the sliding fault blocks a good run.

Pleistocene

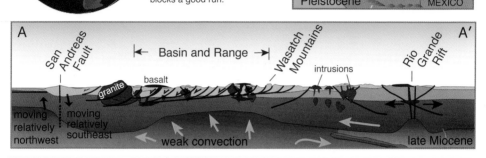

During the first part of the Neogene, Utah was characterized by broad swells of mountains surrounded by basins clogged with sediment eroded from the mountains and erupted from volcanoes. Lakes filled many of the basins. With Basin and Range faulting, Utah began to look like the land we know today.

In Quaternary time, several ice ages occurred, with ice caps and glaciers covering up to 30% of Earth. Sea level was lower because the ice tied up so much water, so land bridges appeared between the northern continents. River down-cutting rapidly exposed millions of years of sedimentary layers and helped carve Utah's spectacular scenery.

23 Ma

Neogene — 5.3 Ma — 2.6 Ma — 0.01 Ma — present

Miocene | Pliocene | Pleistocene | Holocene | Quaternary

23 Ma: Explosive volcanism continues, with thick ash layers accumulating in sedimentary formations such as the Browns Park Formation (24–11 Ma) of northern Utah.

20 Ma: The collapse and spread of volcanics in the huge Marysvale Volcanic Field causes landslides and thrust faulting in southern Utah.

17 Ma: Basin and Range–style extensional faulting begins in Utah.

15 Ma: The Wasatch Fault forms; during the next 15 million years, land west of the fault drops up to 5 miles (8 km) relative to land to the east, creating the Wasatch Front.

15 Ma: Volcanism changes from explosive eruptions of rhyolite and andesite magmas from melted crust to basalt and rhyolite flows fed instead by magmas from the mantle and melted crust. This is a result of the region-wide change from subduction and compression to stretching and rifting, and possibly also passing over, at depth, the remains of the Farallon Plate.

10–5 Ma: The Gulf of California rifts open. Rivers start to drain toward it from the edges of the uplifted Colorado Plateau.

6–5 Ma: Drainages link to become the Colorado River flowing off the Colorado Plateau we know today. The formation of the river is linked to the growth of the Grand Canyon. The particular process and timing of the Colorado drainage evolution and the erosion of the Grand Canyon are hotly debated.

3 Ma: The Isthmus of Panama joins North and South America, resulting in the migration of many animals between the two continents, and many extinctions due to competition between the two populations.

2.5 Ma: Glaciers begin to cover Utah's high mountains, especially the Uinta Mountains.

16,000 years ago: Lake Bonneville is at its largest, covering most of western Utah; its highest level forms the Bonneville Shoreline. Sediment deposited in the lake covers most of western Utah, except for the highest portions of the Basin and Range, which make islands.

Merychippus, *an early horse, had three toes and stood about 35 inches (90 cm) tall.*
—H. Zell photo, Wikimedia Commons

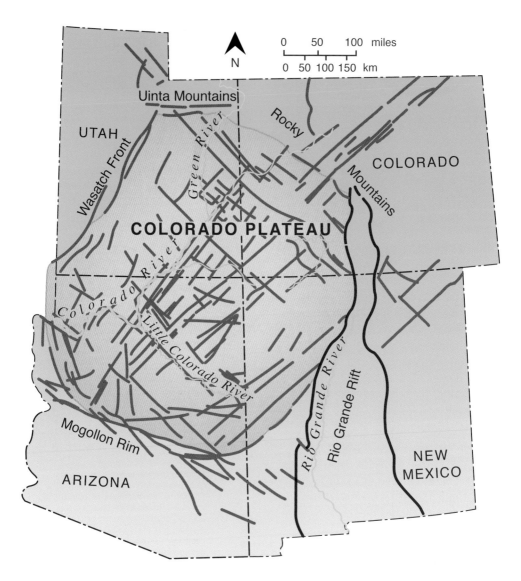

Named not for the state of Colorado but for the river that courses through it, the Colorado Plateau is edged by the Uinta Mountains to the north, the Rockies and the Rio Grande Rift to the east, the Wasatch Front to the west, and the Mogollon Rim to the south. Almost all of it drains into the Colorado River. Precambrian fault zones are shown with brown lines. Note how river channels commonly follow the ancient fault zones.

HIGH, WIDE, AND LONESOME:
THE COLORADO PLATEAU

Southeastern Utah is part of a province known as the Colorado Plateau. It is a special land of flat-lying sedimentary rocks exposed wondrously thanks to an arid climate and deep down-cutting by the Colorado River and its many tributaries. Colorful beds of shale, siltstone, sandstone, and conglomerate are revealed like cake layers, sometimes steeply folded and faulted but never having been metamorphosed. This is a country of Pennsylvanian-age and younger rocks; older rocks are only rarely exposed.

WHERE BRILLIANT COLORS MEET YOUR EYES

The palette of rocks in pink and red, yellow and green, purple and white makes the Colorado Plateau America's most brilliantly colored landscape. Minor constituents give rise to their hues: reds, pinks, and yellows are from tiny particles of iron oxide (the iron has been oxidized), greens and blues from oxygen-poor iron minerals (the iron has been reduced), and lavender from manganese. Then there are the contrasting grays of igneous rocks and the dark iron and manganese coatings of desert varnish. Lichens color surfaces with green and orange blotches and, along with algae and desert varnish, streak cliffs with brown and black.

IN EARTH'S SCULPTURE GARDEN:
WEATHERING AND EROSION

On the Colorado Plateau, in the rain shadow of Utah's High Country, soils are thin and vegetation sparse. When rain does fall, there are few plants to absorb it or to anchor the soil. Cloudbursts quickly run off the land, carrying loose material and filling gullies with flash floods. Though brief, the thick, muddy water of flash floods moves a great deal of sediment. Even car-sized boulders can be carried downstream.

Most of the rocks in the Plateau are fairly porous, allowing rain and snowmelt to sink into them; as a result, shaded rock is often cool and humid, even on hot days. In winter, nighttime temperatures dip below freezing. The freezing water expands, prying loose sand grains and sometimes an outer skin of rock. When the ice thaws during the day, the grains fall away, ready for wind and rain to carry them off. For months each year, this relentless process enlarges alcoves and recesses in the Colorado Plateau's rocks, sculpting some of the region's most spectacular features.

Example of rock weathered by freeze-and-thaw processes in Arches National Park. Field of view is 6 feet (2 m) wide. —Felicie Williams photo

Wind is an important agent of erosion, hammering exposed rock with tiny grains of sand it carries, smoothing surfaces and carrying away loose sediment, sometimes in swirling dust devils. In places, wind leaves behind only a desert pavement of closely spaced pebbles, an armor that slows further erosion.

One other agent of erosion looms large in canyon country: gravity. As softer layers crumble away from the lower portions of canyon walls, stronger layers above them are undermined; eventually, they fall away, like the calving edge of a glacier. If you spend much time in the Plateau, you will hear falling rock. Often it's only small pieces, but occasionally pieces as big as train cars thunder into the canyons below.

Most of the canyons of the Plateau were cut in the last 6 million years, probably a result of a combination of geologic factors, including faulting, continued regional uplift, and the erosional response of the Colorado River's tributaries to its connection with the Gulf of California. The latter factor created a much greater drop in elevation than the river previously had. Streams that earlier wound lazily across broad, gentle slopes surfaced with Tertiary sediments gained new strength due to these changes—rivers with greater elevational relief over their courses, and additional flow, are able to erode landscapes more aggressively. Cutting constantly downward, rivers of the Plateau became trapped in their channels, incising their old meandering paths deep into underlying Mesozoic and Paleozoic rocks.

Many landforms of the Plateau, such as Factory Butte, seen here off Utah 24, are the result of the differential weathering and erosion of hard and soft rock layers. Hard layers (usually sandstone and limestone) form cliffs; soft layers (usually mudstone and shale) form slopes and benches. —Emily Silver photo

THE COLORADO PLATEAU: A RAFT IN A STORMY SEA OF MOUNTAIN BUILDING

Since Precambrian time, mountain building events, or orogenies, have mostly bypassed the Colorado Plateau even though surrounding areas have been repeatedly crumpled and faulted. The continental crust beneath the Plateau is thicker and cooler than that to the west, which may have helped the Plateau stay together. Thicker, cooler rocks are less easily folded and faulted.

Tectonic forces have pushed up several broad uplifts in the Plateau, both during Pennsylvanian to Permian time and some time after mid-Cretaceous time. They have also lifted the entire Colorado Plateau up to 2 miles (3 km).

Widespread marine sedimentary rocks containing fossils of Late Cretaceous age attest to the fact that the Colorado Plateau and adjacent high desert were last at sea level during this period. The cause of the rise of this region is a subject of much research, with geologists investigating many potential factors, all having to do with the forces of plate tectonics and the behavior of the lithosphere and asthenosphere underlying North America.

The timing of the uplift is also uncertain. The Grand Canyon and the deep canyons of Utah's Colorado Plateau appear to have been incised in the last 6 million years or less. But that does not simply suggest that all of the uplift is that recent. Rather, previous faulting and river piracy may have allowed the ancestral river and its system to erode deeply into already high topography. A great roadblock to determining what really happened is that once land rises above sea level, where sediments often contain marine fossils, there is no reliable record of its altitude. Ongoing studies in paleobotany, paleohydrology, paleotemperature, and paleorelief may shed light on the varying elevations of the Plateau, but so far, no strategy has been totally successful.

GETTING TO KNOW UTAH'S PLATEAU COUNTRY

Though the Colorado Plateau has held together persistently over time, within it are many smaller plateaus and basins that have risen and fallen relative to one another. Faults and folds between these plateaus and basins are aligned in two general directions: northeasterly and northwesterly. Old fault zones in the underlying Precambrian rocks that were metamorphosed around 1.65 billion years ago dictate the orientations. These fault zones still become active when tectonic stresses are great. When Precambrian rock on one side of a fault zone is raised higher than that on the other, faults sometimes develop in the overlying sedimentary rocks, but often the layers hold together and drape in monoclines over the deep Precambrian faults.

The oldest sedimentary rock of the region is the Neoproterozoic Uinta Mountain Group, which can be found in the Uinta Mountains. Paleozoic strata are exposed where rocks are domed or faulted upward and in deep canyons. Mesozoic (Triassic, Jurassic, and Cretaceous) sedimentary rocks cover much of the remaining area. Cenozoic rocks appear on the Roan Plateau, the southwestern high plateaus, and in the basins of the Uinta and Henry Mountains. More-recent Quaternary sediments include alluvial fans (many from flash flooding), stream floodplain deposits, a few high glacial moraines, huge boulders and piles of talus at the bases of cliff faces, and a thick layer of windblown sand and silt that surfaces most plateaus.

Here and there across the Colorado Plateau are signs of igneous activity that occurred during Tertiary time: lone mountain ranges that are clustered igneous intrusions, stark volcanic necks (the conduits of former volcanoes), dikes and sills of hard igneous rock that now jut out as ridges and cap mesas, and lava-capped plateaus.

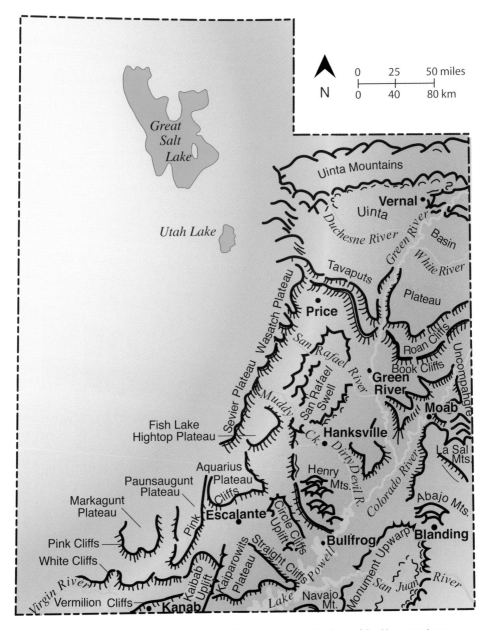

Within Utah's Colorado Plateau, uplifts include the broad anticlines of the Uncompahgre Plateau, San Rafael Swell, Circle Cliffs Uplift, and Monument Upwarp. Along the Plateau's western margin is a string of high plateaus that have some similarities with Utah's High Country, so their highways are described in the following chapter.

Colorado Plateau Stratigraphic Column

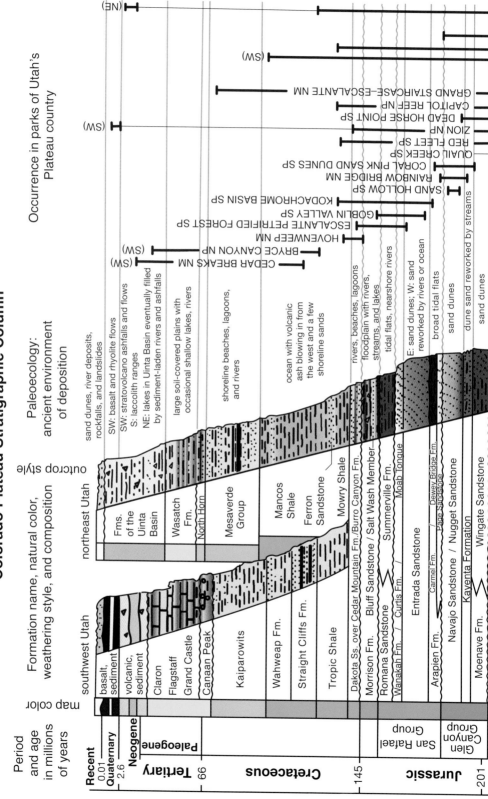

Period and age in millions of years — Recent 0.01, Quaternary, Neogene, Tertiary 2.6, Paleogene, Cretaceous 66, 145, Jurassic, San Rafael Group, Glen Canyon Group, 201

map color

Formation name, natural color, weathering style, and composition

southwest Utah: basalt, sediment; volcanic, sediment; Claron; Flagstaff; Grand Castle; Canaan Peak; Kaiparowits; Wahweap Fm.; Straight Cliffs Fm.; Tropic Shale; Dakota Ss. over Cedar Mountain Fm./Burro Canyon Fm.; Morrison Fm.; Bluff Sandstone / Salt Wash Member; Romana Sandstone; Wanakah Fm. / Curtis Fm.; Summerville Fm.; Moab Tongue; Entrada Sandstone; Carmel Fm.; Dewey Bridge Fm.; Page Sandstone; Arapien Fm.; Navajo Sandstone / Nugget Sandstone; Kayenta Formation; Moenave Fm.; Wingate Sandstone

northeast Utah: Fms. of the Uinta Basin; Wasatch Fm.; North Horn; Mesaverde Group; Mancos Shale; Ferron Sandstone; Mowry Shale

outcrop style

Paleoecology: ancient environment of deposition

- sand dunes, river deposits, rockfalls, and landslides
- SW: basalt and rhyolite flows
- SW: stratovolcano ashfalls and flows
- S: laccolith ranges
- NE: lakes in Uinta Basin eventually filled by sediment-laden rivers and ashfalls
- large soil-covered plains with occasional shallow lakes, rivers
- shoreline beaches, lagoons, and rivers
- ocean with volcanic ash blowing in from the west and a few shoreline sands
- rivers, beaches, lagoons
- floodplain with rivers, streams, and lakes
- tidal flats, nearshore rivers
- E: sand dunes; W: sand reworked by rivers or ocean
- broad tidal flats
- sand dunes
- dune sand reworked by streams
- sand dunes

Occurrence in parks of Utah's Plateau country

- CEDAR BREAKS NM (SW)
- BRYCE CANYON NP (SW)
- HOVENWEEP NM
- ESCALANTE PETRIFIED FOREST SP
- GOBLIN VALLEY SP
- KODACHROME BASIN SP
- SAND HOLLOW SP
- RAINBOW BRIDGE NM
- CORAL PINK SAND DUNES SP
- QUAIL CREEK SP
- RED FLEET SP
- ZION NP (SW)
- DEAD HORSE POINT SP
- CAPITOL REEF NP
- GRAND STAIRCASE-ESCALANTE NM
- (SW)
- (NE)

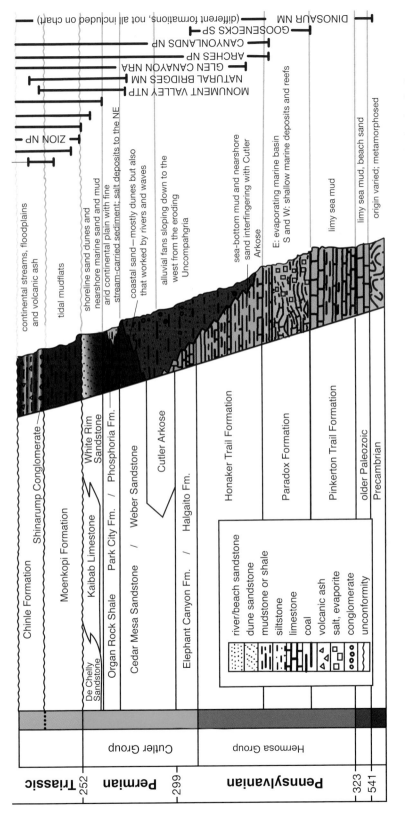

Geologists summarize an area's sedimentary rocks in a stratigraphic column. Generally speaking, red-hued layers were deposited on land, where iron in the sediment was rusted by oxygen into red or yellow shades (an oxidizing environment), while gray-hued ones accumulated where a limited oxygen supply left the iron dark (a reducing environment). Wavy black lines separating formations represent unconformities. This column applies to most areas of the Colorado Plateau, but because the geology of the High Country and Great Basin vary significantly from this sequence, this column cannot be reliably applied in those areas.

ZION NP

DINOSAUR NM (different formations, not all included on chart)
GOOSENECKS SP
CANYONLANDS NP
ARCHES NP
GLEN CANYON NRA
NATURAL BRIDGES NM
MONUMENT VALLEY NTP

continental streams, floodplains and volcanic ash

tidal mudflats

shoreline sand dunes and nearshore marine sand and mud

arid continental plain with fine stream-carried sediment; salt deposits to the NE

coastal sand—mostly dunes but also that worked by rivers and waves

alluvial fans sloping down to the west from the eroding Uncompahgria

sea-bottom mud and nearshore sand interfingering with Cutler Arkose

E: evaporating marine basin
S and W: shallow marine deposits and reefs

limy sea mud

limy sea mud, beach sand

origin varied; metamorphosed

Chinle Formation
Shinarump Conglomerate
Moenkopi Formation
Kaibab Limestone
White Rim Sandstone
De Chelly Sandstone
Organ Rock Shale
Park City Fm. / Phosphoria Fm.
Cedar Mesa Sandstone / Weber Sandstone
Cutler Arkose
Elephant Canyon Fm. / Halgaito Fm.
Honaker Trail Formation
Paradox Formation
Pinkerton Trail Formation
older Paleozoic
Precambrian

river/beach sandstone
dune sandstone
mudstone or shale
siltstone
limestone
coal
volcanic ash
salt, evaporite
conglomerate
unconformity

Triassic
252
Permian
299
Pennsylvanian
323
541

Cutler Group
Hermosa Group

Of What Use Can Such a Desert Be?

The Plateau is a geologist's heaven, for both its clearly exposed rock layers and structures and its bountiful energy resources. The area is ringed by coal beds in Cretaceous and Tertiary layers. Fields of oil and natural gas are trapped in Permian, Cretaceous, and Tertiary rocks. Deposits of radium, vanadium, and uranium occur in fossil river channels in Triassic and Jurassic layers. There is little gold or silver; however, copper is mined in the Lisbon Valley near the Colorado border.

Most people visit this wild country to live for a short time in its amazing landscape, to gaze in awe at its brilliantly colored spires and arches and shadowed purple canyons, and to feel its huge expanse below a wide blue sky. For these reasons, much of the area has been set aside in parks and other preserves. Most of the rest is public land managed by the US Forest Service and the Bureau of Land Management, so it is also accessible. The Colorado Plateau in Utah is a vast country, ready to be explored. It's sunny, hot, and dry, so bring shelter and plenty of water.

Road Guides to the Colorado Plateau

I-70
Colorado State Line—Green River
70 miles (113 km)

Passing into Utah, I-70 rides on a slope of tan Dakota Sandstone. Here, layers dip north off the Uncompahgre Uplift and into the Uinta Basin. Both the Dakota and the underlying Burro Canyon/Cedar Mountain Formation are Cretaceous in age. They can be seen at the milepost-229 viewpoint, which is at the crest of the uplift. The name Burro Canyon is only used east of the Colorado River.

Some highway cuts, particularly those near the turnoff to Westwater, reveal layers of low-grade coal. These started as swampy, forested areas behind the beaches that became Dakota Sandstone. The occasional light gray to white layers within the dark coal are volcanic ash, which preserved fossils of many forest plants that lived at the time. Both the Burro Canyon/Cedar Mountain Formation and Dakota Sandstone contain crossbedded sand deposited by water in river channels. The channels in the Dakota formed along the shore of an advancing Cretaceous sea. River-formed crossbedding is smaller in scale than dune crossbedding, and it is trough shaped.

The highway soon leaves the Dakota and stays on the overlying Mancos Shale all the way to Green River. This soft, fine gray rock was deposited as mud and silt in the Cretaceous Interior Seaway, which covered most of the western interior of North America during Cretaceous time, before the Rocky Mountains rose. To the north are the Book Cliffs. Their sandstone, shale, and coal

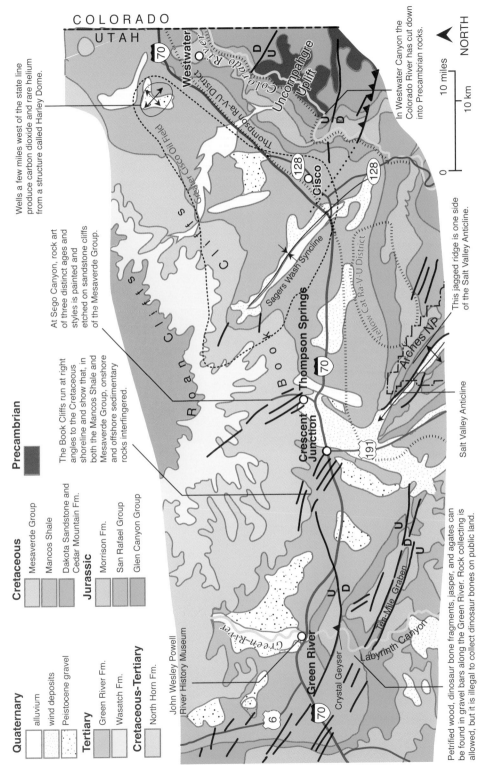

Quaternary
- alluvium
- wind deposits
- Pleistocene gravel

Tertiary
- Green River Fm.
- Wasatch Fm.

Cretaceous-Tertiary
- North Horn Fm.

Cretaceous
- Mesaverde Group
- Mancos Shale
- Dakota Sandstone and Cedar Mountain Fm.

Jurassic
- Morrison Fm.
- San Rafael Group
- Glen Canyon Group

Precambrian

COLORADO
UTAH

NORTH

Wells a few miles west of the state line produce carbon dioxide and rare helium from a structure called Harley Dome.

In Westwater Canyon the Colorado River has cut down into Precambrian rocks.

Uncompahgre Uplift

Colorado River

Westwater

Greater Cisco Oil Field

Thompson Ra-v-U District

Cisco

Sagers Wash Syncline

Yellow Cat Ra-V-U District

Arches NP

This jagged ridge is one side of the Salt Valley Anticline.

Salt Valley Anticline

At Sego Canyon, rock art of three distinct ages and styles is painted and etched on sandstone cliffs of the Mesaverde Group.

The Book Cliffs run at right angles to the Cretaceous shoreline and show that, in both the Mancos Shale and Mesaverde Group, onshore and offshore sedimentary rocks interfingered.

Roan Cliffs

Book Cliffs

Thompson Springs

Crescent Junction

Ten-Mile Graben

Labyrinth Canyon

John Wesley Powell River History Museum

Green River

Green River Fm.

Crystal Geyser

Petrified wood, dinosaur bone fragments, jasper, and agates can be found in gravel bars along the Green River. Rock collecting is allowed, but it is illegal to collect dinosaur bones on public land.

0   10 miles
0   10 km

Geology along I-70 between the Colorado state line and Green River.

beds, also Cretaceous in age, comprise the Mesaverde Group. Hidden above them is a broad bench of Tertiary Wasatch Formation and then, above that, the pale Roan Cliffs of the Green River Formation. Both were mostly deposited in large Tertiary lakes.

The Uncompahgre Uplift, to the south, is a high fault-edged plateau. It was shoved westward and upward, slightly overriding rocks to the west, in the same mountain building event that pushed up the Rockies: the Laramide Orogeny of Cretaceous through Tertiary time. The Colorado River does not quite manage to avoid the uplift's northern end. In Westwater Canyon, the river has carved all the way down into uplifted Precambrian rocks. Wells scattered in the Cisco area obtain oil and natural gas from Jurassic and Cretaceous rocks that are folded and faulted along the uplift's western edge.

Let's take a look at the Book Cliffs to the north. Cliffs in the gently tilted layers of the Plateau erode from the side rather than the top. As weak rock layers on the slopes erode back, harder layers above are undermined and eventually break away. Water from storms carries some of the debris down the usually dry washes. Wind helps by carrying away the small grains—witness the roadside dust storm warnings. Over thousands and even millions of years, the cliffs retreat. The Book Cliffs were once south of where the highway is now!

The cliffs, composed of the Mancos Shale and Mesaverde Group, run at right angles to the Cretaceous shoreline along which the formations were deposited. Since they were deposited along a shoreline, offshore and onshore deposits interfinger in the formations. Uplifts to the west furnished their sand and the silt, and plants that grew in lagoons and swamps later became coal.

A long ridge of red rocks on the skyline southwest of milepost 194 is in Arches National Park. It is composed of the Jurassic Entrada Sandstone, which has weathered into the jagged fins in which the famous arches formed. The low, hilly area north of the ridge is underlain by several formations, including the Morrison.

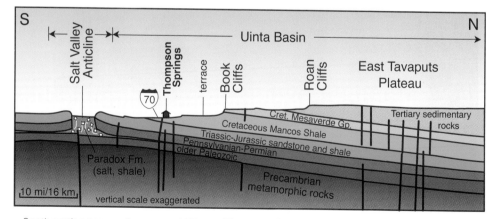

South-north cross section across I-70 near Thompson Springs. I-70 rides on soft beds of Mancos Shale next to the 1,500-foot-tall (460 m) Book Cliffs. The interstate skirts the edge of the Uinta Basin, where Earth's surface bowed down south of the Uinta Mountains during the Laramide Orogeny.

The Salt Wash Member of the Morrison Formation hosts many small deposits of radioactive minerals. Trace amounts of vanadium and uranium came from light-colored feldspar minerals and volcanic ash in the region's rocks. Dissolved by groundwater, they were redeposited as black pitchblende and light-yellow carnotite near and in organic material in river channels. Radium was produced by the decay of the uranium. Radium was mined here starting in the early part of the twentieth century. In the 1930s and through World War II, vanadium was mined for the purpose of strengthening steel. In the 1950s and 1960s, uranium had the limelight because of the Cold War and the proliferation of nuclear weapons.

With a Geiger counter to detect radioactivity, it was easy to locate these small deposits. Mills were built at Moab and Green River, near water, and bought ore from all who brought it in. Closed decades ago, these mills have been dismantled and cleaned up. The radioactive waste from the Moab mill is being moved to a safe, engineered cell dug in the ground near Crescent Junction, away from the Colorado River. There, it will be covered with 9 feet (3 m) of inactive sediment. But now, as the price of uranium increases and Cold War stockpiles shrink, mining interest and activity have been renewed here.

North of the highway the Book Cliffs, formed in layered rocks of the Mesaverde Group, rise above rounded hills of Mancos Shale. One of the shale layers contains more iron, so it weathers to a yellow color. —Felicie Williams photo

The rest area west of milepost 188 is a good place see the Book Cliffs and the Roan Cliffs. The viewpoint is surrounded by hilly badlands of Mancos Shale. Above the shale are cliffs and slopes of the Mesaverde Group, which were shoreline deposits that formed as the sea retreated. A weak layer of siltstone and coal, and then more sandstone, overlie the lower sandstone layers of the Mesaverde. The sandstones were beaches and sandbars, while the coal and siltstone accumulated in marshes and swamps where plants were abundant. Dinosaur trackways found in coal mines near here show that a giant with a three-toed, 15-foot (4.6 m) stride walked these shorelines.

Some of the Mancos Shale hilltops are beveled and wear a topcoat of redbrown gravel. These buttes are remnants of a Pleistocene pediment that sloped downward from the Book Cliffs.

The Green River Formation, which forms the higher-up Roan Cliffs, is famous for its thick layers of oil shale. They hold an estimated 1.8 trillion barrels of oil—about 800 billion barrels of it recoverable—making them the world's richest oil shale deposit. They contain enough oil to supply our country's demand for four hundred years. You can smell the oil in the rock, and the rock will even burn! The oil is tightly held as a waxy hydrocarbon solid known as kerogen in the shale's extremely small pore spaces. The rock must be heated to get the oil to flow out, an expensive process and difficult to do in a way that is environmentally sound. Oil companies have researched ways to extract the oil in place, by fracturing and heating the rock, but so far no method has proven economical.

As you have probably noticed, areas surfaced with Mancos Shale support few plants. There are three reasons for this: the arid climate, with only 5 to 6 inches (13 to 15 cm) of rainfall annually; the high selenium, salt, and alkali content of the ocean-deposited shale; and clay that swells when wet and shrinks when dry, making it hard for plants to stay rooted.

Edged with trees, the Green River flows quietly through the town of Green River. With headwaters in the Wind River Range of Wyoming, it is by far the largest of the Colorado's tributaries, often carrying more water than the Colorado itself. It cuts through the Book Cliffs in Desolation Canyon, named in 1869 by John Wesley Powell, a Civil War veteran and geologist who boldly led the first exploratory trip down the Green and Colorado Rivers. The John Wesley Powell River History Museum east of town has excellent displays on the history and geology of the Green River.

I-70
Green River—Fremont Junction
73 miles (117 km)

There is no better place to look at Permian, Triassic, Jurassic, and Cretaceous rocks than the San Rafael Swell, the long, low rise west of Green River. Pushed eastward between 65 and 56 million years ago during the Laramide Orogeny, it rose into an asymmetrical anticline 100 miles (160 km) long from north to south and about 45 miles (72 km) wide.

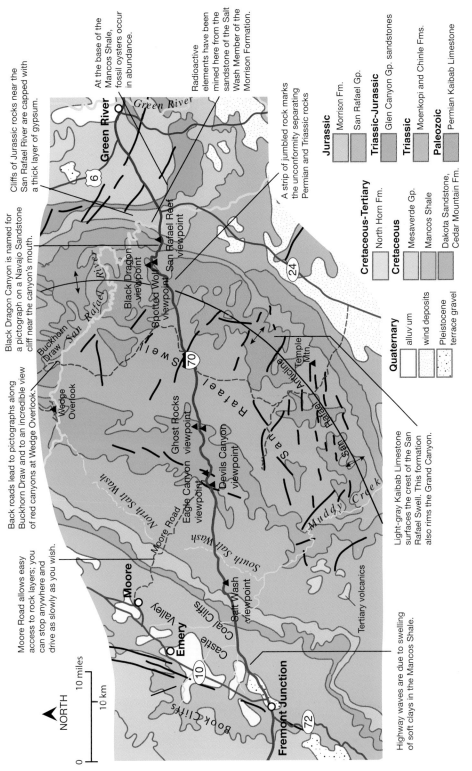

At the base of the Mancos Shale, fossil oysters occur in abundance.

Radioactive elements have been mined here from the sandstone of the Salt Wash Member of the Morrison Formation.

A strip of jumbled rock marks the unconformity separating Permian and Triassic rocks

Cliffs of Jurassic rocks near the San Rafael River are capped with a thick layer of gypsum.

Black Dragon Canyon is named for a pictograph on a Navajo Sandstone cliff near the canyon's mouth.

Back roads lead to pictographs along Buckhorn Draw and to an incredible view of red canyons at Wedge Overlook.

Moore Road allows easy access to rock layers; you can stop anywhere and drive as slowly as you wish.

Light-gray Kaibab Limestone surfaces the crest of the San Rafael Swell. This formation also rims the Grand Canyon.

Highway waves are due to swelling of soft clays in the Mancos Shale.

Jurassic
Morrison Fm.
San Rafael Gp.

Triassic-Jurassic
Glen Canyon Gp. sandstones

Triassic
Moenkopi and Chinle Fms.

Paleozoic
Permian Kaibab Limestone

Cretaceous-Tertiary
North Horn Fm.

Cretaceous
Mesaverde Gp.
Mancos Shale
Dakota Sandstone, Cedar Mountain Fm.

Quaternary
alluvium
wind deposits
Pleistocene terrace gravel

NORTH

0   10 miles
0   10 km

Green River
Green River
6

San Rafael River
Buckhorn Draw
Wedge Overlook

San Rafael Reef viewpoint
Black Dragon viewpoint
Spotted Wolf viewpoint

San Rafael Swell

70

San Rafael Anticline

Temple Mtn

Muddy Creek

24

Ghost Rocks viewpoint
Eagle Canyon viewpoint
Devils Canyon viewpoint

Moore Road
North Salt Wash
South Salt Wash

Salt Wash viewpoint

Moore
Emery
10
Castle Valley
Coal Cliffs
Castle Cliffs

Book Cliffs
Fremont Junction
72

Tertiary volcanics

Geology along I-70 between Green River and Fremont Junction.

The Swell exposes around 7,800 feet (2,380 m) of sedimentary layers. As I-70 crosses the San Rafael Swell, it travels through older and older rock layers to the summit, and then back through the same layers down the other side.

I-70 heads west from Green River on soft Cretaceous Mancos Shale: dark-gray marine shale that contains coal and crossbedded nearshore sandstone. The shale also contains fossil mollusks and ammonites. A prominent bed of oyster shells near its base, at about milepost 150, forms a line of low, rounded knolls. The oysters indicate a shallow environment, such as a shoal, and are sometimes so abundant that they are quarried for road material.

Near the Utah 24 interchange, watch for pastel-colored shales and sand-stones of the Cretaceous Cedar Mountain and Jurassic Morrison Formations. Both formations were deposited on river floodplains. The beds that are soft shades of green were buried relatively quickly, and the decaying of abundant plant material in them used up oxygen, thereby preventing the oxidation (rust-ing) of the iron minerals they contain. Other beds remained in contact with the atmosphere, so the iron in them oxidized to red and purple. All the beds con-tain a large proportion of volcanic ash, which was blown into the region from the volcanoes whose roots are now the Sierra Nevada of California and Nevada. Dinosaur fossils are found in these formations all over the West.

Near the bridge over the San Rafael River there are castle-like, dark-red siltstone cliffs of Jurassic Summerville Formation capped with a massive bed of gypsum. The siltstone and gypsum were deposited on tidal flats along the shore of a desert-bordered sea that stretched from here to northern Canada.

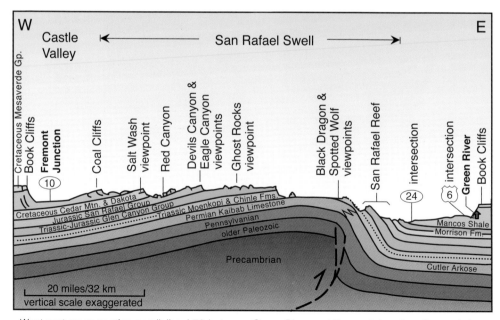

West-east cross section paralleling I-70 between Green River and Fremont Junction. Precambrian rocks far beneath the surface rose eastward along a fault, causing the San Rafael Swell's asymmetry.

The gypsum layer, as well as thin gypsum veins that penetrate the siltstone, was left by the evaporation of seawater in isolated embayments. The evaporating sea must have left salt deposits too, but the more soluble salt has long since washed away.

The eastern flank of the San Rafael Swell looms high as you approach it. The Swell is edged with a formidable sandstone barrier—the San Rafael Reef—jutting up 1,200 feet (370 m) high. The Reef made the entire anticline relatively inaccessible before I-70 was constructed. Even now, the interstate is the only highway to cross it.

From the viewpoint near milepost 144, you can see a quick succession of the Reef's Jurassic rocks. The viewpoint is on the pale greenish-tan, sandy Curtis Formation, which was deposited in fairly shallow ocean water. Below it, an angular unconformity, or erosional surface, separates the Curtis from red beds of the Entrada Sandstone, deposited on coastal flats in a desert climate. Much of it was dune sand, but it graded westward into tidal flats. Often found as an erosion-resistant cliff former, here it is quite soft and erodes into a wide racetrack valley. Beneath the Entrada, dark-red Carmel Formation beds record more tidal flats, in which sand and silt were deposited. Below the Carmel Formation is the Page Sandstone, also deposited in sand dunes. Another unconformity separates the Page from the Navajo Sandstone.

I-70 next slices through the colossal cockscombs of yellow-white Page and Navajo Sandstones and then through red sharp-edged cliffs of Triassic-Jurassic Wingate Sandstone; it is dark and shiny with desert varnish. All three of these formations are eolian (wind-deposited) and thus marked by the broad, sweeping crossbeds of ancient sand dunes. They formed along the southeastern shore of a Jurassic sea, at latitudes similar to today's Sahara. The continent's drift and rotation have since carried these former dunes far to the north. Between the Navajo and the Wingate are thin layers of flat, stream-deposited sandstone and siltstone of the Kayenta Formation.

West of the reef is another racetrack valley, this one in the dark, brick-red Triassic Chinle and Moenkopi Formations, though here the lower Moenkopi beds have been bleached a yellowish color. Both formations are composed of easily eroded mudstone and siltstone with occasional ledges of limestone. The Chinle also includes sandstone and, at its base, a conglomerate that represents a period of higher-energy erosion. Uranium is found in many places in the Swell in the Chinle Formation. It is concentrated in sandstone deposited in former river channels, and in the conglomerate, both places where there had been abundant decaying organic material such as leaves and wood.

Climbing toward the crest of the San Rafael Swell, I-70 emerges from the Triassic red beds onto light-gray to buff Permian Kaibab Limestone, which is well exposed around the rest areas west of milepost 140. It makes slabs of rock, is fairly sandy, and was probably deposited near the shore. Farther west it contains less sand. From the Black Dragon viewpoint the Permian White Rim Sandstone, the oldest rock exposed on the Swell, is visible in the canyons. Both it and the Kaibab extend south into Arizona, where they form the rim of the Grand Canyon; there the White Rim is called the Coconino Sandstone. The White Rim is prominent in Canyonlands as well, where nonmarine Triassic

Jurassic Navajo Sandstone
Jurassic Kayenta Fm.
Triassic-Jurassic Wingate Ss.
Triassic Chinle Fm.
Triassic Moenkopi Fm.
Permian Kaibab Limestone

West of the Navajo outcrops, an angular cliff mostly composed of Triassic-Jurassic Wingate Sandstone towers over ledges and slopes of older Triassic rock. —Felicie Williams photo

Permian Kaibab Limestone surfaces much of the crest of the San Rafael Swell. It is a fairly sandy limestone here, its eastern outcrops having been deposited near the shore. E. D. McKee's rock hammer for scale. —Felicie Williams photo

rocks overlie it; the ocean that deposited the Kaibab did not extend that far to the east.

About 1 mile (1.6 km) west of the Devils Canyon and Eagle Canyon viewpoints, visible only along the eastbound lanes, is an interesting band of rock containing chunks of Kaibab Limestone, sandstone, and chert, oriented in all directions and embedded in sand. This deposit is probably a karst surface that developed when limestone, exposed to the atmosphere by a drop in sea level, dissolved and caved in, creating the jumbled rock.

West of the summit, we descend back through the same rock layers. Since the rocks dip more gently on the west side of the San Rafael Swell, the outcrop bands are considerably wider. In order, watch for the dark-red Triassic Moenkopi and Chinle Formations, covered in places with fine, pink windblown sand; the massive Wingate Sandstone with its sharp-edged cliffs; the Kayenta Formation's light-colored, finely bedded sandstone; and the dramatic yellow-white knobs of the Navajo and Page Sandstones, both marked with long crossbedding.

Next comes the Jurassic San Rafael Group, named for its superb exposures on the Swell: The dark-red Carmel Formation siltstone and the Entrada Sandstone together form the wide Red Valley. From the Salt Wash viewpoint, hatted sentinels weathered in the pale-greenish sandstone of the Curtis Formation gaze eastward. The Curtis Formation also forms a gently dipping cuesta near milepost 107. West of here the Summerville Formation is again recognized by

In Eagle Canyon, dark red Triassic beds are overlain by the three sandstones of the Glen Canyon Group: cliffs of Wingate, shelves of Kayenta, and a rounded metropolis of Navajo. In the far distance across Castle Valley, Cretaceous rocks form the Book Cliffs. Taken from the Eagle Canyon viewpoint. —Felicie Williams photo

its castlelike bluffs and superabundant gypsum. Fabulous trackways of ptero-saurs have been found in the Summerville on this side of the Swell.

Castle Valley and its badlands expose the weak but colorful Jurassic Mor-rison and Cretaceous Cedar Mountain Formations. In recent years, more than eighty new fish, amphibians, lizards, dinosaurs, birds, and mammals have been discovered in the Cedar Mountain Formation on this side of the Swell. Creta-ceous Dakota Sandstone caps a west-dipping cuesta at the bridge across Muddy Creek. In roadcuts west of the bridge there are coal layers in the Mancos Shale; they are thicker here than on the east side of the Swell. As you get closer to Fremont Junction, scattered dark boulders of weathered volcanic rock, eroded from high lava-covered plateaus to the southwest, herald the beginning of the faulted transition zone separating the Colorado Plateau from Basin and Range country.

US 6/US 191
Green River—Price
57 miles (92 km)

From its junction with I-70, US 6/US 191 heads northward into a wide, empty valley floored with Cretaceous Mancos Shale, which it follows all the way to Price. The broad anticline of the San Rafael Swell rises to the west, with steeply dipping Jurassic sandstone seemingly rolling off its eastern flank. Deep below, less-flexible Precambrian rocks are faulted. The Swell was pushed up during the Laramide Orogeny, when rock west of here was forced eastward up a long, curving fault plane.

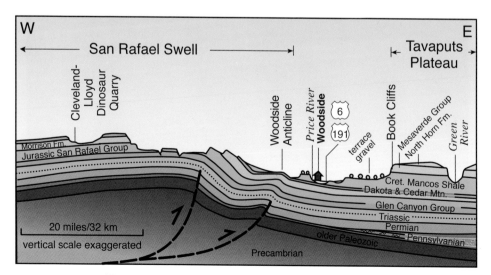

West-east cross section across US 6/US 191 near Woodside.

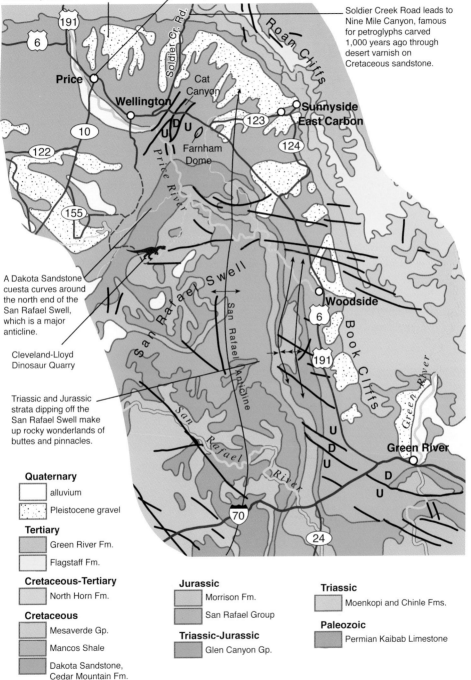

Beveled hills of Mancos Shale are covered with reddish-brown Pleistocene gravel. The mustard-yellow parts of the shale are colored by oxidized iron minerals.

Among the many exhibits at the Utah State University Eastern Prehistoric Museum are dinosaur footprints found in local coal mines.

NORTH

0   10 miles

10 km

Soldier Creek Road leads to Nine Mile Canyon, famous for petroglyphs carved 1,000 years ago through desert varnish on Cretaceous sandstone.

191

6

Soldier Cr. Rd.

Roan Cliffs

Price

Cat Canyon

Wellington

U   D   u

123

Sunnyside
East Carbon

10

Farnham Dome

124

122

155

Price River

San Rafael Swell

San Rafael Anticline

Woodside

6

Book Cliffs

191

A Dakota Sandstone cuesta curves around the north end of the San Rafael Swell, which is a major anticline.

Cleveland-Lloyd Dinosaur Quarry

Triassic and Jurassic strata dipping off the San Rafael Swell make up rocky wonderlands of buttes and pinnacles.

San Rafael River

Green River

Green River

U
D
U

D
U

70

24

Quaternary

☐ alluvium

▦ Pleistocene gravel

Tertiary

☐ Green River Fm.

☐ Flagstaff Fm.

Cretaceous-Tertiary

☐ North Horn Fm.

Cretaceous

☐ Mesaverde Gp.

☐ Mancos Shale

☐ Dakota Sandstone,
  Cedar Mountain Fm.

Jurassic

☐ Morrison Fm.

☐ San Rafael Group

Triassic-Jurassic

☐ Glen Canyon Gp.

Triassic

☐ Moenkopi and Chinle Fms.

Paleozoic

☐ Permian Kaibab Limestone

Geology along US 6/US 191 between Green River and Price.

To the east are the Book Cliffs, with sandstone layers of the Cretaceous Mesaverde Group above the Mancos Shale. Still higher and out of sight from most parts of the highway are the pale Roan Cliffs, composed of Tertiary (Paleocene and Eocene) lake deposits. The successive layers of the Mesaverde show that the Cretaceous Interior Seaway was receding from this area when its sediments were deposited: the lowest layers are sandy beach and sandbar deposits; the middle layers contain coal derived from plant material, which was deposited in coastal swamps behind beaches; and the top layers consist of floodplain sediments, deposited in rivers headed for the departed ocean.

Hydrocarbons are abundant in this region. Oil and gas are produced mostly from Cretaceous and Tertiary rocks in the Uinta Basin, which is to the northeast behind the Book and Roan Cliffs. Coal is mined from sandstones of the Mesaverde Group near East Carbon and Sunnyside

The Uinta Basin includes layers of sandstone with abundant asphalt, or tar; these hydrocarbon-rich beds are from 10 to 300 feet (3 to 90 m) thick. Called tar sands, these layers have sparked interest since they were discovered over one hundred years ago, but extracting the tar has not been profitable. A current proposal involves dissolving the tar in a citrus-based solvent. The Green River Formation of the Roan Cliffs contains a vast untapped reserve of oil shale, but it also has not proven profitable. The oil is so thick and waxy that it must be heated above 640°F to 700°F (340°C to 370°C) before it will flow out of the shale. The tar sand and oil shale deposits draw considerable attention each time oil prices rise.

Above slopes of Mancos Shale, sandstone ledges of the Mesaverde Group form the Book Cliffs. The strata dip very gently northeast, into the Uinta Basin. View is to the northeast. —Felicie Williams photo

The Tertiary Green River Formation was deposited between 52 and 45 million years ago in a large freshwater lake between rising mountain ranges. Volcanic ash helped fill the lake, preserving many fossils, including insects, fish, leaves, and even birds and feathers. From top to bottom: Chlorocyphidae, *unidentified flower and leaf, and* Gosiutichthys. —Photos courtesy of the Virtual Fossil Museum, www.fossilmuseum.net

Wells in a few places in this broad valley, including near Woodside, produce carbon dioxide from the Navajo Sandstone and older rocks. It is used to make dry ice and in the oil-drilling industry.

Curving northwest of Woodside, the road crosses several terraces of pinkish gravel above Mancos Shale. The terraces are remnants of a Pleistocene pediment, an erosional slope west of the Book Cliffs topped with gravel. Streams carried the gravel here from the east during the wetter Pleistocene. Since then, erosion has cut down through the graveled surface.

As the highway gradually curves around the north end of the San Rafael Swell, notice the rock layers dipping downward along the edge of the Swell. Just before Wellington the road crosses Farnham Dome, a small anticline from which significant amounts of carbon dioxide have been produced. Like the San Rafael Swell, the dome formed during the Laramide Orogeny.

Price lies near the western edge of the Colorado Plateau. Beyond Price, the faulted Wasatch Plateau marks the beginning of the transition zone separating the Basin and Range from the Colorado Plateau.

US 40
Heber City—Duchesne
70 miles (113 km)

Starting in Utah's High Country, US 40 crosses the edges of the Uinta Mountains and the Wasatch Range and passes onto the Colorado Plateau. Thrust faulting reached this region during the Sevier Orogeny; the Charleston-Nebo Thrust Fault lies beneath the valley just south of Heber City, coming to the surface near Strawberry Reservoir to the east. The east-west-oriented Uinta Anticline, which forms the core of the Uinta Mountains, lies just north of Heber City; the anticline formed during the Laramide Orogeny.

When tectonic pressure from the west eased following the Sevier and Laramide Orogenies, the thrust sheets that had been forced eastward—crumpling the crust and forming mountains in the process—slid back a little way to the west. This decreased the pressure beneath the surface. Since deeply buried rock melts below a certain pressure, some crust melted, and the resulting large pockets of magma rose upward. Some of the magma crystallized as large stocks in the central Wasatch Range, but remnants of volcanic flows are also preserved in the low area between the Uinta Mountains and the Wasatch. Within the last 17 million years, the long north-south Heber Valley has dropped downward due to Basin and Range extension.

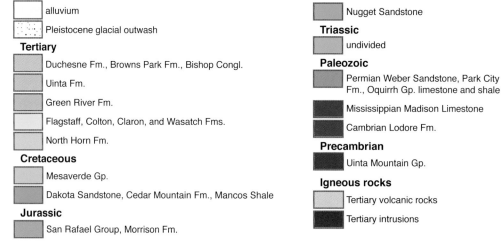

Quaternary
- alluvium
- Pleistocene glacial outwash

Tertiary
- Duchesne Fm., Browns Park Fm., Bishop Congl.
- Uinta Fm.
- Green River Fm.
- Flagstaff, Colton, Claron, and Wasatch Fms.
- North Horn Fm.

Cretaceous
- Mesaverde Gp.
- Dakota Sandstone, Cedar Mountain Fm., Mancos Shale

Jurassic
- San Rafael Group, Morrison Fm.

Triassic-Jurassic
- Nugget Sandstone

Triassic
- undivided

Paleozoic
- Permian Weber Sandstone, Park City Fm., Oquirrh Gp. limestone and shale
- Mississippian Madison Limestone
- Cambrian Lodore Fm.

Precambrian
- Uinta Mountain Gp.

Igneous rocks
- Tertiary volcanic rocks
- Tertiary intrusions

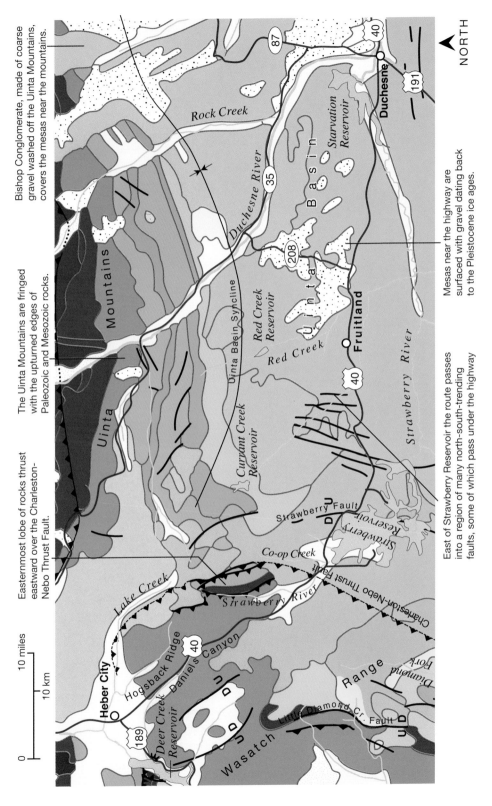

Bishop Conglomerate, made of coarse gravel washed off the Uinta Mountains, covers the mesas near the mountains.

The Uinta Mountains are fringed with the upturned edges of Paleozoic and Mesozoic rocks.

Easternmost lobe of rocks thrust eastward over the Charleston-Nebo Thrust Fault.

Mesas near the highway are surfaced with gravel dating back to the Pleistocene ice ages.

East of Strawberry Reservoir the route passes into a region of many north-south-trending faults, some of which pass under the highway.

NORTH

10 miles

10 km

Duchesne

Starvation Reservoir

Duchesne River

Rock Creek

Uinta Mountains

Uinta Basin

Uinta Mountains

Red Creek Reservoir

Currant Creek Reservoir

Uinta Basin Syncline

Red Creek

Fruitland

Strawberry River

Strawberry Fault

Strawberry Reservoir

Co-op Creek

Lake Creek

Strawberry River

Charleston-Nebo Thrust Fault

Heber City

Hogsback Ridge

Daniels Canyon

Deer Creek Reservoir

Wasatch Range

Diamond Fork

Little Diamond Cr. Fault

Geology along US 40 between Heber City and Duchesne.

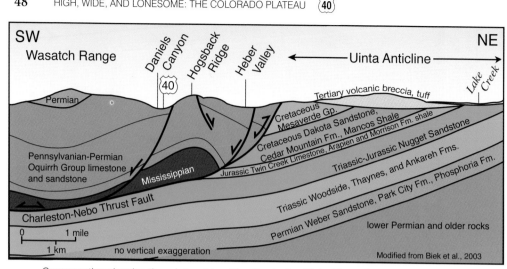

Cross section showing the relationship of the Charleston-Nebo Thrust Fault (active during the Sevier Orogeny), the Uinta Anticline (uplifted during the Laramide Orogeny), and the Basin and Range down-dropping of Heber Valley (faults active within the last 17 million years).

East of Heber Valley, US 40 heads up Daniels Canyon between massive ridges of dark-gray Oquirrh Group siltstone, sandstone, and limestone of Pennsylvanian-Permian age. These strata are well east of their birthplace, having been pushed here above the Charleston-Nebo Thrust Fault during the Sevier Orogeny. The rocks are sliced by many lesser faults and alternate with Triassic rocks also brought eastward by the thrust faulting. Vegetation and numerous rockslides and landslides make the rocks hard to see.

Just east of the pass above Daniels Canyon, the hills to the north comprise the last prominent lobe of Triassic, Pennsylvanian, and Permian rocks that were thrust eastward along the Charleston-Nebo Thrust Fault. Winding downhill toward Strawberry Reservoir, we enter the Colorado Plateau proper.

The reservoir, completed in 1916, lies at the east portal of an extensive tunnel and aqueduct system called the Central Utah Project. The system conveys water from the Strawberry and Duchesne Rivers westward through the mountains to the cities of the Wasatch Front.

Between Strawberry Reservoir and Duchesne, US 40 follows a surface of Tertiary Uinta Formation gravel, sand, and silt washed south off the Uinta Mountains. These strata, as would be expected, coarsen northward toward their source. As streams and rivers exit mountains and enter flatter terrain, they lose energy. They drop the largest sediments first from the load they carry.

The Uinta Mountains are the erosional remnant of a long east-west-trending anticline pushed up during the Laramide Orogeny. Paleozoic and Mesozoic sedimentary rocks flank the mountains; irregular ridges at the base of the mountains are the sedimentary rocks' upturned edges. Harder beds form cuestas and hogbacks; softer ones have eroded into racetrack valleys that encircle the range.

Some of the mesas between the highway and the Uinta Mountains are capped with gravel and sand that washed off the mountains in Pleistocene time, and the flat, smooth-surfaced valleys near Fruitland are glacial outwash

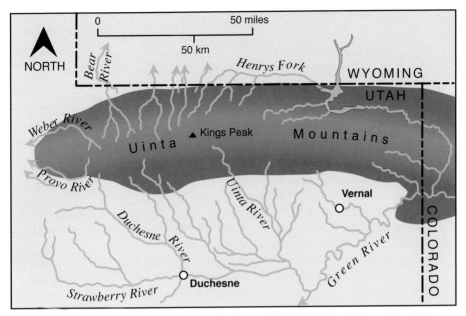

Except for water diverted to the west by the Central Utah Project (not pictured here), the abundant outflow from the south flank of the Uinta Mountains feeds into the Green River.

Co-op Creek valley, north of Strawberry Reservoir, formed as a long, broad outwash plain like many seen today in Alaska, where braided streams of glacial meltwater carry abundant sediment. Splays of the Charleston-Nebo Thrust Fault lie beneath the valley floor. —Lucy Chronic photo

plains, broad valleys that abundant glacial meltwater cut down through both the glacial sand and gravel and the Uinta Formation; the valleys are glutted by sediment, so they are very flat. During the Pleistocene the Uinta Mountains had the largest glaciated area in Utah, with extensive glaciers in the mountains.

After passsing the Strawberry Reservoir, US 40 enters the Uinta Basin, which is south of the Uinta Mountains. The basin is a major oil and gas province. Numerous wells dot the area between Fruitland and Duchesne. The hydrocarbons originate in Tertiary shale that is particularly rich in organic matter.

Starvation Reservoir, outside of Duchesne, is also part of the Central Utah Project. Because the Uinta Basin is more than 1,000 feet (300 m) higher than the Great Basin west of the mountains, its water flows westward by gravity through the tunnels and aqueducts of the Central Utah Project, with enough energy to turn electricity-generating turbines along the way.

Near milepost 83 there are good views of gently sloping mesas northeast of Duchesne, stretching like long fingers from the mountains. The highest ones are capped with Oligocene Bishop Conglomerate. The lower ones, of which there are two levels near Duchesne, are capped with younger Pleistocene gravel.

US 40
Duchesne—Colorado State Line
88 miles (142 km)

East of Duchesne, the land gains the wide-open feeling of the tectonically peaceful Colorado Plateau. This section of US 40 crosses the Uinta Basin, where Earth's crust sank downward south of the Uinta Anticline during the Laramide Orogeny. The basin is filled with up to 22,000 feet (6,700 m) of Tertiary sedimentary rock. Much of it accumulated in Lake Uinta, which occupied the basin for a great part of Tertiary time. Fine-grained lake deposits interfinger with the coarse alluvium shed from the Uinta Mountains. The lake deposits include parts of the North Horn, Flagstaff, Wasatch, Green River, and Uinta Formations. A series of soil and alluvial deposits that filled the last of the basin compose the upper Uinta and Duchesne River Formations. The Uinta Basin is one of the primary sources of oil and gas in Utah; wells and drilling equipment are a common sight along this route.

At Duchesne US 40 crosses from the valley of the Strawberry River into the Duchesne River drainage. Both rivers drain the south side of the Uinta Mountains. The Strawberry and part of the Duchesne are diverted into Starvation Reservoir and thence through tunnels and aqueducts of the Central Utah Project.

Terraces sloping downward from the Uintas are capped with thick layers of Pleistocene gravel scoured from the mountain summits when they were cloaked with glaciers. The highway crosses the Duchesne River at Myton and ascends some of these terraces; road cuts expose the cobbly gravels. Near the Uintas, the gravels contain cobbles; however, terraces farther south—away from the mountains—are finer grained.

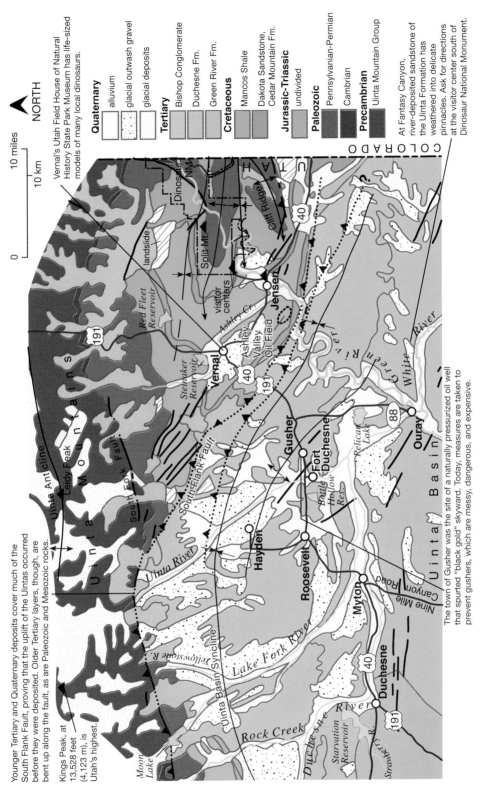

Quaternary
alluvium
glacial outwash gravel
glacial deposits

Tertiary
Bishop Conglomerate
Duchesne Fm.
Green River Fm.

Cretaceous
Mancos Shale
Dakota Sandstone, Cedar Mountain Fm.

Jurassic-Triassic
undivided

Paleozoic
Pennsylvanian-Permian
Cambrian

Precambrian
Uinta Mountain Group

NORTH

10 miles
10 km

Younger Tertiary and Quaternary deposits cover much of the South Flank Fault, proving that the uplift of the Uintas occurred before they were deposited. Older Tertiary layers, though, are bent up along the fault, as are Paleozoic and Mesozoic rocks.

Kings Peak, at 13,528 feet (4,123 m), is Utah's highest.

Vernal's Utah Field House of Natural History State Park Museum has life-sized models of many local dinosaurs.

At Fantasy Canyon, river-deposited sandstone of the Uinta Formation has weathered into delicate pinnacles. Ask for directions at the visitor center south of Dinosaur National Monument.

The town of Gusher was the site of a naturally pressurized oil well that spurted "black gold" skyward. Today, measures are taken to prevent gushers, which are messy, dangerous, and expensive.

Geology along US 40 between Duchesne and the Colorado state line.

The Uinta Mountains owe their east-west orientation to events that occurred 770 to 740 million years ago, when an east-west extensional basin formed here. The basin filled with as much as 28,000 feet (8,500 m) of sediment, which became the Precambrian Uinta Mountain Group, not to be confused with the Tertiary Uinta Formation that is found in the Uinta Basin, south of the mountains.

During the Laramide Orogeny, long after the sediment had turned to stone, pressure from the west-southwest forced the great block of rock up into a single large anticline, which is about 160 miles (260 km) long from east to west and 30 to 40 miles (50 to 64 km) wide. It has faults on its north and south sides. The top of the Precambrian surface gradually rose 3.4 miles (5.5 km) above sea level, while in the Uinta Basin it ultimately bowed down 5 miles (8 km) below sea level.

Soon after the range formed, it was virtually buried by layers and layers of mud, sand, and cobbles that washed off its peaks. In Miocene time, erosion began to eat away at these soft sediments, and they were eroded off the great fold's crest, exposing its hard core of the Uinta Mountain Group sandstone, shale, and conglomerate.

North of Roosevelt, fine pinkish sandstone and siltstone of the Tertiary Uinta Formation show beneath the cap of Pleistocene gravel. These beds contain middle Eocene–age fossils that record the origins of most modern mammal groups. Here, the beds dip toward the mountains, off a small anticline, the crest of which roughly parallels the mountain front. This little anticline is an oil trap and is at the eastern end of a large oil field. The town of Hayden, north of Roosevelt, is named after the geologist who, between 1867 and 1879, led the first geological and geographical surveys of the western territories.

Oil wells in the Uinta Basin produce what geologists call "Uinta Basin black wax crude oil." It is warm and liquid but solidifies after about four hours of cooling. Crude oil, direct from the well, is a blend of many chemical compounds,

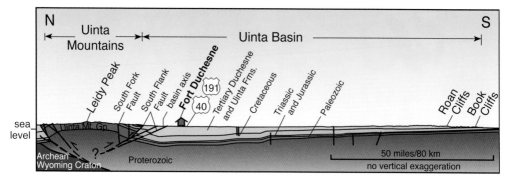

North-south cross section across the Uinta Mountains and the Uinta Basin. For many of the Colorado Plateau highways in this book, the accompanying cross sections are drawn like cartoons, with a lot of vertical exaggeration so the different formations can be seen. This cross section, with no vertical exaggeration, loses that detail but shows well the contrast between the fairly flat-lying Colorado Plateau and the Uinta Mountains.

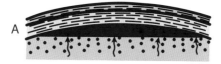

Oil migrates upward in permeable rocks such as sandstone or cavernous limestone, floating above groundwater until it is (A) trapped in anticlines, (B) along faults, or (C) by impermeable rock, usually shale. Areas of permeable rock act as natural reservoirs.

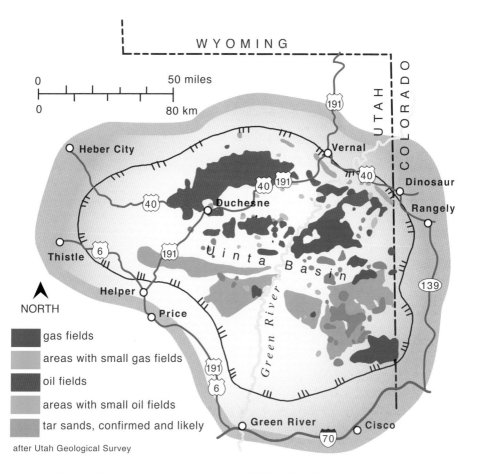

Hydrocarbons of the oil, gas, and tar sand fields of the Uinta Basin are mostly produced from Tertiary rocks.

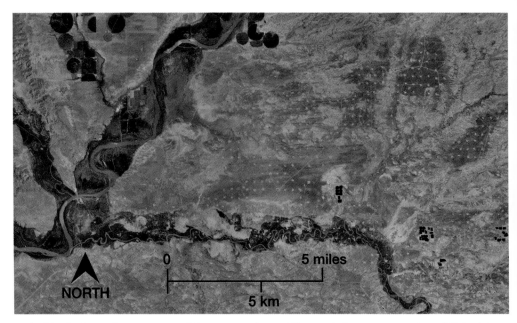

In this air photo of an area east of Ouray, natural gas drilling pads form a pattern of small light dots. Folds, faults, and impermeable beds trap gas the same way they do oil. A natural gas boom began in the Uinta Basin in 1995, fueled by a surge in price and demand as power plants converted from coal to gas. —USDA-FSA-APFO photo

Production operations on the ground. —Felicie Williams photo

only a few of which are useful without refining. During refining, large oil molecules are broken apart by high temperature and pressure and by chemical cracking and then regrouped to make fuels, oils, greases, and chemicals.

Badlands in soft Tertiary siltstones have developed due to the arid climate and the volcanic ash content of the rock, which weathers into swelling clay that discourages vegetation from taking root. Watch highway cuts for faults and small folds. Most of the pinkish rocks belong to the Uinta Formation; they include massive crossbedded sandstone, conglomerate, and easily eroded mudstone layers.

East of Vernal, sedimentary rocks are steeply folded and faulted along the southern flank of the Uinta Mountains. One fold is called Split Mountain because the Green River has cut a sheer-walled passage through it. At one time Tertiary deposits covered this part of the Uintas. The river carved downward into them and established its course. By the time the river reached the harder, older rocks, it could not escape its canyon to find an easier way around the anticline.

The south flank of Split Mountain is surfaced with pale Pennsylvanian-Permian Weber Sandstone. The sweeping crossbeds of the Weber prove it originated as a field of windblown sand dunes. The Ashley Valley Oil Field near Dinosaur National Monument produces oil from the Weber Sandstone. This field was discovered just after World War II, making it the Uinta Basin's oldest oil field.

Roadcuts near the river crossing at Jensen expose the erosional surface, or unconformity, between Cretaceous sedimentary rocks and their coarse gravel covering. The highway continues east on top of this gravel. To the north, Cliff Ridge, another anticline southeast of Split Mountain, rises steeply, its face marked with chevrons of Permian and Triassic rock.

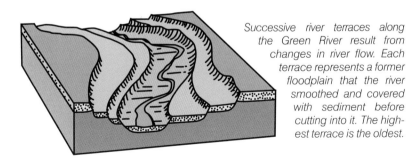

Successive river terraces along the Green River result from changes in river flow. Each terrace represents a former floodplain that the river smoothed and covered with sediment before cutting into it. The highest terrace is the oldest.

At milepost 164 the highway passes from drab shale onto the colorful Cedar Mountain and Morrison Formations, Cretaceous and Jurassic layers famous for holding the area's dinosaur bones. The Uinta Mountains continue eastward beyond the Colorado border, with Weber Sandstone surfacing their southern slope in a third faulted anticline called Blue Mountain. A viewpoint near milepost 166 contains exhibits on the Uinta Mountains.

US 89
Arizona State Line—Kanab
64 miles (103 km)

Leaving Arizona just north of Wahweap, US 89 curves northwestward on a wide bench eroded in soft Jurassic Carmel Formation, which sits above the pale cliffs of Navajo Sandstone that wall Lake Powell and the Glen Canyon Dam. Above the Carmel is the Entrada Sandstone. Its cliffs form the high mesas and buttes just above this end of Lake Powell. Like the similar Navajo Sandstone, the Entrada started as Jurassic sand dunes.

Between mileposts 4 and 5, younger rocks appear to the northeast. The lowest, yellowish-brown ledges are Cretaceous Dakota Sandstone, with a whitish slope of Jurassic Morrison Formation below. Above the Dakota are gray slopes of Tropic Shale and light-brown cliffs of the Straight Cliffs Formation, also Cretaceous in age. They edge the Kaiparowits Plateau and together make up the extensive Gray Cliffs of Utah's Grand Staircase. Above them are higher buttes of Tertiary sedimentary rocks: the Pink Cliffs, out of sight from this part of the highway.

Closely spaced vertical joints give some of these rocks a stockade-like appearance. As the shale at the bottom of the stack erodes, it undermines the cliffs. Blocks and slabs of sandstone break off along the joints. With successive rockfalls the cliffs gradually retreat northward.

Patchy sand dunes appear along parts of the route. Their pale, uniform grains of sand, which sometime drift onto the highway, are recycled from the far older, lithified dunes of the Navajo and Entrada Sandstones.

The highway climbs steadily from Lake Powell. Near the bend in the road between mileposts 17 and 18, it crosses a drainage divide between streams draining toward Lake Powell and those draining west toward the Paria River.

Red and gray layers of the Carmel Formation that edge the highway indicate there were cyclic conditions of deposition: gray layers were deposited in oxygen-depleted conditions, perhaps in locations where abundant decaying organic debris used up the oxygen, and red layers were deposited in areas with abundant oxygen. Gypsum layers are also present; they precipitated from ocean water that became saturated with the mineral during periods of intense evaporation.

Shortly after crossing the Paria River, the highway crosses the East Kaibab Monocline, which is actually the steep eastern face of a huge anticline—the Kaibab Uplift—the crest of which trends northeast to southwest. The monocline drapes over a Precambrian fault, which was reactivated by tectonic pressure during the Laramide Orogeny. Younger beds overlying the fault folded due to the pressure until they were nearly vertical. These form the Cockscomb, which the road winds through. The Grand Canyon cuts right across the Kaibab Uplift, where the relief of the fold—about 5,000 feet (1,500 meters)—and the underlying fault can be seen.

The Navajo Sandstone is well exposed at the big road cut between mileposts 24 and 25, where the highway slices through the monocline. The route then crosses successively older, colorful, steeply tilted rock layers.

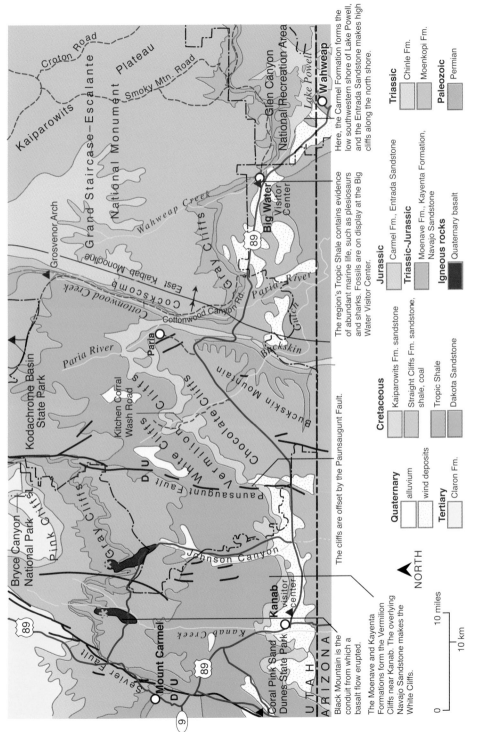

Geology along US 89 between the Arizona state line and Kanab.

Black Mountain is the conduit from which a basalt flow erupted.

The Moenave and Kayenta Formations form the Vermilion Cliffs near Kanab. The overlying Navajo Sandstone makes the White Cliffs.

The cliffs are offset by the Paunsaugunt Fault.

The region's Tropic Shale contains evidence of abundant marine life, such as plesiosaurs and sharks. Fossils are on display at the Big Water Visitor Center.

Here, the Carmel Formation forms the low southwestern shore of Lake Powell, and the Entrada Sandstone makes high cliffs along the north shore.

NORTH

0   10 miles

    10 km

Quaternary
alluvium
wind deposits

Tertiary
Claron Fm.

Cretaceous
Kaiparowits Fm. sandstone
Straight Cliffs Fm. sandstone, shale, coal
Tropic Shale
Dakota Sandstone

Jurassic
Carmel Fm., Entrada Sandstone

Triassic–Jurassic
Moenave Fm., Kayenta Formation, Navajo Sandstone

Igneous rocks
Quaternary basalt

Triassic
Chinle Fm.
Moenkopi Fm.

Paleozoic
Permian

Near milepost 25 the Carmel and underlying Navajo Sandstone swing up in the giant Cockscomb, which defines the East Kaibab Monocline. —Lucy Chronic photo

At milepost 26, US 89 turns north along a valley eroded in tilted red sandstone and siltstone of the Triassic Moenkopi Formation. The streams here have cut deep gullies, or arroyos, in their older valley floors. Arroyos appear to be related to long-term climate cycles. In the West, arroyos formed between 1890 and 1939 and also around AD 1200.

Farther north along the valley, Triassic and Jurassic rocks suddenly swoop upward, the most resistant ones forming the Cockscomb. Farther west the layers level out as they cross Buckskin Mountain, the crest of the monocline. Part of the mountain's surface is covered with windblown sand. Beneath it is erosion-resistant, tan Kaibab Limestone, the youngest Paleozoic rock here. This limestone surfaces the Kaibab Uplift and, to the south, forms both rims of the Grand Canyon. The highway closely follows the contact between the Kaibab Limestone and the red Moenkopi Formation around the north end of Buckskin Mountain.

As can be seen in some of the small canyons along US 89, the contact between the Moenkopi and Chinle Formations is irregular, showing that there was a period of erosion between them. Jointing in the near-vertical rocks is a result of stress caused by tectonic folding during Cretaceous and Tertiary time.

From milepost 30, look north to distant cliffs of Tertiary sedimentary rock. These are the Pink Cliffs of the Grand Staircase. Northwestward are several small peaks, one a grayish volcanic neck, the other two erosional remnants of Navajo Sandstone raised high by the Kaibab Uplift. Below the Pink Cliffs are more red cliffs, these composed of the Kayenta and Moenave Formations, which rise above colorful Chinle Formation slopes. These rocks continue westward as the Vermilion Cliffs, the long line of promontories that guard Utah's southern frontier. Looking southwest from milepost 44, the Mt. Trumbull volcano, elevation 8,028 feet (2,447 m), and several smaller volcanoes appear in the far distance.

North of Buckskin Mountain, a side road leads to the old movie set of Paria, where stunning exposures of Triassic Chinle Formation rise to a crown of Triassic-Jurassic Moenave Formation sandstone. The Chinle is a terrestrial deposit with plentiful volcanic ash, and often petrified wood. —Lucy Chronic photo

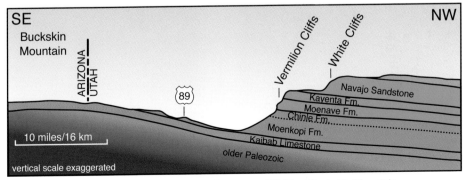

Southeast-northwest cross section across US 89 at Buckskin Mountain.

In the valley between mileposts 45 and 46, US 89 crosses the Paunsaugunt Fault. Rocks to its west have dropped; because the beds dip slightly to the south, the outcrops of the Vermilion Cliffs and Chocolate Cliffs of the Moenkopi Formation appear father to the south, much closer to the highway. This is a good opportunity to see the red Moenave beds: siltstone deposited in slow streams, ponds, and lakes, and sandstone deposited in swifter creeks and dune fields. White cliffs of Navajo Sandstone can be seen to the north from milepost 78.

US 163
Arizona State Line—Bluff
45 miles (72 km)

US 163 enters Utah in the western part of Monument Valley. Here, on the crest of a wide anticline—the Monument Upwarp—Permian through Triassic sedimentary rocks form some of Utah's most iconic scenery. The highway rests on a surface of hard, resistant Permian Cedar Mesa Sandstone that is generally covered by windblown sand. Tidal-flat mudstone and gypsum beds in the formation can be seen best near Monument Pass. Northwest of Mexican Hat, they grade gradually into dune-deposited sandstone.

Monument Valley, along with all the land south of the San Juan River, is within the Navajo Reservation. Monument Valley Tribal Park Visitor Center has great views, and tours of the valley can be arranged there (guides are required).

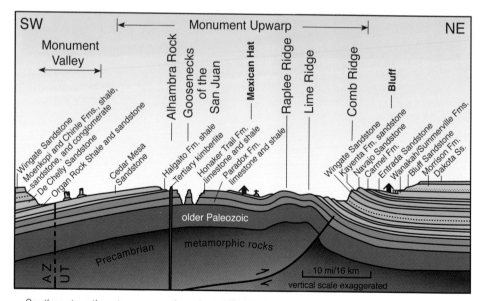

Southwest-northeast cross section along US 163 between the Arizona state line and Bluff. Between 70 and 40 million years ago, during the Laramide Orogeny, the Monument Upwarp rose when its section of crust, pushed eastward, rode upward on a deep fault. Its eastern edge crumpled, forming Raplee and Lime Ridges (both anticlines) and the Mexican Hat Syncline.

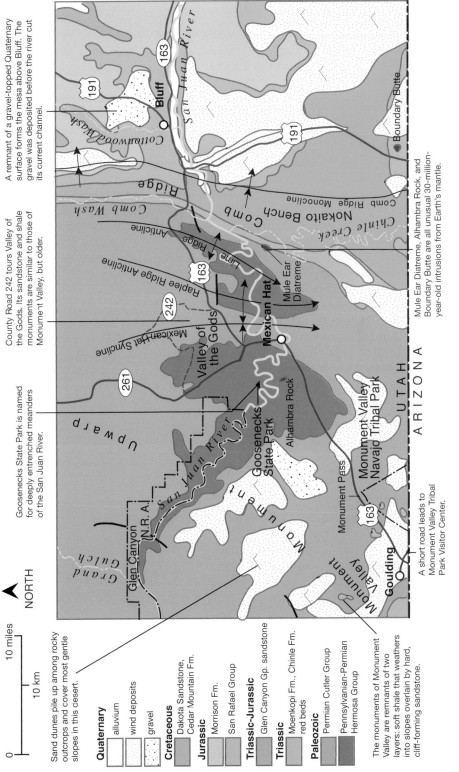

NORTH

0　10 miles

0　10 km

A remnant of a gravel-topped Quaternary surface forms the mesa above Bluff. The gravel was deposited before the river cut its current channel.

County Road 242 tours Valley of the Gods. Its sandstone and shale monuments are similar to those of Monument Valley, but older.

Goosenecks State Park is named for deeply entrenched meanders of the San Juan River.

Sand dunes pile up among rocky outcrops and cover most gentle slopes in this desert.

Mule Ear Diatreme, Alhambra Rock, and Boundary Butte are all unusual 30-million-year-old intrusions from Earth's mantle.

The monuments of Monument Valley are remnants of two layers: soft shale that weathers into slopes overlain by hard, cliff-forming sandstone.

Quaternary
alluvium
wind deposits
gravel

Cretaceous
Dakota Sandstone, Cedar Mountain Fm.

Jurassic
Morrison Fm.
San Rafael Group

Triassic-Jurassic
Glen Canyon Gp. sandstone

Triassic
Moenkopi Fm., Chinle Fm. red beds

Paleozoic
Permian Cutler Group
Pennsylvanian-Permian Hermosa Group

San Juan River

163
191
Bluff
Cottonwood Wash
Comb Ridge
Comb Wash
Comb Ridge Monocline
Nokaito Bench
Chinle Creek
191
Boundary Butte
242
261
Raplee Ridge Anticline
Mexican Hat Syncline
Lime Ridge Anticline
163
Valley of the Gods
Mexican Hat
Mule Ear Diatreme
Monument Upwarp
Grand Gulch
Glen Canyon N.R.A.
San Juan River
Goosenecks State Park
Alhambra Rock
Monument Valley Navajo Tribal Park
Monument Pass
Monument Valley
Goulding
163
UTAH
ARIZONA
A short road leads to Monument Valley Tribal Park Visitor Center.

Geology along US 163 between the Arizona state line and Bluff.

Reddish-brown Organ Rock Shale below the tall monuments is silty, sandy sediment washed westward by rivers from an ancient mountain range. The sheer cliffs above, composed of De Chelly Sandstone, were dunes in a field that covered much of northeast Arizona, northwest New Mexico, and a little of southern Utah. Now well cemented, the sandstone tends to break along vertical joints. Some of the monuments wear layered caps composed of red Triassic Moenkopi Formation, deposited by rivers on a westward-sloping coastal plain, and overlying Chinle Formation, laid down in rivers and lakes. The Chinle's lowest member, the Shinarump Conglomerate, forms strong ledges.

Alhambra Rock juts up northeast of Monument Pass, an errant monument with a very unusual origin. It is one of several diatremes in the region. It formed about 30 million years ago when a gas-driven vortex of material blasted upward from below Earth's crust. It contains pieces of Earth's mantle and of the rocks it punched through. Possibly moving faster than the speed of sound, the material probably followed old Precambrian faults. The rock is kimberlite, the type of rock that diamonds occur in, though none have been found here.

East of Alhambra Rock, the hogbacks of Comb Ridge mark the east side of the Monument Upwarp. The rocks of Comb Ridge are mostly Jurassic and therefore younger than those of Monument Valley. Below Comb Ridge a sealed cell contains radioactive waste. Uranium ore was processed here; it was mined from the Moss Back and Shinarump Conglomerates of the Chinle Formation around Mexican Hat and Monument Valley between 1957 and 1963.

The entrenched meanders of Goosenecks State Park. —Felicie Williams photo

This region's main river, the San Juan, rises in the San Juan Mountains of southwest Colorado and is a tributary of the Colorado. Before Navajo Dam was built upstream, the San Juan carried an average of 34 million tons (30.8 million metric tons) of sediment per year to the Colorado. During a 1911 flood the river rose 50 feet (15 m) above its low-water mark downstream from Mexican Hat. Now, the river flows well enough for rafting in spring and early summer. In fall and winter it may be almost dry.

Four miles (6 km) north of Mexican Hat, Utah 261 leads west to Goosenecks State Park. The goosenecks of the park are the deeply incised, or entrenched, meanders of the San Juan River. Meanders like this develop when a river that once looped lazily across a gentle surface later cuts downward while staying in its meandering course. The entrenched meanders of Goosenecks State Park are often used in geology textbooks as the classic example. The 1,000-foot (300 m) gorge exposes Pennsylvanian rocks: the upper half is interbedded limestone and shale of the Honaker Trail Formation, while the lower portion is Paradox Formation, here marine limestone beds that formed in shallow water with reefs. The Paradox rocks here were deposited outside of the Paradox Basin. (See the US 191: Monticello—I-70 road guide in this chapter and Arches National Park in the final chapter, Something Special, for more about the Paradox Basin.)

Northeast of Mexican Hat, US 163 climbs across red-brown Halgaito Formation shale onto the Raplee Ridge and Lime Ridge Anticlines. The anticlines are surfaced with marine limestone of the Honaker Trail Formation. Road cuts

The Mexican Hat, lower right, is a remnant of Permian Halgaito Shale northeast of town. Behind it, flatirons of the Halgaito and underlying Honaker Trail Formation rise across Raplee Ridge. —Felicie Williams photo

show its many individual layers thickening and thinning with the reef-like accumulations of limy algae and animal shells that settled on the ocean floor during Permian time.

East of the crest of the Lime Ridge Anticline, about 15 miles (24 km) from Mexican Hat, the highway crosses younger rocks again: red-brown Halgaito Formation shale and then Cedar Mesa Sandstone, with a massive bed of gray gypsum near its lower slope. Comb Wash is a long valley eroded in the relatively soft, colorful Permian Organ Rock Shale and Triassic Moenkopi and Chinle Formations. The long line of cuestas and hogbacks here illustrates differential weathering on a grand scale: Resistant Triassic-Jurassic Wingate Sandstone forms the cliffs of the lower part of Comb Ridge. Between it and the imposing vertical cliff of Comb Ridge's crest is a shelf of thin Kayenta Formation beds, which are more easily eroded. The crest is resistant Navajo Sandstone. Its sweeping crossbedding records countless individual sand dunes.

Even younger rocks, of Jurassic age, appear east of Comb Ridge: the red-tinted San Rafael Group, which includes the Carmel Formation, Entrada Sandstone, and Summerville Formation, and still farther east the light-colored Bluff Sandstone, forming cliffs on both sides of the San Juan River. Its crossbedded layers represent a small dune field. Locally, the Bluff Sandstone is the lowest bed of the Morrison Formation.

Comb Ridge edges the Monument Upwarp. The ridge's sharp lines are shaped by mass wasting: witness the pile of car-sized boulders on the edge of the highway. —Felicie Williams photo

Approaching the town of Bluff, US 163 descends across successive flat-topped, gravel-covered terraces. The terraces formed when Cottonwood Wash and the San Juan River repeatedly cut down through their floodplains. Gravel on the highest terrace was carried here by a river that was swollen with glacial meltwater 660,000 years ago. Since then, the San Juan has cut down 380 feet (115 m).

US 191
Arizona State Line—Monticello
69 miles (111 km)

From the state line, US 191 passes gradually upward through younger and younger Jurassic formations. First it crosses a broad plain of Navajo Sandstone covered with active sand dunes, and then it climbs onto the red Carmel Formation. To the east, dark intrusive rock forms Boundary Butte. Low cliffs north of the butte are Bluff Sandstone. North of the San Juan River crossing, the highway rises through the Carmel Formation, with its distinctive crinkly outcrops, and onto terraces of gravel deposited by a Pleistocene river.

Bluff lies at the confluence of Cottonwood Wash and the San Juan River. The wash is typical of streams in the arid Southwest: it is usually dry or just barely

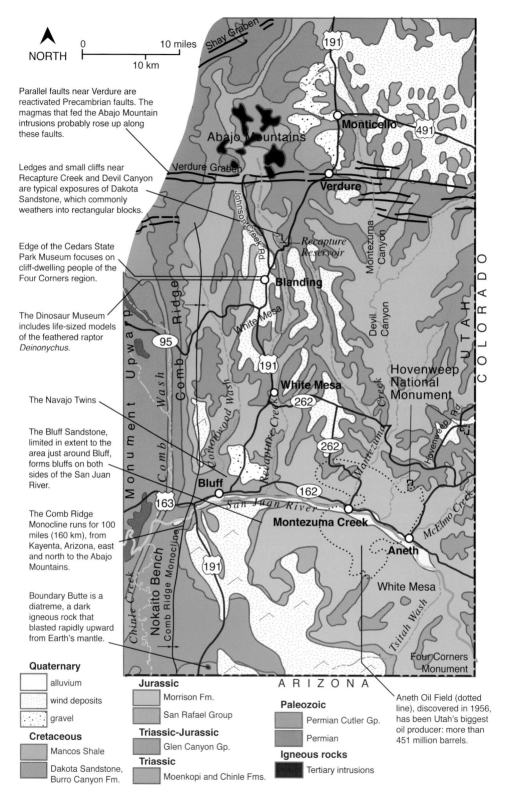

NORTH

0   10 miles

10 km

Parallel faults near Verdure are reactivated Precambrian faults. The magmas that fed the Abajo Mountain intrusions probably rose up along these faults.

Ledges and small cliffs near Recapture Creek and Devil Canyon are typical exposures of Dakota Sandstone, which commonly weathers into rectangular blocks.

Edge of the Cedars State Park Museum focuses on cliff-dwelling people of the Four Corners region.

The Dinosaur Museum includes life-sized models of the feathered raptor *Deinonychus*.

The Navajo Twins

The Bluff Sandstone, limited in extent to the area just around Bluff, forms bluffs on both sides of the San Juan River.

The Comb Ridge Monocline runs for 100 miles (160 km), from Kayenta, Arizona, east and north to the Abajo Mountains.

Boundary Butte is a diatreme, a dark igneous rock that blasted rapidly upward from Earth's mantle.

Shay Graben

Abajo Mountains

Monticello

491

Verdure Graben

Verdure

Johnson Creek Rd.

Recapture Reservoir

Montezuma Canyon

Blanding

Comb Ridge

Monument Upwarp

95

White Mesa

191

Devil Canyon

Wash

Cottonwood Wash

White Mesa

262

Hovenweep National Monument

Hovenweep Rd.

Montezuma Creek

Comb

Bluff

262

Redhouse Creek

163

San Juan River

162

Montezuma Creek

Aneth

McElmo Creek

Nokaito Bench

Comb Ridge Monocline

191

White Mesa

Chinle Creek

Tsitah Wash

Four Corners Monument

Aneth Oil Field (dotted line), discovered in 1956, has been Utah's biggest oil producer: more than 451 million barrels.

A R I Z O N A

U T A H   C O L O R A D O

Quaternary

alluvium

wind deposits

gravel

Cretaceous

Mancos Shale

Dakota Sandstone, Burro Canyon Fm.

Jurassic

Morrison Fm.

San Rafael Group

Triassic-Jurassic

Glen Canyon Gp.

Triassic

Moenkopi and Chinle Fms.

Paleozoic

Permian Cutler Gp.

Permian

Igneous rocks

Tertiary intrusions

Geology along US 191 between the Arizona state line and Monticello.

flowing. Overloaded with sand and gravel, its channel shifts with every storm, making braided patterns in its bed. The San Juan's channels shift as well, but over a longer time frame; its gravelly islands persist for many years. Its headwaters are above Navajo Dam in the San Juan Mountains.

From 1891 to 1893 Bluff saw a small gold rush. Some 1,200 prospectors washed sand and gravel along the San Juan River looking for gold that the river had carried here from north of Durango, Colorado. One operation produced $3,000 worth of gold in one month—quite a fortune at the time. Most of the gold, though, was too fine grained for profitable mining.

Winter frost and rain, as well as wind, are responsible for much of the erosion in this area. Wind erosion is strongest near the ground, where gusts pick up particles of sand and hurl them against nearby rock. Erosion works slowly. William Henry Jackson photographed the Navajo Twins north of Bluff in 1875 for an early government survey of the western territories. Geologists on the survey team were sure the Twins would soon fall. But comparison of today's scene with the Jackson photographs shows little change. Terraces tell us river down-cutting here proceeds at an average rate of about 0.5 inch (1.3 cm) per one hundred years. The rate of erosion occurring outside of river channels is even less.

North of the Navajo Twins the route climbs through a narrow canyon onto a Bluff Sandstone bench. To the northeast are outlying mesas of pink and

Erosion along vertical joints carved the Navajo Twins in soft Summerville Formation shale and its cap of more-resistant Bluff Sandstone. The Summerville layers were tidal-flat mudstones deposited after a period of erosion, whereas the Bluff Sandstone was a small regional dune field. —Felicie Williams photo

yellowish-gray Morrison Formation, the youngest of the Jurassic layers. Pale Navajo Sandstone appears to the west, swooping upward to form jagged Comb Ridge, the steep eastern edge of the Monument Upwarp. (See the US 163: Arizona State Line—Bluff road guide for more on these features.)

The high country north of Bluff overlooks much of the Four Corners region, the only place in the nation where four states (Arizona, Colorado, New Mexico, and Utah) meet. To the southeast are the mesas and tablelands that surround the Four Corners. Sleeping Ute Mountain appears to the east, in Colorado. To the north are the Abajo Mountains. These island-like ranges are clusters of small laccoliths. The intrusion that forms Sleeping Ute Mountain is 72 million years old, coinciding with the Laramide Orogeny.

Several small igneous intrusions called laccoliths (pink rock) compose the Abajo Mountains. The laccoliths formed as magma was injected between sedimentary layers, lifting and doming the layers above. Remnants of the domed layers surround the mountains and appear between the peaks, whereas the peaks themselves are igneous.

The rainbow-colored Morrison Formation shows up near the junction of US 191 and UT 262 (Hovenweep Road) and along the south edge of the town of White Mesa. It was deposited as fine soil, mud, and clay in river floodplains, with a generous lacing of volcanic ash. Winding river channels that crossed the floodplains left lens-shaped deposits of resistant sandstone that cover the Morrison in places.

Hovenweep National Monument preserves several ancestral Puebloan ruins: towers, kivas (square-walled underground rooms), and cliff dwellings scattered across the Utah-Colorado border 20 miles (32 km) east of US 191. The road leading to the monument (Utah 262 and Hovenweep Road) crosses valley after valley of the muddy Morrison Formation edged by short cliffs of Cretaceous Burro Canyon and Dakota sandstones. The ancestral Puebloans built with Dakota Sandstone, which breaks naturally into rough, blocky pieces. They built near springs, which are common at the base of the Dakota Sandstone. Rain and snowmelt sink into the porous sandstone but are stopped by the clay below it, flowing out where the contact is exposed in the canyons. Ancestral Puebloan dwellings were also built in Montezuma Canyon, just southeast of Monticello.

North of the Utah 262 turnoff, US 191 climbs through the Morrison Formation and the river- and ocean-deposited sands, conglomerates, and clays of the Burro Canyon Formation to the top of White Mesa, which is capped with hard, resistant Dakota Sandstone. The irregular erosional contact (unconformity)

The rainbow-colored Morrison Formation covered by thick sandstone of the Cretaceous Burro Canyon Formation. —Felicie Williams photo

between the colorful Morrison and the drab Burro Canyon is mostly covered here, but it is visible in big road cuts north of Blanding.

On the mesas around Blanding and Monticello, reddish windblown soil conceals the Dakota Sandstone. This fine, uniform soil has made the area famous for its pinto beans, which grow well in it.

Near the little community of Verdure the highway crosses a graben, a narrow slice of land that has dropped down about 200 feet (60 m) between two parallel east-west faults. Just before Monticello, the highway passes onto Cretaceous Mancos Shale, surfaced by fans of Pleistocene gravel eroded from the Abajo Mountains.

The Monticello uranium district was, for a time, Utah's most productive, and it still produces ore. To form these deposits, groundwater dissolved trace elements from sediment it flowed through, particularly from volcanic ash. The water reached areas that contained plant debris and deposited uranium, vanadium, and radium minerals and small amounts of chromium, copper, selenium, barium, and molybdenum. In the Monticello area, these minerals are found in river-channel sandstones in the lower part of the Morrison Formation and conglomerate in the Triassic Chinle Formation. Since the deposits formed in stream channels, they are usually long, slender, and winding.

Southeast of Monticello a smooth-topped storage cell holds radioactive waste. Monticello's mill processed ore from 1942 to 1959. The sandy mill tailings, still containing a small amount of radioactive minerals, were used in construction projects all over town. They also blew around in the wind. Between 1989 and 2004, 424 contaminated sites were cleaned up. For those who grew up with the mill, there are ongoing efforts to establish the harm done to them by radiation.

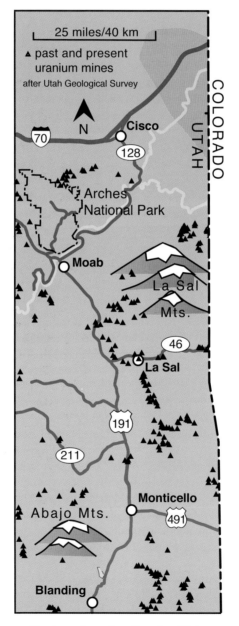

Uranium mines of southeastern Utah.

US 191
Monticello—I-70
86 miles (138 km)

North of Monticello, US 191 runs on a surface of gravel composed of rounded igneous rocks washed from the nearby Abajo Mountains. Beneath is the Mancos Shale, which was deposited in a shallow Cretaceous sea. The crumbly, clayey Mancos makes a weak and often bumpy roadbed. Below the shale, visible in deep gullies, is the Dakota Sandstone, which formed on beaches and sandbars as the sea flooded in from the east. Below it is conglomerate, shale, and sandstone of the Burro Canyon Formation, which was deposited by eastward-flowing rivers that preceded the sea's encroachment.

About 10 miles (16 km) north of town, the highway descends through these layers and through the river-floodplain clays, soils, and sandstones of the Jurassic Morrison Formation. Castle-like outcrops along the west edge of the valley are the older Jurassic Summerville Formation, a fine siltstone deposited on a tidal flat. Watch for small faults that offset the Summerville beds.

This section of US 191 crosses the Paradox Basin, a long-buried basin that formed southwest of a range called Uncompahgria, which was part of the Pennsylvanian-Permian-age Ancestral Rocky Mountains. The Paradox Basin was a shallow, nearly landlocked sea during Pennsylvanian time. Dozens of cycles of flooding and evaporation saturated the water with evaporites—salt, potash, and gypsum—that settled out of the water, forming deposits up to 4,000 feet (1,220 m) thick.

Church Rock is an erosional remnant of Jurassic Entrada Sandstone capped by the lighter Moab Tongue of the Curtis Formation and sitting on brick-red siltstone of the Carmel and Dewey Bridge Formations. These rocks make splendid scenery in Canyon Rims Recreation Area. In the background are the igneous La Sal Mountains. —Felicie Williams photo

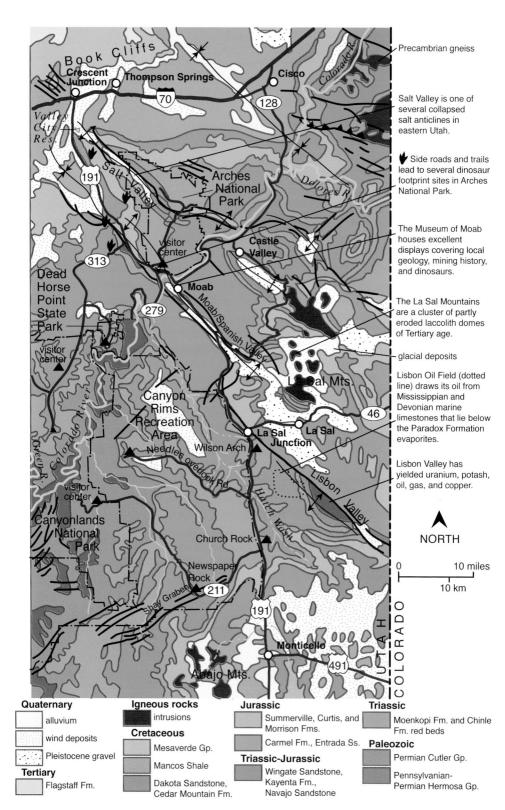

Precambrian gneiss

Salt Valley is one of several collapsed salt anticlines in eastern Utah.

Side roads and trails lead to several dinosaur footprint sites in Arches National Park.

The Museum of Moab houses excellent displays covering local geology, mining history, and dinosaurs.

The La Sal Mountains are a cluster of partly eroded laccolith domes of Tertiary age.

glacial deposits

Lisbon Oil Field (dotted line) draws its oil from Mississippian and Devonian marine limestones that lie below the Paradox Formation evaporites.

Lisbon Valley has yielded uranium, potash, oil, gas, and copper.

NORTH

0   10 miles

10 km

Book Cliffs

Crescent Junction
Thompson Springs
Cisco
70
128
Colorado R.
Valley City Res.
191
Salt Valley
Arches National Park
Delores R.
visitor center
Castle Valley
Moab
313
Dead Horse Point State Park
279
visitor center
Moab/Spanish Valley
La Sal Mts.
46
Canyon Rims Recreation Area
Green R.
Colorado River
La Sal Junction
La Sal
visitor center
Needles overlook Rd
Wilson Arch
Lisbon Valley
Hatch Wash
Canyonlands National Park
Church Rock
Newspaper Rock
Shay Graben
211
191
UTAH
COLORADO
Monticello
491
Abajo Mts.

Quaternary

alluvium

wind deposits

Pleistocene gravel

Tertiary

Flagstaff Fm.

Igneous rocks

intrusions

Cretaceous

Mesaverde Gp.

Mancos Shale

Dakota Sandstone, Cedar Mountain Fm.

Jurassic

Summerville, Curtis, and Morrison Fms.

Carmel Fm., Entrada Ss.

Triassic-Jurassic

Wingate Sandstone, Kayenta Fm., Navajo Sandstone

Triassic

Moenkopi Fm. and Chinle Fm. red beds

Paleozoic

Permian Cutler Gp.

Pennsylvanian-Permian Hermosa Gp.

Geology along US 191 between Monticello and I-70.

Evaporites weigh less than most rock and can flow slowly, so as soon as other layers covered them, the evaporites of the Paradox Basin began to rise. They rose above a series of long northwest-trending faults, lifting overlying rocks into long, parallel anticlines. Along the crest of each anticline, the salt weathered away and the overlying rocks collapsed, also weathering away in time. This left valleys where the crests of the anticlines used to be. Northwest of Moab, the highway follows the Moab Valley (Spanish Valley), one of these faulted anticlines that became a valley.

Side roads lead westward into Canyon Rims Recreation Area, Canyonlands National Park, and Indian Creek State Park (Newspaper Rock) and to the Needles and Anticline Overlooks. The remarkable sculpted rocks of Canyonlands

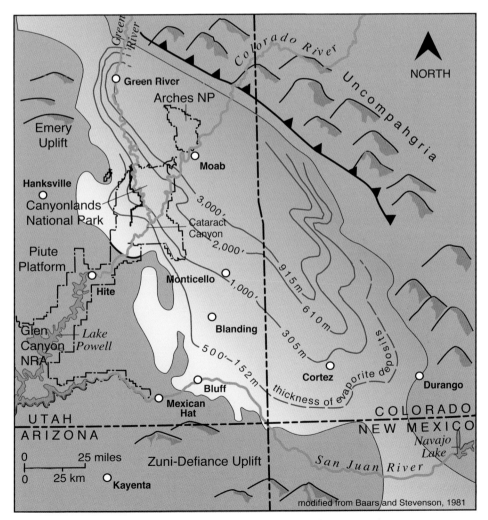

The Paradox Basin formed southwest of Uncompahgria, a mountain range that was roughly where the gentle rise of the Uncompahgre Uplift is now.

and Arches National Parks, covered in the final chapter, Something Special, partly owe their shape to their position over the mobile salt layers of the Paradox Basin.

Several of the region's salt anticlines have trapped upward-migrating oil and gas. Wells in Lisbon Valley, east of US 191, produce oil, natural gas, and carbon dioxide. The carbon dioxide is reinjected into the ground to force more oil upward. A great deal of uranium and vanadium has also come from the Jurassic Morrison and Triassic Chinle Formations of Lisbon Valley. Copper has also been mined nearby, along the faults that edge the Lisbon Valley salt anticline.

Wherever the sandy Salt Wash Member of the Morrison Formation is exposed in eastern Utah, it is dotted with old prospect pits and small mines. Almost all are inactive. Vanadium and uranium (and its decay product radium) are found in black pitchblende and powdery bright-yellow carnotite. Ore occurs in fossil river channels that also contain a lot of carbon. Fossilized trees are some of the richest ore, with more than 5.5 percent uranium oxide and 8 percent vanadium oxide.

Ore in the region was mined first for radium, starting around 1898; it was used for research and medicine and, because it glows in the dark, in paint for clock and instrument panels. Radium was difficult to recover, but at $70,000 to $100,000 per gram, it was worth the effort. From 1930 through World War II, vanadium was wanted for steel. After 1942, the old tailings were reprocessed for uranium, and more mines were found to build the US government stockpile.

The Entrada Sandstone is an especially good conduit for water, or aquifer. Some of its water comes to the surface at Kane Springs Rest Area, about 7 miles (11 km) north of La Sal Junction.

Carnotite. —Felicie Williams photo

The Moab area is particularly interesting geologically because of the way the underlying salt formed the Moab Valley Salt Anticline and, as a result of the anticline's erosion and collapse, the cliffs that surround town. To the west, rounded orange knobs of Jurassic Navajo Sandstone appear above straight cliffs of Wingate Sandstone. Below them are slopes of Triassic and Permian shale. To the east, the Navajo Sandstone, and above it the Entrada Sandstone, form gentler bluffs, with rubble of other formations along their base. Rising gypsum has broken through along both edges of the valley floor, making soft, hummocky hills. Potash, deposited with the gypsum and salt, is mined and concentrated a few miles downriver from Moab for use as fertilizer.

Looking northwest up the Moab Valley Salt Anticline, layers dip away to east and west. At the far end of the valley there are intensely broken rocks. These were originally about 2,000 feet (610 m) higher, before the evaporites dissolved and the top of the anticline collapsed. —Felicie Williams photo

Just north of Moab, the Atlas Mill operated from 1956 until 1984, processing uranium ore from some three hundred mines. Its 16-million-ton (14.5-million-metric-ton) tailings pile was unlined and on the bank of the Colorado River. A poisonous plume flowed into the river from the pile until remediation wells were drilled. The Department of Energy is currently moving the tailings by train to a site 30 miles (48 km) to the north, near Crescent Junction.

The turnoff to Arches National Park comes soon after the valley narrows. A little farther north, Utah 313 leads west to Dead Horse Point State Park and the Island in the Sky part of Canyonlands National Park. (See the final chapter, Something Special, for more information on Arches and Canyonlands National Parks.)

Near the Utah 313 junction the highway begins to follow a syncline rimmed by prominent bluffs of Entrada Sandstone below soft, colorful rocks of the Jurassic Morrison and Cretaceous Cedar Mountain Formations. North of the airport, Dakota and Cedar Mountain sandstones form hogbacks to the east. To the west, a resistant layer in the Mancos Shale forms a long ridge.

US 191
Helper—Duchesne
47 miles (76 km)

The section of US 191 between Helper and Duchesne crosses from the edge of the Uinta Basin to its center, climbing through the Cretaceous Book Cliffs and Tertiary Roan Cliffs and onto the Tavaputs Plateau before dropping down through Indian Canyon to Duchesne.

Numerous changes in the type of rock occur in this region, often within a single formation. This is because the sediments were deposited between the mountains that were forming during the Sevier Orogeny, whose uneroded roots are exposed in the Wasatch Range, and the slightly younger Uinta Mountains, which were uplifted during the Laramide Orogeny. Each range had its complement of a basin bowed down by the weight of the mountains that were thrust up next to it. Each range shed sediment into its basin. Fans of coarse sediment were deposited next to the mountains. Away from the mountains, these fans graded into fine sediment, which was often deposited in large bodies of water. Coarser material was also deposited during times of increased uplift and faulting.

The basin east of the Sevier Thrust Belt held the Cretaceous Interior Seaway. Its eastward-receding shoreline became the Mesaverde Group, the rock we start our journey on at Helper.

Castle Gate, where US 191 turns northeast, was named for two sandstone promontories on either side of the Price River. They looked like an opening gate with turrets on either side as one approached the promontories while traveling up the canyon. One promontory was removed to widen US 6. Here, the Mesaverde Group includes the Castlegate Sandstone, the prominent cliff-forming, crossbedded river sandstone that formed the promontories, and below it the coal-bearing Blackhawk Formation. The Blackhawk's thick layers of coal have been mined since the late 1800s; it is one of Utah's most productive sources of coal.

Castle Gate was the site of one of Utah's worst coal mining disasters. In 1924, 172 men and boys were killed when inadequately watered-down coal dust ignited and exploded. This and a similar disaster nearby resulted in much better safety laws for coal mines.

US 191 climbs from Castle Gate through younger and younger layers. In places the rocks have been baked red by burning coal lit by lightning or forest fires. The Castlegate Sandstone is at road level by milepost 254; above it are the Price River and North Horn Formations, which are primarily river deposits with a few lake-deposited shales and limestones. Angular unconformities in the North Horn Formation give evidence of both the last phases of the Sevier Orogeny and the beginning of the Laramide Orogeny.

Nearby wells are part of a coal bed–methane development, a relatively new technology that recovers methane from underground coal beds—in this case, from the Blackhawk Formation.

Soon after the junction with Emma Park Road, US 191 crosses onto the Green River Formation, which forms the Roan Cliffs and Gray Head Peak to

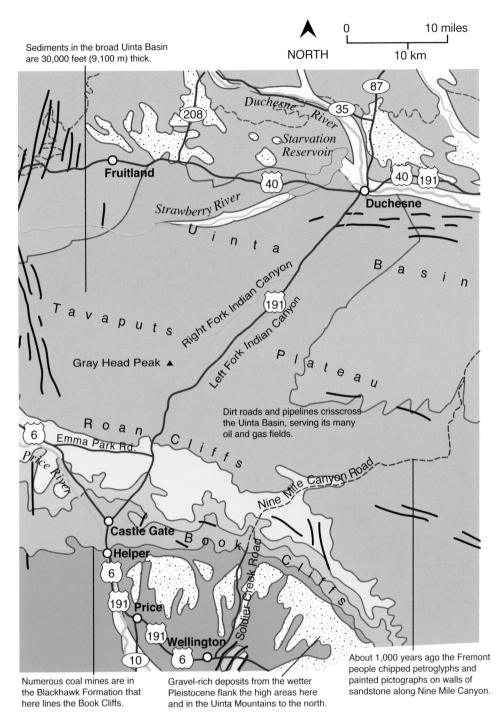

Sediments in the broad Uinta Basin are 30,000 feet (9,100 m) thick.

NORTH

0   10 miles

10 km

208

Duchesne River

35

87

Starvation Reservoir

40

40   191

Fruitland

Strawberry River

Duchesne

U i n t a

B a s i n

T a v a p u t s

Right Fork Indian Canyon

191

Left Fork Indian Canyon

Gray Head Peak ▲

P l a t e a u

Dirt roads and pipelines crisscross the Uinta Basin, serving its many oil and gas fields.

6

R o a n

C l i f f s

Emma Park Rd.

Price River

Nine Mile Canyon Road

Castle Gate

B o o k

Helper

C l i f f s

6

Soldier Creek Road

191   Price

191   Wellington

10   6

Numerous coal mines are in the Blackhawk Formation that here lines the Book Cliffs.

Gravel-rich deposits from the wetter Pleistocene flank the high areas here and in the Uinta Mountains to the north.

About 1,000 years ago the Fremont people chipped petroglyphs and painted pictographs on walls of sandstone along Nine Mile Canyon.

Quaternary

☐ alluvium

☐ glacial outwash gravel

Tertiary

☐ Uinta and Duchesne Fms.

☐ Green River Fm.

☐ Colton and Wasatch Fms.

Cretaceous-Tertiary

☐ North Horn Fm.

Cretaceous

☐ Mesaverde Gp.

☐ Mancos Shale

Geology along US 191 between Helper and Duchesne.

the north. The Green River's immense thickness (up to 6,000 feet, or 1,830 m) accumulated gradually in Lake Uinta as the rising Uinta Mountains warped the Uinta Basin downward. Distant volcanic eruptions added layers of fine ash to the formation. It contains many fossils and is especially noted for its beautifully preserved freshwater fish and insects.

Coal in the Blackhawk Formation was deposited in swamps along the edge of the Cretaceous Interior Seaway. It contains abundant evidence of life, including dinosaur footprints, trees, and other swamp vegetation. —Lucy Chronic photo

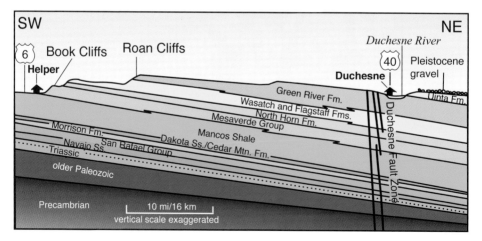

This cross section, essentially parallel to US 191, shows how the sedimentary rock layers that were deposited in this region slope downward toward the center of the Uinta Basin.

Cliffs of Green River Formation near Duchesne give a feel for the great amount of sediment deposited in Lake Uinta. Where the formation is thickest, it contains some 6 million varves (inset), pairs of sediment layers that are only millimeters thick. Varves develop due to seasonal changes in the amount of organic matter and sediment that settled out of lakes. They are common in lake deposits. —Lucy Chronic and Felicie Williams photos

The Green River Formation is also a rich oil reservoir, providing much of the oil produced in the Uinta Basin and containing one of the world's largest reserves of oil shale. Along Indian Canyon, new oil development is evident. Around milepost 278, look for an increasing number of white layers in roadcuts; these are lime-rich mudstones (marlstones) and evaporites that were deposited as Lake Uinta evaporated and shrank in size.

Utah 9
I-15—Mount Carmel Junction
57 miles (92 km)

As Utah 9 leaves I-15, it travels eastward across the Virgin Anticline, an oval-shaped fold about 12 miles (19 km) long that was bowed up by pressure from the west during the Sevier Orogeny. The top, or axis, of the anticline has since eroded away, leaving a rock-rimmed valley, the northern part of which is occupied by Quail Creek State Park and its reservoir. This and several of the other

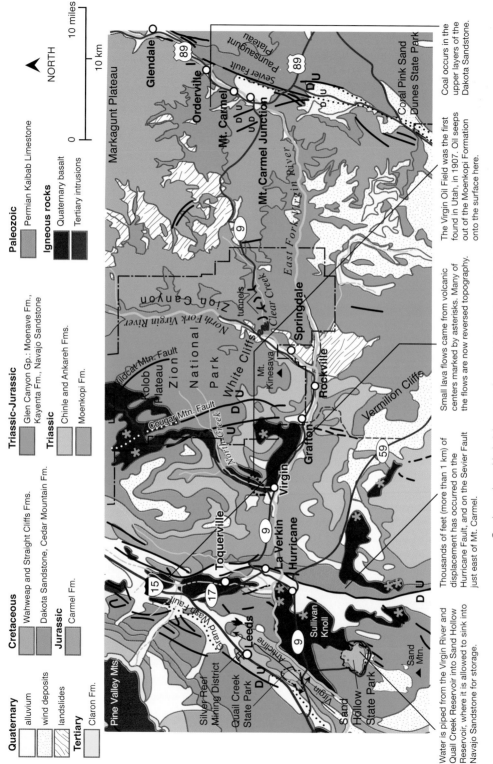

Quaternary

alluvium

wind deposits

landslides

Tertiary

Claron Fm.

Cretaceous

Wahweap and Straight Cliffs Fms.

Dakota Sandstone, Cedar Mountain Fm.

Jurassic

Carmel Fm.

Triassic-Jurassic

Glen Canyon Gp.: Moenave Fm., Kayenta Fm., Navajo Sandstone

Triassic

Chinle and Ankareh Fms.

Moenkopi Fm.

Paleozoic

Permian Kaibab Limestone

Igneous rocks

Quaternary basalt

Tertiary intrusions

NORTH

Pine Valley Mts

Markagunt Plateau

Paunsaugunt Plateau

Glendale

Orderville

Mt. Carmel

Mt. Carmel Junction

East Fork Virgin River

Sevier Fault

Coral Pink Sand Dunes State Park

Zion Canyon

North Fork Virgin River

Wildcat Mtn. Fault

Kolob Plateau

Zion National Park

White Cliffs

Cougar Mtn. Fault

North Creek

Mt. Kinesava

Clear Creek

tunnels

Springdale

Rockville

Grafton

Virgin

Vermilion Cliffs

Toquerville

La Verkin

Hurricane

Leeds

Silver Reef Mining District

Quail Creek State Park

Sand Hollow State Park

Sand Mtn.

Sullivan Knoll

Grand Wash Fault

Anticline

Virgin

Coal occurs in the upper layers of the Dakota Sandstone.

The Virgin Oil Field was the first found in Utah, in 1907. Oil seeps out of the Moenkopi Formation onto the surface here.

Small lava flows came from volcanic centers marked by asterisks. Many of the flows are now reversed topography.

Thousands of feet (more than 1 km) of displacement has occurred on the Hurricane Fault, and on the Sevier Fault just east of Mt. Carmel.

Water is piped from the Virgin River and Quail Creek Reservoir into Sand Hollow Reservoir, where it is allowed to sink into Navajo Sandstone for storage.

Geology along Utah 9 between I-15 and Mount Carmel Junction.

nearby anticlines form natural traps for oil, which has been found in small quantities only 475 to 800 feet (145 to 244 m) down, below impervious rock layers.

Triassic rocks are well exposed in roadcuts across the anticline. The cross-bedded, pebbly Shinarump Conglomerate forms a prominent cuesta that completely rings the fold. Below it is the deep-red Moenkopi Formation, with a candy-stripe pattern of sandstone and gypsum layers. Permian Kaibab Limestone forms the core of the anticline beneath the windblown soil of its central valley.

East of the anticline many basalt flows are now reversed topography: they originally flowed down long slender valleys, but over time softer sediments eroded around the flows and left them as long slender ridges. The flow near milepost 5 is 258,000 years old. It came from the small volcanic cone to the southeast, near milepost 7. At the Virgin River bridge, near milepost 11, the river has cut a deep canyon through another flow—this one 353,000 years old. It erupted from Sullivan Knoll (also known as Volcano Mountain) and flowed down the river, damming it and forcing it to make a new channel to the north. Pillow basalt at the base of the flow formed when the lava flowed into the water.

The Hurricane Fault, which the highway crosses a little east of the Utah 17 turnoff, is a major normal fault that runs from Cedar City to south of the Grand Canyon. It marks the western edge of the transition zone between the Basin and Range country and the Colorado Plateau. The fault is active; a 1992 earthquake here measured magnitude 5.8, but bigger earthquakes—up to magnitude 7—could happen at any time. It and several other normal faults formed in the transition zone as the crust thinned and was pulled westward due to tectonic extension. The faults have caused the blocks of crust between them to stair-step down from the Colorado Plateau to the much lower Great Basin. Such faulting greatly increases the erosional energy of the rivers by giving them elevated topography to course down. Many of the amazing erosional features seen in Utah's spectacular parks are the result of faulting caused by tectonic extension.

The volcanic flows are handy indicators of tectonic movement over time: east of the Hurricane Fault, the same 353,000-year-old-lava flow exposed at bridge level (and west of the fault) is 250 feet (76 m) higher, so rocks on the west side are slipping around 8 inches (20 cm) per 1,000 years. Other parts of the fault are slipping at different rates; farther north, an offset lava flow slips at a rate close to 22 inches (56 cm) per 1,000 years. All together, the layers indicate there have been thousands of feet of displacement along this fault.

Hot springs in the gorge of the Virgin River produce sulfurous steam that can often be smelled from the bridge. Groundwater heated deep beneath the surface finds easy passage upward along the permeable fault zone.

Looking north as the highway rises east of Hurricane, the rock beds dip downward (to the west) into the fault zone. For about 1 mile (1.6 km) the road crosses several slivers of the faulted Triassic Moenkopi Formation and Permian Kaibab Limestone; the large faults in Utah are often zones of faulting, with several related and connected faults cutting the rocks into slivers. Some of the limestone was highly polished and finely grooved by fault movement. Geologists call such polished surfaces slickensides.

The limestone east of Hurricane has a rough karst surface, showing that it was exposed to air in middle to late Permian time before being filled in and covered with Triassic conglomerate and sandstone. This boundary coincides with the greatest extinction known, when more than 90 percent of all marine species disappeared.

The country soon takes on the character of the Colorado Plateau, a land of intricately sculpted but generally flat-lying sedimentary rocks. As individual plateaus erode, they become mesas, buttes, and then pinnacles; there are plenty of examples of this process visible from Utah 9.

Utah 9 approaches Zion National Park while crossing Triassic badlands composed of the chocolate-red Moenkopi Formation and rainbow-colored Chinle Formation. Large landslides are common in the soft and clayey Chinle. Zion's great cliffs and towers rise high as the road continues eastward. At milepost 24 the mountain straight ahead is Mt. Kinesava. Watch for columnar jointing in a lava flow north of the highway. Based on the rate of river down-cutting that has occurred in volcanic flows in and near the park, geologists believe Zion was carved in the last 2 million years (see also the final chapter, Something Special). The town of Springdale has long been known for its many rock and mineral shops.

The grandeur of Zion Canyon is revealed soon after Springdale. The canyon's massive Navajo Sandstone cliffs show large-scale crossbedding that formed in a huge Jurassic dune field. Below the Navajo, with close-spaced horizontal bedding, are pinker stream deposits of the Kayenta Formation, and then steep slopes of red Moenave Formation siltstone and sandstone followed by the Chinle.

Utah 9 zigzags up above Clear Creek and out of Zion Canyon across steep Moenave and Kayenta Formation slopes. Then it tunnels through Navajo Sandstone cliffs, which were, until this highway was built, a formidable barrier.

Long after the sediments of the Plateau had been buried and lithified, the land rose and numerous vertical joints became pathways for water. Water weakens the Plateau's rock by dissolving some of the calcium carbonate that cemented its grains together. Helped by plant roots seeking moisture, joints widen into narrow defiles and secret canyons. Some of the little valleys on the

East of the tunnels, large-scale crossbeds show beautifully in the Navajo Sandstone. Fine sand, blown up the gentle windward slopes of dunes, was deposited in these layers on the dune's steeper leeward slopes, where the wind slowed. Horizontal bands between sets of crossbeds contain silt and sand deposited in flat areas between the dunes. —Lucy Chronic photo

surface of the Plateau don't drain at all. Rainfall and snowmelt just sink into the rock, percolating downward and ultimately feeding springs in deep canyons.

Continuing eastward, the highway gradually passes into younger rock layers. A little west of milepost 54 the Carmel Formation, cut by small faults, appears near the highway. Farther east, the road crosses hummocky landslides with greenish-brown soil derived from Cretaceous layers higher up.

The White Cliffs of Utah's Grand Staircase soon appear to the east. The Navajo Sandstone and the Carmel Formation are much higher in the Grand Staircase than they are in canyons below and south of Utah 9. Separated from us by the Sevier Fault, another of the transition zone normal faults, the cliffs are part of the Paunsaugunt Plateau. Displacement along the Sevier Fault here amounts to several thousand feet.

We descend toward Mount Carmel Junction through the soil-covered, soft Carmel Formation. The valley of the East Fork of the Virgin River lies right along the Sevier Fault.

Utah 10
Price—Fremont Junction
69 miles (111 km)

From Price Utah 10 follows Castle Valley southwest between the San Rafael Swell and the Wasatch Plateau. Price lies on recently deposited river gravels that surface a broad valley eroded in soft Cretaceous Mancos Shale. Long tongues of coarse gravel and sand of Pleistocene age extend out from the surrounding Book Cliffs, topping mesas around and south of town. West of town, the Book Cliffs, with capstones of Cretaceous Mesaverde Group sandstone, hide much of the higher Wasatch Plateau behind them.

South of Price, the highway climbs onto one of the Pleistocene terraces, and you can see the low arch of the San Rafael Swell farther south. A 100-mile-long (160 km), 45-mile-wide (72 km) anticline of Paleozoic and Mesozoic rocks, the Swell was pushed eastward and upward during the Laramide Orogeny by pressure generated from subduction on the continent's west coast. The Swell's western slope, which we'll be seeing on this route, is more gently inclined than its eastern slope. Most uplifts in the Colorado Plateau have gentle slopes like this on the western side and steep ones on the eastern side.

South of milepost 57, a road leads east to the Cleveland-Lloyd Dinosaur Quarry, a National Natural Landmark. The quarry is in the Jurassic Morrison Formation, like the famous quarry at Dinosaur National Monument. Here, paleontologists have unearthed skeletons of at least seventy individual dinosaurs that died at a waterhole during a drought, possibly poisoned by bad water. Two-thirds of the bones are from *Allosaurus*; others are from *Camarasaurus*, *Ceratosaurus*, *Stegosaurus*, and *Camptosaurus*. Many are those of young animals. Volcanic ash overlying the dinosaur layer indicates the layer is about 147 million years old. Discovered in the early 1900s, the site has added significantly to our knowledge of dinosaurs.

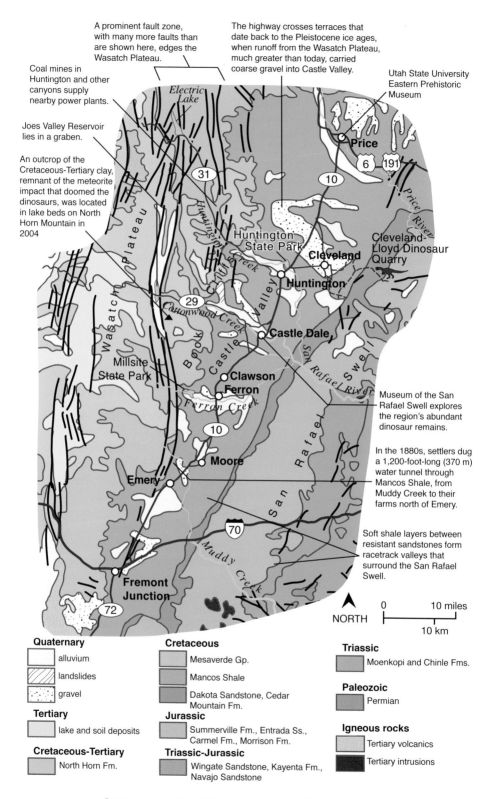

A prominent fault zone, with many more faults than are shown here, edges the Wasatch Plateau.

The highway crosses terraces that date back to the Pleistocene ice ages, when runoff from the Wasatch Plateau, much greater than today, carried coarse gravel into Castle Valley.

Coal mines in Huntington and other canyons supply nearby power plants.

Utah State University Eastern Prehistoric Museum

Joes Valley Reservoir lies in a graben.

An outcrop of the Cretaceous-Tertiary clay, remnant of the meteorite impact that doomed the dinosaurs, was located in lake beds on North Horn Mountain in 2004

Electric Lake

Price

31

6   191

10

Huntington State Park

Cleveland

Cleveland-Lloyd Dinosaur Quarry

Huntington

Plateau

Wasatch

Huntington Creek

Cottonwood Creek

29

Castle Dale

Castle Valley

San Rafael River

San Rafael Swell

Millsite State Park

Book

Clawson

Ferron

Ferron Creek

10

Museum of the San Rafael Swell explores the region's abundant dinosaur remains.

Moore

Rafael

San

In the 1880s, settlers dug a 1,200-foot-long (370 m) water tunnel through Mancos Shale, from Muddy Creek to their farms north of Emery.

Emery

70

Muddy Creek

Soft shale layers between resistant sandstones form racetrack valleys that surround the San Rafael Swell.

Fremont Junction

72

NORTH

0   10 miles

10 km

Quaternary

- alluvium
- landslides
- gravel

Tertiary

- lake and soil deposits

Cretaceous-Tertiary

- North Horn Fm.

Cretaceous

- Mesaverde Gp.
- Mancos Shale
- Dakota Sandstone, Cedar Mountain Fm.

Jurassic

- Summerville Fm., Entrada Ss., Carmel Fm., Morrison Fm.

Triassic-Jurassic

- Wingate Sandstone, Kayenta Fm., Navajo Sandstone

Triassic

- Moenkopi and Chinle Fms.

Paleozoic

- Permian

Igneous rocks

- Tertiary volcanics
- Tertiary intrusions

Geology along Utah 10 between Price and Fremont Junction.

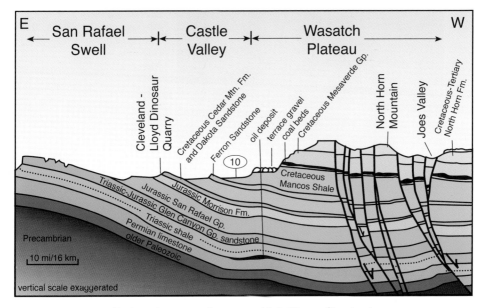

East-west cross section across Utah 10 near Cleveland.

South of the quarry turnoff, Utah 10 crosses one terrace after another, dropping between them to Mancos Shale on the valley floor. Wherever the highway crosses this shale, the pavement is bumpy because the shale contains clay that shrinks and swells depending on its moisture content.

Utah 10 passes many areas where salt encrusts the surface. Both naturally occurring water and that from irrigation dissolves salt from the Mancos Shale—a naturally salty marine deposit. The water later evaporates, leaving the salt behind and eventually making the land useless for agriculture.

Farther south the rocks are gently folded into several small domes, which are traps for gas and oil. Between mileposts 41 and 40, a dome east of the highway trapped oil between the grains in Cretaceous Ferron Sandstone and, deeper down, in cavities and pore spaces of Permian limestone.

Big power plants near Castle Dale and Huntington burn low-sulfur, low-ash coal mined nearby in the Ferron Sandstone. The mines are underground, and their portals are in the canyons to the west. Their proximity to the plants keeps transporation costs low, and thus electricity rates as well.

South of Ferron, Utah 10 converges with faults that signal the beginning of the transition zone that separates the Basin and Range from the Colorado Plateau. The faults disappear under volcanic rocks of the Fish Lake Hightop Plateau, which is to the southwest across I-70. The plateau is part of the Marysvale Volcanic Field, where Tertiary and Quaternary lava, ash, and breccia cover a region 70 miles (113 km) across. Dark lava boulders washed from this region rest near Utah 10 south of milepost 3.

Crystallizing salt has swelled the base of a fencepost along Utah 10. —Felicie Williams photo

Castle Dale is named for the castellated turrets on the sides of some of the Pleistocene terraces that edge the valley (lower center). Erosion shaped them into steep-sided mesas that are often isolated from the mountains they once flanked. The high cliffs of sandstone, siltstone, and coal above are part of the Mesaverde Group. Pink areas on these cliffs mark places where coal layers have burned and baked surrounding rock. —Felicie Williams photo

Utah 24
Fruita—I-70 near Green River
84 miles (135 km)

East of Fruita and the Capitol Reef National Park visitor center, Utah 24 follows the Fremont River through its narrow canyon past younger and younger rocks. Each layer bends downward along a prominent monocline called the Waterpocket Fold. (See the final chapter, Something Special, for more on Capitol Reef National Park.)

East of milepost 86, the Fremont River made a tight loop until Utah 24 was cut across the loop's neck. A new river channel was made next to the highway, complete with a waterfall. Along the river, boulders darkened by desert varnish are igneous rock from high country to the west.

Emerging from the canyon, Utah 24 enters a painted desert of red, green, and purple rocks of the Jurassic Morrison and Cretaceous Cedar Mountain Formations. These layers contain abundant clay that swells when wet and shrinks when dry, so few plants manage to grow on them. The cuesta east of these desert badlands is capped with tan Cretaceous Dakota Sandstone. The Dakota is not always present because much of it was eroded away as this region was uplifted during mid-Cretaceous time.

In these cliffs of Wingate Sandstone, rounded deep hollows called tafoni begin forming when rainwater dissolves the cement around sand grains and then freezes, expanding and prying grains loose. Wind whirls away the loosened grains, abrading the stone. —Lucy Chronic photo.

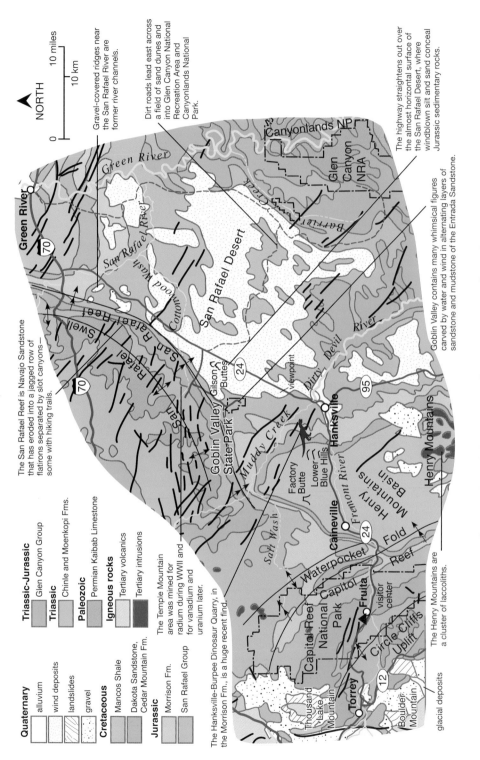

Quaternary

- alluvium
- wind deposits
- landslides
- gravel

Cretaceous

- Mancos Shale
- Dakota Sandstone, Cedar Mountain Fm.

Jurassic

- Morrison Frm.
- San Rafael Group

Triassic-Jurassic

- Glen Canyon Group

Triassic

- Chinle and Moenkopi Frms.

Paleozoic

- Permian Kaibab Limestone

Igneous rocks

- Tertiary volcanics
- Tertiary intrusions

The Temple Mountain area was mined for radium during WWII and for vanadium and uranium later.

The Hanksville-Burpee Dinosaur Quarry, in the Morrison Frm., is a huge recent find.

The San Rafael Reef is Navajo Sandstone that has eroded into a jagged row of flatirons separated by slot canyons—some with hiking trails.

Gravel-covered ridges near the San Rafael River are former river channels.

Dirt roads lead east across a field of sand dunes and into Glen Canyon National Recreation Area and Canyonlands National Park.

The highway straightens out over the almost horizontal surface of the San Rafael Desert, where windblown silt and sand conceal Jurassic sedimentary rocks.

Goblin Valley contains many whimsical figures carved by water and wind in alternating layers of sandstone and mudstone of the Entrada Sandstone.

The Henry Mountains are a cluster of laccoliths.

NORTH

0   10 miles

10 km

Green River

San Rafael River

San Rafael Swell

San Rafael Reef

Cottonwood Wash

Green River

San Rafael Desert

Barrier Creek

Canyonlands NP

Glen Canyon NRA

Gilson Buttes

Goblin Valley State Park

viewpoint

Dirty Devil River

Hanksville

Muddy Creek

Factory Butte

Lower Blue Hills

Caineville

Fremont River

Henry Mountains

Henry Mountains Basin

Salt Wash

Waterpocket Fold

Capitol Reef National Park

Capitol Reef

Fruita

visitor center

Circle Cliffs Uplift

Torrey

Thousand Lake Mountain

Boulder Mountain

glacial deposits

Geology along Utah 24 between Fruita and I-70 near Green River.

Farther east are gray hills of Mancos Shale. Look for dark coal layers and white bentonite. The cliff north of milepost 102 sparkles with gypsum crystals. Fossil oysters are common near mileposts 111 and 112, in the lowest layers of the Mancos Shale and the Dakota Sandstone below it. In places, benches of Mancos rich in oyster shells have been quarried for road material.

Utah 24 crosses the Caineville Anticline, the crest of which is west of Caineville, and then a gentle syncline—the Henry Mountains Basin—between Caineville and Hanksville. Near Hanksville, roadcuts again display colorful Morrison Formation shale and below it castellated brown bluffs of the Summerville Formation, which was deposited on Jurassic tidal flats.

The bridge north of Hanksville is at the confluence of the Fremont River and Muddy Creek. Together they become the Dirty Devil River, so named by John Wesley Powell in 1869. Where it flowed into the Colorado, Powell and his men found the water "exceedingly muddy" with "an unpleasant odor," and when one of the men said it was a "dirty devil," the name stuck. The Dirty Devil is overloaded with sediment, so it shifts course often, creating a braided channel.

Roadcuts north of the bridge between mileposts 119 and 120 expose Jurassic Entrada Sandstone. Here the rock is fairly well cemented, so it forms cliffs.

At the viewpoint near milepost 123, an informational display identifies topographic features. Thousand Lake Mountain and Boulder Mountain rise high to the west and southwest, both capped with Tertiary volcanic rocks. The Henry Mountains to the south are a cluster of igneous intrusions called laccoliths. In Oligocene time magma pushed up and between the sedimentary layers but did

The subtle grays of Mancos Shale seem rare in Utah's red rock country. In the Lower Blue Hills, a sandstone cap protects the soft shale. Once the cap crumbles, the shale will quickly weather to rounded, wavelike hills. —Felicie Williams photo

not erupt at the surface. The sedimentary rocks that were domed upward in this process are now eroded away; their upturned edges ring the mountains. Goblin Valley is to the west, and some of its stone goblins are just visible.

Farther north along the road, west of the highway near the Goblin Valley turnoff, Gilson Butte and Little Gilson Butte are decoratively carved remnants of Entrada Sandstone.

Goblin Valley State Park is intriguing, fun, and photogenic. The brick-red Entrada Sandstone here was deposited on tidal flats that had dunes and river channels. Above it, the pale green-white Curtis Formation was a near-shore marine deposit. The dark-red Summerville Formation was deposited in nearshore tidal flats, the overlying Morrison Formation in river floodplains. Together these rocks record an advance (to the east) and retreat (to the west) of a sea during Jurassic time. Just before the park entrance, a dirt road leads west to Little Wildhorse Canyon and several trails in slot canyons.

North of the turnoff to Goblin Valley State Park, Utah 24 parallels a stark hogback of Navajo Sandstone. Known as the San Rafael Reef, it will be in sight the rest of the way to I-70. The Reef is the eastern edge of the San Rafael Swell, a section of crust that was pushed up with the Wasatch Range and the Rockies around 60 million years ago.

The reef made the vast and uninhabited San Rafael Swell inaccessible until the uranium boom of the 1950s and '60s. When uranium ore was found in sandstone of the Chinle Formation, dirt roads proliferated. Many mines were developed. The mines are now inactive, and many have been sealed. Though it

In Goblin Valley, the Entrada Sandstone has eroded into hundreds of weird and amusing figures because of the rock's alternating hard and soft layers of mudstone, sandstone, and siltstone. —Felicie Williams photo

can be tempting to explore the old mines in this backcountry haven, they are generally unsafe and often radioactive. Pennsylvanian through Triassic rocks in the area contain many small occurrences of elaterite, a tar-like, brown hydrocarbon that seeps to the surface; so far the substance has not proven economical to mine.

The San Rafael Reef is the eastern edge of the San Rafael Swell. The word reef *emphasizes how dangerous it was for early settlers to pass.* —Felicie Williams photo

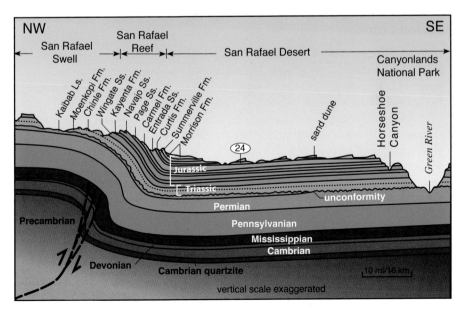

Northwest-southeast cross section across Utah 24. Precambrian rock beneath the Swell is probably offset by a fault, though the fault does not reach the surface. Sedimentary rocks have absorbed the displacement in uncountable small faults.

Small dunes edge much of the highway through the San Rafael Desert, their sand derived mostly from Jurassic rocks. Wind carries sediment loosened by water and frost and sandblasts grains loose from the rock as well. The dune field spreads far, from the Dirty Devil River to the San Rafael River, and from the Reef eastward to the edge of Canyonlands National Park. Strong wind lifts dust and fine sand high, sometimes carrying it far enough to the east to color snow on the Rockies.

Two types of sand dunes are common along the San Rafael Swell: longitudinal dunes with crests that parallel the wind, and parabolic dunes that resemble boomerangs. The arms of parabolic dunes are held in place by plant roots, while the wind blows sand out from their centers.

As Utah 24 descends toward the San Rafael River, sinuous gravel-covered ridges become visible to the west. They were once river channels. Their gravel became cemented with caliche, a limy crust that grows as alkaline groundwater evaporates. Eventually, the rivers found adjacent softer shale to dig into and left their cemented gravel beds high and dry. Look for gravel-filled channels in roadcuts as Utah 24 approaches I-70.

Utah 95
Hanksville—Blanding
126 miles (203 km)

Utah 95 leaves Hanksville on a plain of windblown silt and sand. Desert plants anchor small dunes. The occasional sculpted outcrops are of Entrada Sandstone and, beneath it, dark-red Carmel Formation siltstone. Ahead the gentle Colorado Plateau is broken by the Henry Mountains.

A few narrow mesas, including those near mileposts 2, 3, and 4, are the remains of a Pleistocene pediment. They are protected from erosion by stream gravel composed of hard gray igneous rock. It washed down from the Henry Mountains during the Pleistocene, when glaciers carved deeply into the peaks, and is partly cemented by white caliche. Younger streams have cut lower than the pediment, isolating the mesas.

The core of the Henry Mountains is composed of diorite porphyry intrusions of Tertiary age. Diorite has a larger percentage of dark-colored minerals than granite; *porphyry* refers to the texture of large crystals—here white plagioclase and black, spindly hornblende—embedded in a finer-grained groundmass.

Watch for small folds in Jurassic rocks near the Garfield County line, as well as Entrada Sandstone goblins. Farther south, the highway goes up and down through the Jurassic formations. Funnel-like depressions and small offsets in the bedding of the Jurassic Entrada Sandstone and underlying dark-red Carmel

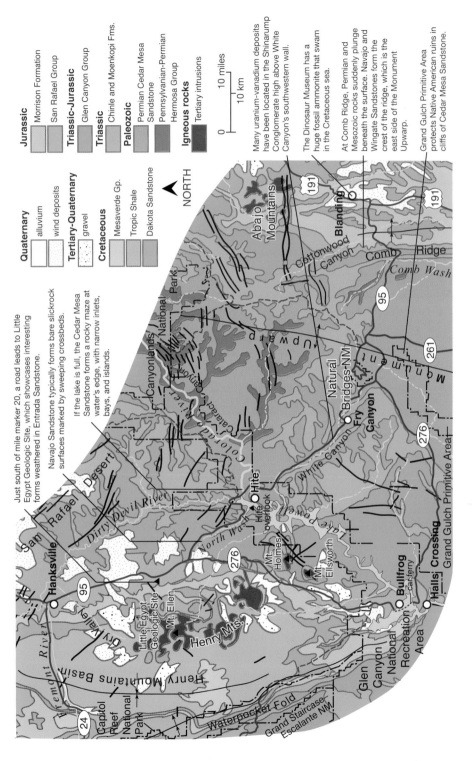

Geology along Utah 95 between Hanksville and Blanding.

Quaternary
- alluvium
- wind deposits

Tertiary-Quaternary
- gravel

Cretaceous
- Mesaverde Gp.
- Tropic Shale
- Dakota Sandstone

Jurassic
- Morrison Formation
- San Rafael Group

Triassic-Jurassic
- Glen Canyon Group

Triassic
- Chinle and Moenkopi Fms.

Paleozoic
- Permian Cedar Mesa Sandstone
- Pennsylvanian-Permian Hermosa Group

Igneous rocks
- Tertiary intrusions

NORTH

0   10 miles
0   10 km

Just south of mile marker 20, a road leads to Little Egypt Geologic Site, which showcases interesting forms weathered in Entrada Sandstone.

Navajo Sandstone typically forms bare slickrock surfaces marked by sweeping crossbeds.

If the lake is full, the Cedar Mesa Sandstone forms a rocky maze at water's edge, with narrow inlets, bays, and islands.

Many uranium-vanadium deposits have been located in the Shinarump Conglomerate high above White Canyon's southwestern wall.

The Dinosaur Museum has a huge fossil ammonite that swam in the Cretaceous sea.

At Comb Ridge, Permian and Mesozoic rocks suddenly plunge beneath the surface. Navajo and Wingate Sandstones form the crest of the ridge, which is the east side of the Monument Upwarp.

Grand Gulch Primitive Area protects Native American ruins in cliffs of Cedar Mesa Sandstone.

Jurassic and Cretaceous rocks ring the Henry Mountains. Beyond the cliffs in the foreground, the beds rise upward. The Morrison Formation is only sandstone here; its usual colored shale is absent because this area was near the sloping edge of the basin in which it was deposited, and thus the basin received slightly coarser sediment from a nearby highland. —Felicie Williams photo

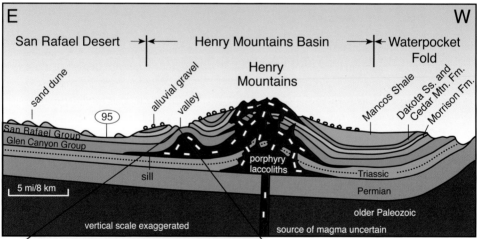

The Henry Mountains formed when magma flowed upward and between sedimentary layers, doming and stretching them but not ever erupting at the surface. Inset: Sedimentary rocks adjust to stresses like this by micro-fracturing—breaking along very small shears (lines in the rock) that spread throughout the rock.

Formation are common over much of southeast Utah. These features formed while the sediment was still soft, as the weight of the developing, heavy Entrada dunes on still-wet Carmel beds forced the water in them to whirl upward through the Entrada dunes.

Less than 2 miles (3 km) south of the Utah 276 turnoff, lighter beds of Jurassic dune sandstone appear below the Carmel Formation. The upper 100 feet (30 m) is Page Sandstone, which is separated from the Navajo Sandsone below by an unconformity. Both weather into rounded slopes. About 3 miles (5 km) south of the turnoff, the Kayenta Formation, red river-deposited sandstone and siltstone, appears below the Navajo, followed by darker Triassic-Jurassic Wingate Sandstone, another dune deposit. The Navajo, Kayenta, and Wingate make up the Glen Canyon Group.

The typical Wingate Sandstone surface is deeply pitted by weathering. Less-weathered rock is smoother where slabs have broken away. Blue-black patina—desert varnish—develops on the sandstone as wind-borne clay combines with manganese and iron oxides leached from the sandstone by repeated soaking and drying. Lichens, mosses, algae, and fungi growing where water flows down the cliffs after rain make vertical black streaks. Together, these coatings mask the crossbedding in the Wingate.

Beginning at Hog Spring, the colorful blue, purple, and red Chinle Formation shows up below the Wingate Sandstone. Deposited on a wide Triassic floodplain dotted with forests, shallow ponds, and marshes, the Chinle contains mudstone, sandstone, conglomerate, and lots of volcanic ash. Its multicolored layers preserve ancient soil horizons. Hard angular ledges near and at the formation's base, close to the Glen Canyon National Recreation Area sign, are the Moss Back and Shinarump Conglomerates, hosts to many uranium deposits.

Below the Chinle, the Triassic Moenkopi Formation's dark-red beds are products of a coastal plain, where mud, silt, and sand, washed from highlands to the south and east, accumulated in well-ordered layers. Hite overlook is in the Moenkopi Formation.

White rocks that edge the lake at Hite are Cedar Mesa Sandstone, an erosion-resistant nearshore and dune sandstone that weathers into rounded, potholed surfaces. Its crossbeds formed under the easterly winds of the tropics during Permian time.

Lake Powell's upper end is always murky from sediment carried in by the Colorado and Dirty Devil Rivers. The sediment is gradually building a delta into the lake. Finer sediment travels farther and settles in deeper, stiller water. Upstream, the whitewater stretch of the Colorado River in Cataract Canyon has challenged river runners ever since John Wesley Powell's first trip in 1869.

Southeast of the bridges the strata rise eastward onto the Monument Upwarp. Several faults offset the Cedar Mesa Sandstone and overlying Organ Rock Shale and White Rim Sandstone, all of Permian age. The White Rim's layers, deposited on the shoreline of an ocean, soon grade eastward into Cutler Arkose, which was shed westward from the Permian mountain range of Uncompahgria.

About 0.5 mile (0.8 km) past the Hite overlook, roadcuts expose Moenkopi Formation mudstone marked by carrot-like wedges—former mud cracks that filled with sand. —Felicie Williams photo

The area southeast of Hite has a lot of soft windblown soil colored by hematite from the Moenkopi Formation. Near the highway, the White Creek swings in looping meanders inherited from a time before canyon cutting began—when it still flowed in more easily eroded sediments; it now flows some 60 feet (18 m) below the top surface of the Cedar Mesa Sandstone.

The roadbed holds to the Cedar Mesa Sandstone surface over the crest of the Monument Upwarp. The view to the south shows the low dome of this immense anticline, which was pushed up during the Laramide Orogeny around 70 million years ago. In the distance ahead, in Colorado, is Sleeping Ute Mountain. The Carrizo Mountains are to the south, in Arizona, and to the north the Abajo Mountains. All have igneous intrusions for cores.

As the road dives into Comb Wash, we see the same Triassic and Jurassic strata we saw between Hanksville and Lake Powell. Soft Permian and Triassic shales have weathered to form the valley. Resistant Glen Canyon Group sandstones form the amazing Comb Ridge. East of the ridge, the younger Jurassic layers make a succession of valleys and hogbacks until, near milepost 118, the road climbs onto a tableland of Cretaceous Dakota Sandstone. Canyons reveal weak purple and green mudstones of the Cedar Mountain and Morrison Formations below the Dakota. A veneer of windblown silt and sand hides the rocks near the junction with US 191.

Flame-colored crossbeds in Cedar Mesa Sandstone are colored by hematite. The entire sandstone originally contained hematite. Long after it had formed, probably during the Laramide Orogeny or the Tertiary igneous episode, groundwater moved through the more porous areas of the rock, dissolving and carrying away hematite.
—Felicie Williams photo

Comb Ridge's Navajo Sandstone barrier curves southwestward to Kayenta, Arizona, defining the eastern, steeper flank of the Monument Upwarp. —Felicie Williams photo

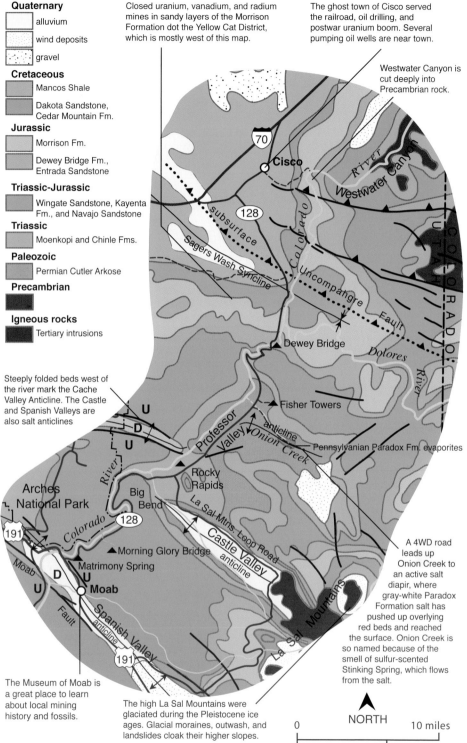

Quaternary
- alluvium
- wind deposits
- gravel

Cretaceous
- Mancos Shale
- Dakota Sandstone, Cedar Mountain Fm.

Jurassic
- Morrison Fm.
- Dewey Bridge Fm., Entrada Sandstone

Triassic-Jurassic
- Wingate Sandstone, Kayenta Fm., and Navajo Sandstone

Triassic
- Moenkopi and Chinle Fms.

Paleozoic
- Permian Cutler Arkose

Precambrian

Igneous rocks
- Tertiary intrusions

Closed uranium, vanadium, and radium mines in sandy layers of the Morrison Formation dot the Yellow Cat District, which is mostly west of this map.

The ghost town of Cisco served the railroad, oil drilling, and postwar uranium boom. Several pumping oil wells are near town.

Westwater Canyon is cut deeply into Precambrian rock.

Steeply folded beds west of the river mark the Cache Valley Anticline. The Castle and Spanish Valleys are also salt anticlines

Pennsylvanian Paradox Fm. evaporites

A 4WD road leads up Onion Creek to an active salt diapir, where gray-white Paradox Formation salt has pushed up overlying red beds and reached the surface. Onion Creek is so named because of the smell of sulfur-scented Stinking Spring, which flows from the salt.

The Museum of Moab is a great place to learn about local mining history and fossils.

The high La Sal Mountains were glaciated during the Pleistocene ice ages. Glacial moraines, outwash, and landslides cloak their higher slopes.

Cisco

Dewey Bridge

Fisher Towers

Professor Valley

Onion Creek anticline

Rocky Rapids

Arches National Park

Big Bend

La Sal Mtns. Loop Road

Castle Valley anticline

Morning Glory Bridge

Matrimony Spring

Moab

Moab Fault

Spanish Valley anticline

La Sal Mountains

Colorado River

Westwater Canyon

Uncompahgre Fault

Dolores River

Sagers Wash Syncline

subsurface

UTAH | COLORADO

0   10 miles

NORTH

10 km

Geology along Utah 128 between Cisco and Moab.

Utah 128
Cisco—Moab
44 mi (71 km)

Utah 128 parallels I-70 south of the Book Cliffs, passing through a nearly flat valley of Cretaceous Mancos Shale and crossing part of the Greater Cisco Oil and Gas Field. Hydrocarbons are trapped here in small faulted anticlines in Jurassic and Cretaceous rocks. The road follows the Sagers Wash Syncline southeast. Note how beds slope toward the highway on both sides of the valley. This syncline marks the southwest flank of the Uncompahgre Plateau, which was pushed up around 70 million years ago by tectonic pressure brought on by subduction along the west coast.

Low, buff sandstone cliffs at milepost 33 are composed of the Dakota Sandstone and Cedar Mountain Formation; their sediments represent an environment in which beaches of the encroaching Cretaceous sea were overriding the river channel gravels that preceded them. Below them is the Morrision Formation, a thick sequence of soft multicolored shale deposited on a broad Late Jurassic floodplain. Both the Morrison and the Cedar Mountain are known for their dinosaur fossils, which are particularly abundant in ancient river channels and water holes. Between mileposts 32 and 31 the road passes the Salt Wash Member, which was deposited in sand dunes and rivers. At milepost 31, Utah 128 crosses into red cliff-forming Entrada Sandstone, which was laid down in a widespread dune field during Late Jurassic time.

At Dewey Bridge, the Jurassic Dewey Bridge Formation comes to light under the Entrada Sandstone. A 40-foot-thick (12 m) sequence of dark-red, soft, interbedded sandstone and mudstone deposited on a tidal flat, the Dewey Bridge Formation is the time equivalent of the Carmel Formation elsewhere

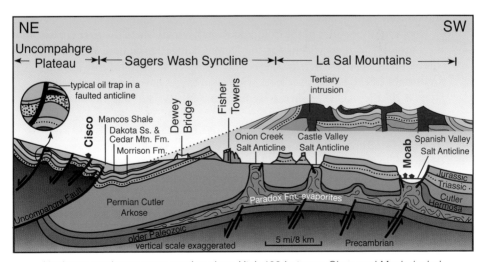

Northeast-southwest cross section along Utah 128 between Cisco and Moab, including the La Sal Mountains, which are projected in the distance behind the actual path of the cross section.

The foundation of the historic Dewey Bridge rests on the Dewey Bridge Formation, with cliffs of Entrada Sandstone and Morrison Formation above. The bridge, built in 1916, burned in 2008. —Felicie Williams photo

in Utah. Almost everywhere its beds are wavy; they were contorted while still soft, probably due to a combination of the loading on of heavy Entrada sand overhead, the growth and weathering of thin layers of gypsum among the beds, and possibly earthquake shaking.

Notice how the valley of the Colorado River narrows where the river crosses the Entrada and then widens for the Dewey Bridge Formation. Downriver, this becomes a pattern: narrow canyons have formed where hard cliff-forming sandstones extend below river level; wide open spaces have formed where soft layers at river level have washed away, undermining the cliffs and causing massive pieces to fall from them.

South of the Dewey Bridge, Utah 128 curves into a canyon of Navajo Sandstone; the sandstone has orange-pink cliffs with joints and long crossbeds that developed in dunes. Near the boat ramp just past the bridge, the road descends into the darker-orange, stream-deposited Kayenta Formation sandstone, which is interbedded with siltstone and occasionally conglomerate. Below the Kayenta, the red Wingate Sandstone was a field of dunes in Triassic to Jurassic time. Near milepost 29 the canyon cuts into Triassic shales and sandstones: the dark-red Chinle Formation over the similar but slightly darker Moenkopi Formation. In places, a thin white bed marks their contact. Look for a conglomerate composed of pebbles at the base of the Moenkopi as the canyon widens out into Professor Valley. Below it is the Permian Cutler Arkose.

The purplish-brown Cutler Arkose is a gritty, coarse rock interbedded with a few beds of windblown sandstone and, near its base, limestone. Its coarseness, high feldspar content (feldspar breaks down more rapidly than other minerals), and poorly rounded, poorly sorted grains show it wasn't carried far from its source. The Cutler is an amazing 16,000 feet (4,900 m) thick here! It was eroded from a high mountain range to the northeast called Uncompahgria. The range was pushed up by forces related to a tectonic collision along the east coast that helped form the supercontinent Pangaea. Thrust southwestward, Uncompahgria weighed down the land west of it, forming the Paradox Basin.

The Paradox Basin filled with evaporite deposits that mixed with minor amounts of other sediments and became the Paradox Formation. Buried by younger sediments, the evaporites, which weighed less than the overlying sediments, flowed upward through some of the overlying heavier sediments and domed others. This formed several anticlines, which are commonly called salt anticlines.

West of milepost 18, the beds are extremely folded and faulted. This is the crest of the Cache Valley Anticline. Its folded and faulted form leads northwest to the Salt Valley Anticline in Arches National Park. Southeast, it curves across the valley and becomes the Onion Creek Anticline. This salt anticline, along with several others, provided pathways for the magmas that formed the La Sal Mountains. Hard quartzite forming the Rocky Rapids on the Colorado River is

The Fisher Towers, east of Utah 128, are composed of Cutler Arkose. The tallest tower, capped by more-resistant sandstone layers of the Moenkopi Formation, is 900 feet (274 m) high—three football fields end to end. Hiking trails lead around their base. —Felicie Williams photo

Morning Glory Bridge, in Navajo Sandstone, is a 2.2 mile (3.5 km) hike from the Negro Bill parking lot. It is called a bridge because it spans a creek, though it formed like an arch from a spalled and weathered rock fin. Note the strong parallel jointing that helped form the huge rock fin. Spanning 243 feet (74 m), it is the world's seventh longest known free-standing natural arch, 9 feet (3 m) longer than Rainbow Bridge. —Felicie Williams photo

West of Moab, cliffs of Navajo Sandstone darkened with desert varnish rise behind the Colorado River, which is red with mud after a thunderstorm. —Felicie Williams photo

probably Moenkopi Formation sandstone, bleached and cemented by quartz-rich fluids that escaped from the magma along the anticline.

Utah 128 has crossed into rock of the younger Moenkopi Formation by milepost 17 and into that of the Chinle Formation at milepost 13. Utah 128 quickly crosses through the Wingate, Kayenta, and Navajo sandstones as it approaches the US 191 junction. About 0.2 mile east of the junction, on the south side of the road, Matrimony Spring marks the base of the Navajo. Rainwater trickles down through the porous sandstone until it reaches a silty layer, which it flows along and out of the cliff face. The steep canyon ends suddenly when the highway crosses into another faulted salt anticline, the wide Moab Valley (Spanish Valley). At the junction, turn left (south) onto US 191 into Moab.

Utah 276
Natural Bridges—Utah 95 via Halls Crossing at Lake Powell
92 miles (148 km)

A treasure of a scenic byway, Utah 276 crosses the western side of the Monument Upwarp, ferries across Lake Powell, and threads through the Henry Mountains. The car ferry runs during daylight hours only and is closed for part of winter.

Utah 276 heads southwest from Utah 95 several miles northwest of the turnoff to Natural Bridges National Monument. The highway drifts gradually downward through piñons, junipers, and windblown sand that cover a surface of Permian Cedar Mesa Sandstone, itself made of windblown sand 270 million years ago. The long Red House Cliffs, west of the road, are the Permian Organ Rock Shale and Triassic Moenkopi and Chinle Formations. The Organ Rock and Moenkopi were once tidal mudflats; their bedding surfaces still have ancient ripple marks, mud cracks, and raindrop impressions. White gypsum layers distinguish parts of the Moenkopi Formation. Near milepost 84, light-colored sandstone was a river channel. Above, a thick conglomerate of the Chinle Formation contains gritty river and floodplain sediment deposited on a low, northwest-sloping plain. Its varicolored Petrified Forest Member is banded with ancient soil layers.

Just west of the Clay Hills Pass the road leaves the Chinle Formation shale and enters an area with thick cliff-forming beds of dark red-brown dune-deposited Wingate Sandstone. It then continues down the slope of the Monument Upwarp on the Wingate and overlying, horizontally bedded Kayenta Formation, passing scattered weathered knobs of lighter-hued Navajo Sandstone. Around milepost 56, the highway enters a large field of orange sand dunes, the fine sand of which was derived from weathered Mesozoic rock. The orange color is iron oxide derived—with the original sand of the Mesozoic rocks—from even older igneous and metamorphic rocks of Precambrian age.

Lake Powell is named for geologist and ethnographer John Wesley Powell. In 1869 Powell captained a team of ten men on four specially built dories in the first descent and survey of the canyons of the Green and Colorado Rivers. Powell

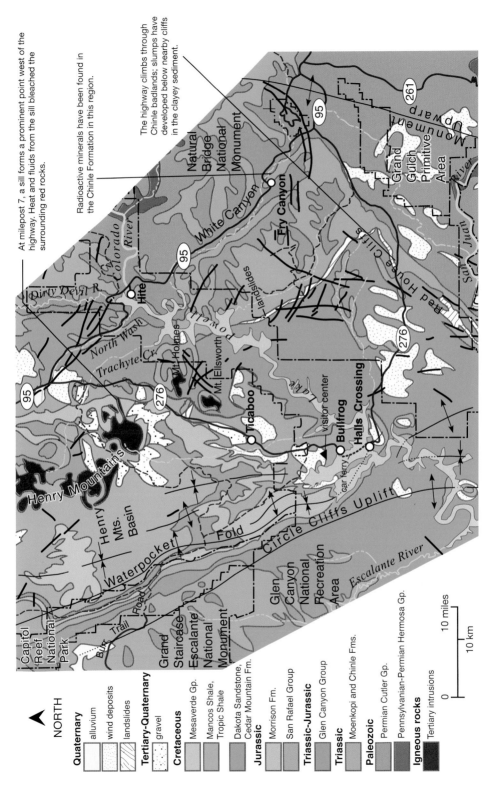

At milepost 7, a sill forms a prominent point west of the highway. Heat and fluids from the sill bleached the surrounding red rocks.

Radioactive minerals have been found in the Chinle Formation in this region.

The highway climbs through Chinle badlands; slumps have developed below nearby cliffs in the clayey sediment.

NORTH

Quaternary
alluvium
wind deposits
landslides

Tertiary-Quaternary
gravel

Cretaceous
Mesaverde Gp.
Mancos Shale, Tropic Shale
Dakota Sandstone, Cedar Mountain Fm.

Jurassic
Morrison Fm.
San Rafael Group

Triassic-Jurassic
Glen Canyon Group

Triassic
Moenkopi and Chinle Frms.

Paleozoic
Permian Cutler Gp.
Pennsylvanian-Permian Hermosa Gp.

Igneous rocks
Tertiary intrusions

0   10 miles
0   10 km

Geology along the Utah 276 loop.

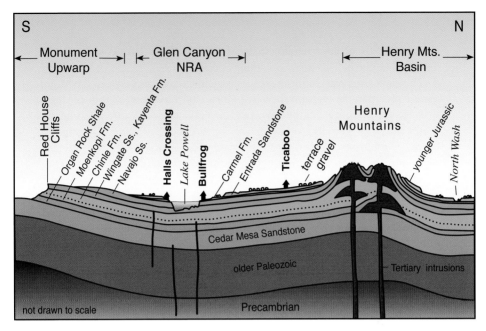

S   N

Monument Upwarp ⟷ ⟷ Glen Canyon NRA ⟷ ⟷ Henry Mts. Basin

Red House Cliffs
Organ Rock Shale
Moenkopi Fm.
Chinle Fm.
Wingate Ss., Kayenta Fm.
Navajo Ss.
Halls Crossing
Lake Powell
Bullfrog
Carmel Fm.
Entrada Sandstone
Ticaboo
terrace gravel
Henry Mountains
younger Jurassic
North Wash

Cedar Mesa Sandstone

older Paleozoic

Tertiary intrusions

not drawn to scale

Precambrian

North-south cross section along Utah 276 through Halls Crossing and the Henry Mountains.

A dune field on the southeast side of Lake Powell. Tracks of beetles and tail-dragging lizards cross the wind-rippled surface. Plants stabilize many of the dunes. —Felicie Williams photo

also mapped the geology, naming many of the formations. Lake Powell is further discussed in the final chapter, Something Special.

The Halls Crossing–Bullfrog area is a syncline that rests between the Monument Upwarp to the east and the Circle Cliffs Uplift to the west. Younger Jurassic rocks are present here: the dark-red Carmel Formation, a tidal-flat deposit found near the waterline of Lake Powell, and rounded cliffs of orange-red Entrada Sandstone. To the northeast the syncline broadens into the Henry Mountains Basin, with even younger rocks in its center.

Behind the Bullfrog visitor center the Entrada Sandstone records a change in environment: long wind-deposited dune crossbeds change to horizontally bedded layers that were deposited in and/or were reworked by an advancing ocean.

Gold fever reached Halls Crossing in 1883, and a dredge was laboriously brought in. Dredges capture gold by digging up river gravel and then washing it through a long sluicebox with ladderlike riffles. The much heavier gold settles to the bottom, where it is caught in the riffles. The dredge proved incapable of capturing the river gravel's fine "flour" gold. Surface tension held the tiny particles of flour gold on top of the water, so it was carried off when the gravel was washed. The small amount of gold in the Henry Mountains was quickly mined out.

North of Bullfrog, Utah 276 climbs onto a surface of Pleistocene gravel. Many of the rounded pieces are hard igneous rock from the Henry Mountains. To the northwest, east-sloping flatirons disappearing into the distance mark the Waterpocket Fold.

The intrusive rocks of the Henry Mountains are mostly porphyry, with two distinct grain sizes. The large crystals of white feldspar and black hornblende grew over time while the magma was deep in the Earth. As the magma rose upward, the rest of the magma chilled quickly, forming the small mineral grains surrounding the larger ones. —Felicie Williams photo

As the Henry Mountains magma intruded sedimentary rock layers, they buckled and were faulted. The layers then eroded into a pediment that was overlain by gravel washed off the mountains during Pleistocene time. Creeks have cut down significantly since then, leaving the old gravel deposits high on the sides of valleys. —Felicie Williams photo

The Henry Mountains are laccoliths, a term coined in a study of these mountains by geologist G. K. Gilbert in 1877. Laccoliths form when magma intrudes between layers of sedimentary rock, forming sills. Some sills swell, doming up overlying sedimentary layers.

North of the pass at the south end of the Henry Mountains, the highway rides on the deeply gullied Page and Navajo Sandstones, identical dune deposits separated by an unconformity. Younger rocks crop out between the highway and the mountains: red-orange Entrada Sandstone capped by white Curtis Formation, a shoreline deposit. Over them is the dark red-brown, horizontally bedded Summerville Formation, a tidal-flat deposit, followed by pale cliffs of Jurassic sandstone—the Morrison Formation's Salt Wash Member. Finally, these rocks are topped with thin Cedar Mountain and Dakota sandstones, both shoreline deposits, and a little gray marine Mancos Shale; all three are Cretaceous in age.

Utah 276 climbs and then levels out on a dune-covered surface of Entrada Sandstone. Cliffs of the Summerville Formation are now closer to the road. Utah 276 descends into the typically contorted beds of the brilliant red Carmel Formation and ends at Utah 95.

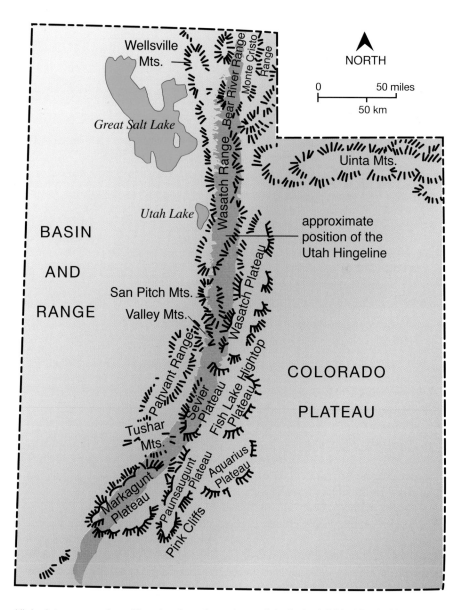

High plateaus, smothered in volcanic rocks and part of the faulted, folded Rocky Mountains, make up Utah's High Country. This land records multiple periods of mountain building.

UTAH'S BACKBONE: THE HIGH COUNTRY

Utah's high mountains, paralleling the Utah Hingeline, the transition between the stable Colorado Plateau and the more easily folded Basin and Range, run like a backbone down the center of the state. The mountains come in many shapes and varieties. Geologically and geographically they compose three basic regions:

- The Uinta Mountains, a range of the Rocky Mountains in the northeast corner of the state, is one big anticline, with faults along both edges. The range is unusual in that it trends east to west, whereas most ranges of the Rockies trend north to south. The Uintas extend westward almost to the Wasatch Range, but their underlying anticlinal structure, though intensely faulted, continues on through the Wasatch.

- The Wasatch Range experienced two episodes of mountain building after Precambrian time that superimposed whole formations, folds, and faults onto each other. The range was shoved up as a series of overlapping thrust sheets during the Sevier Orogeny and then was later sliced by normal faults during Basin and Range extension.

- The High Country south of the Wasatch Range bears the simplicity of the Colorado Plateau. Basin and Range faults have divided this land into long north-south-trending plateaus, tilting them gently eastward in the process. These tablelands are partly blanketed by great piles of volcanic material— lava flows, breccia, and ash—that was in place before Basin and Range faulting began. The erosion-resistant volcanic rocks help the plateaus maintain their present high elevations.

MOUNTAIN BUILDING DRAMATICALLY EXPOSED

Utah's mountains are at the center of the world's understanding of thrust faulting and extensional (normal) faulting. Here, where plant cover is minimal and uplift and massive erosion have exposed the rocks well, the evidence of Earth's movement is dramatic indeed. Highways through these mountains often follow deep canyons, whose walls record thrust faults and normal faults, anticlines and synclines, ancient soil horizons, and episodes of erosion. Wells drilled through the layers add to our understanding. What follows are the main events that helped shape these mountains and the rocks and structural features that record the events. In all cases, the compass directions given are for modern-day Utah rather than those for the particular age in which a structural feature formed.

- **2,700 to 1,650 million years ago:** Hard, intensely metamorphosed Precambrian schist and gneiss are exposed in Utah's northern mountains. The rocks record long episodes of deposition, uplift, and erosion followed by continental collisions that caused their metamorphism and reset their radiometric age to 1,650 million years ago. One mountain building episode involved two early continents colliding along an east-west suture line that runs across the state approximately where the fault along the north side of the Uinta Mountains lies today.

- **1,650 to 770 million years ago:** A great unconformity, a gap in time with no record, is all there is to represent this time frame in Utah; either no sediments were being deposited or those that were deposited were subsequently eroded away. During this time, previous mountain ranges were eroded down to near sea level.

- **770 to 740 million years ago:** At the beginning of this period of time, Utah was part of a larger supercontinent called Rodinia. The continent started to rift west of Utah and parallel to the Utah Hingeline (a rough north-south line that crosses the state not far from I-15). The western part of the supercontinent would become Australia and Antarctica, and the eastern part would become North America. As the rift opened in Nevada, the ocean inundated Utah. The part of the crust west of the Utah Hingeline but east of the rift stretched and thinned. At the same time, an extensional basin formed, caused by north-south tectonic stress related to the rifting. The

Gneiss of the Farmington Canyon Complex was once sedimentary rock. It was metamorphosed at high temperature and pressure during mountain building 1,650 million years ago. —Lucy Chronic photo

tension caused the basin to sink south of a long east-west-trending normal fault that ruptured at the 1,650-million-year-old suture between the two Precambrian continents mentioned above. The extensional basin lay roughly where the Uinta Mountains are today. Rivers flowed west and south off of higher surrounding land into the basin, and up to 28,000 feet (8,500 m) of river, alluvial fan, delta, and shallow marine deposits filled it.

- **740 to 251 million years ago:** During late Precambrian and Paleozoic time, almost all of Utah was tectonically quiet. Mostly marine sediment accumulated across the state. Thicker and more continuous sequences of sediment accumulated over the stretched and thinned crust west of the Utah Hingeline, as the land there slowly subsided in a hingelike manner, with thicker sediments accumulating the farther west they were. In the late Paleozoic, Uncompahgria—a mountain range—rose along the state's eastern edge and then eroded down.

- **251 to 145 million years ago:** Mountain building to the west encroached on the region and the ocean receded. Mesozoic layers reflect these developments: sediment deposited on land replaced that deposited in oceans; sedimentary layers grew coarser, indicating they were deposited closer to their highland sources by powerful rivers; ash from distant volcanoes arrived; and vast dune fields developed behind the mountains because the mountains blocked moisture from westerly storms.

- **145 to 125 million years ago:** The Sevier Orogeny, due to pressure from the subduction of the Farallon Plate on the west coast, moved east into Utah. In much of central Utah this resulted in a period of erosion and the development of an unconformity spanning most of Cretaceous time and into the beginning of Paleogene time.

- **125 to 50 million years ago:** The Sevier Orogeny peaked in Utah. Thick sedimentary layers west of the Utah Hingeline were thrust eastward in enormous sheets, one after another. Entire ranges—including the Wasatch Range—were forced upward into tremendous anticlines by tectonic compression. A large basin developed east of the hingeline, in front of the uplifted mountains.

- **70 to 40 million years ago:** The wave of mountain building reached the crust east of the Utah Hingeline during the Laramide Orogeny. When compression from the subduction of the Farallon Plate reached the Colorado Plateau, the crust and overlying sedimentary layers behaved differently than they had east of the Utah Hingeline. Instead of huge thrust sheets moving east over adjacent sediments, mountain blocks were uplifted along almost vertical faults, and the sedimentary layers were warped into anticlines and synclines, with the anticlines often forming directly over the faults. The San Rafael Swell, Circle Cliffs Uplift, and other uplifts, which had begun to rise during the Sevier Orogeny, were warped upward even more. The mass of the Uinta Mountain Group, rocks that had been deposited in the Uinta extensional basin, rose wholesale, partly along the old faults—especially the fault along the northern side of the Uinta Mountains—that had developed

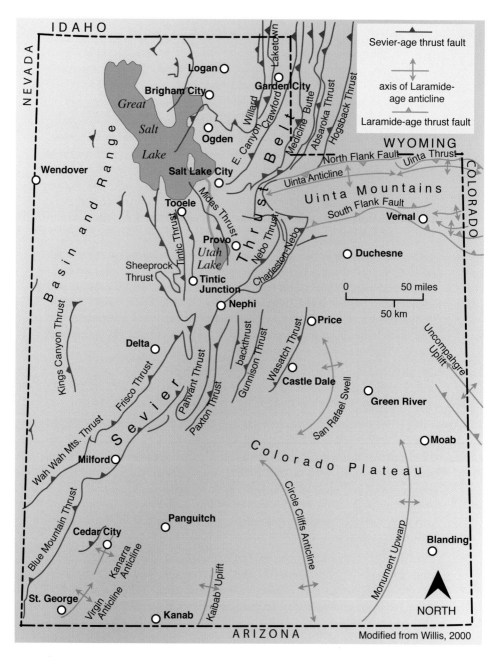

Thrust faults developed throughout western and central Utah during the Sevier and Laramide Orogenies. In the western desert, more thrust faults are present, but they are buried; only short sections are exposed in the isolated ranges. The triangles on the faults are on the blocks of crust that were thrust over rock beneath the faults.

during the Precambrian. Basins flexed down between the uplifts and filled with sediment eroded from the mountains.

- **40 to 30 million years ago:** Mountain building faded as subduction ended on the west coast. The continent relaxed as if letting out its breath. Rock layers slid back westward a little as the subduction-caused tectonic compression was relieved.

- **40 to 20 million years ago:** As subduction ended, the Farallon Plate began to sink beneath the continent. Bodies of magma rose from the newly opened wedge of asthenosphere above the Farallon Plate, incorporating melted crust along the way. These magmas intruded the crust along the many faults left from mountain building, built bodies of magma like the granite batholiths in the Wasatch Range, and produced volcanic eruptions that blanketed southern Utah. Hot fluids escaping the intrusions left valuable minerals.

- **17 million years ago to present:** Tectonic extension increased across Utah, causing the crust to stretch westward. This happened for two reasons: gradual uplift raised the entire Plateau up to 2 miles (3 km) relative to lands west of it, causing tension between the two different elevations; and northward shear intensified along the developing San Andreas Fault, which, in essence, pulled the crust in a northwesterly direction. Normal faults developed throughout the Basin and Range region, forming the many small ranges of western Utah, the Wasatch Front, and the high plateaus as blocks of crust dropped downward along the faults. The deep faults tapped reservoirs of basaltic magma in the Earth's mantle, leading to volcanic eruptions of dark lavas and cinder cones.

Erosion whittled down the mountains and carved deep canyons. Pleistocene time brought glaciers to the High Country and more water to increase the erosional power of streams and rivers. Of course, geologic processes continue today as they have in the past. Helped along by heavy winter snows and summer rains, landslides and wintertime avalanches are frequent. Slides sometimes dam valleys, and the lakes that form later break through them and cause floods. Many slides—old and new—can be seen from the highways in the High Country; some were even initiated by highway construction.

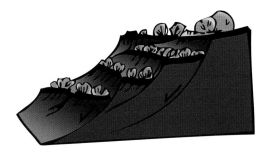

Small landslides and slumps, their horseshoe-shaped scarps showing where they broke away from slopes above, have damaged many highway cuts. Often several slide surfaces form cat steps, or small terraces, each still covered with soil, grass, and shrubbery.

MANY USES FOR THESE MOUNTAINS

Water from snowmelt and rain is the most important commodity provided by Utah's mountains. This water is needed for irrigation and all other uses because there is so little rainfall in Utah. Dams in the mountain valleys also produce hydroelectric power.

Mines and drill holes pepper the mountains and produce many things, including gold, silver, hydrocarbons, and building materials. Today, the mountains are crisscrossed by roads that lead to both current mines and old mining ghost towns.

Utah's mountains provide a beautiful and varied playland. Ski areas, mountain bike and hiking trails, and pristine wilderness areas draw millions of visitors and residents alike. But more importantly, Utah's mountain wildernesses preserve life in many forms and are the reason for that life in the first place.

ROAD GUIDES FOR UTAH'S HIGH COUNTRY

I-70
Fremont Junction—Cove Fort
89 miles (143 km)

Fremont Junction lies in Castle Valley, a long valley eroded in Cretaceous Mancos Shale below the Book Cliffs. From Fremont Junction, I-70 soon begins to climb past the cliffs, thick yellow-gray sandstones alternating with beds of slope-forming shale. The layers belong the Cretaceous Mesaverde Group.

The older layers of the Mesaverde Group contain several black coal seams. Coal forms when peat in marshes and swamps is compacted by the weight of overlying sediment and is warmed by Earth's heat. Here, the interfingering of sandstone, shale, and coal was caused by the come-and-go retreat of the Cretaceous Interior Seaway. Some of the coal, probably ignited by lightning, has burned and oxidized surrounding rocks to a light brick-red. The Mesaverde Group is about 2,000 feet (610 m) thick here. It is topped by thick layers of the North Horn Formation, siltstone and sandstone deposited across the Cretaceous-Tertiary time boundary. The North Horn accumulated along the front of the mountains of the Sevier Orogeny and in the adjacent basin. On top of these layers is the Flagstaff Formation, soft pink layers that were soils and lake sediments during Tertiary time.

The area is cut into slivers by numerous north-south normal faults; the highest mountains are faulted slivers. Faults parallel many ridges. In roadcuts, the offsets along faults are marked by changes in color. North-south stretches of the highway run parallel to faults.

Landslides are common in these soft rocks. Watch for their hummocky rubble. Some slides are quite near the road. In places, the slumping has even buckled the highway. The region's faults increase the likelihood that slides will

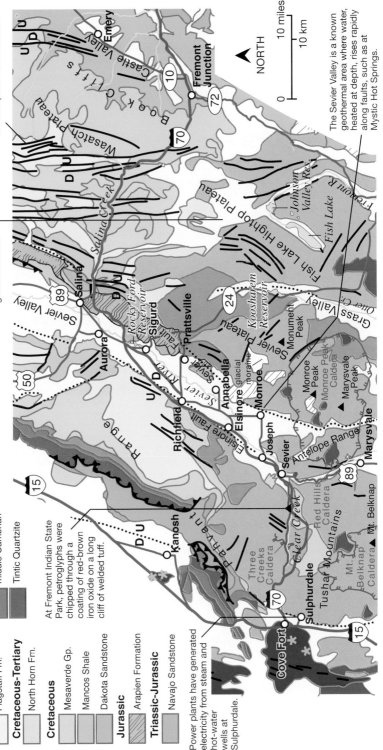

Geology along I-70 between Fremont Junction and Cove Fort.

Quaternary

- alluvium
- landslides

Tertiary

- Sevier River Fm.
- Green River Fm.
- Flagstaff Fm.

Cretaceous-Tertiary

- North Horn Fm.

Cretaceous

- Mesaverde Gp.
- Mancos Shale
- Dakota Sandstone

Jurassic

- Arapien Formation

Triassic-Jurassic

- Navajo Sandstone

Paleozoic

- Pennsylvanian-Permian
- Mississippian
- Devonian
- Silurian
- upper Cambrian
- middle Cambrian
- Tintic Quartzite

Igneous rocks

- Quaternary basalt
- Miocene volcanic rocks
- Oligocene volcanic rocks
- Tertiary basalt
- Tertiary intrusions

Scores of north-south-trending faults have been mapped in this area; only a few can be shown at this scale. The steep slopes of barely cemented sediment make the area prone to landslides.

Some Quaternary deposits on the Fish Lake Hightop Plateau are glacial moraines. Others are landslides, stream gravels, and glacial outwash.

The Sevier Valley is a known geothermal area where water, heated at depth, rises rapidly along faults, such as at Mystic Hot Springs.

At Fremont Indian State Park, petroglyphs were chipped through a coating of red-brown iron oxide on a long cliff of welded tuff.

Power plants have generated electricity from steam and hot-water wells at Sulphurdale.

occur: springs along the faults keep the rocks and soil wet and lubricated, so movement along the faults easily triggers landslides.

Around milepost 70, I-70 begins to follow the path of Salina Creek, passing into Salina Canyon. Roadcuts between mileposts 70 and 69 again expose coal-bearing beds of the Mesaverde Group. Former stream channels of crossbedded sandstone thread through them, which is typical of sediments deposited in a shoreline environment.

In the big roadcut near milepost 58, tilted Cretaceous sandstone and shale are overlain by younger lime-rich sandstone that was deposited in a Paleocene

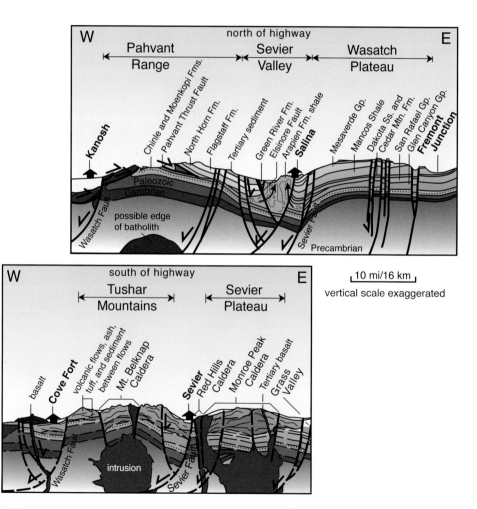

Between Fremont Junction and Cove Fort, I-70 crosses the transition zone between the Colo-rado Plateau and the Basin and Range. In the transition zone, huge curved slices of rock along the edge of the Plateau slid down and westward along faults during the extensional tectonic episode that formed the Basin and Range. Faults developed in sedimentary and mid-Tertiary volcanic rocks alike, and younger basalt and rhyolite magmas rose along some of the faults.

Contorted layers near milepost 63 reveal a moment in time captured in the North Horn Formation. Newly deposited layers settled into those below them, creating the wavy pattern sandwiched between smoother layers. —Felicie Williams photo

In the lower part of Salina Canyon, a striking angular unconformity (dashed line) divides vertical Jurassic layers from near-level Tertiary ones. The lower layers were bent by thrust faulting and back thrusting during the Sevier Orogeny and then were eroded nearly level before the overlying Tertiary layers were deposited. —Felicie Williams photo

lake. A minute or so farther west, yellowish-white Jurassic Arapien Formation shale appears lower down in the canyon walls.

At the mouth of Salina Canyon, shortly before the junction with US 89, the Sevier Fault, a normal fault, abruptly cuts off rocks of Tertiary age. West of the fault the Jurassic Arapien Formation erodes into barren badlands of an unusual whitish color that are banded here and there with red or yellow. The Arapien formed in an evaporating arm of the Jurassic-age Sundance Sea, so the formation contains thick layers of salt and gypsum. Very few plants tolerate both the salt and the selenium in these layers, which explains the barrenness of the hills.

Arapien Formation evaporites are mined for salt, gypsum, and very fine clays from weathered volcanic ash. The salt, for which the town of Salina is named, is used for road deicing, livestock, table salt, and a health supplement; the gypsum is mostly used for drywall board. The clays—montmorillonite, bentonite, and fuller's earth—are used in refining and decolorizing oils, as health supplements, and in cosmetics and skin care. The bentonite is also used in the process of drilling for oil and gas.

After exiting Salina Canyon, I-70 heads in a southerly direction through the upper portion of Sevier Valley, which is a graben. The Sevier River flows north here and for most of its length through Utah's High Country. Since about 1870, most of its water has been diverted for irrigation. In years of plentiful rain it flows to the Sevier Desert. In the wet years of 1983–87 and 2011, it reached all the way to the normally dry Sevier Lake.

The Pahvant Range to the west consists of tilted Paleozoic through Tertiary strata, including the silty, lime-rich soil and lake deposits that make pink and white bands along the mountain front. The Paleozoic rocks, which were thrust up during the Sevier Orogeny, crop out on the west side of the mountains, out of view. Terraces of Quaternary age stretch into the Sevier Valley from the mountains. They are covered by alluvial fans and floodplain deposits that are older than those present on the valley floor.

The Sevier Fault, a Basin and Range normal fault, runs along the valley's eastern edge. It slices this side of the Colorado Plateau for roughly 300 miles (480 km). Land west of it has dropped down, but the amount is variable. The fault is active: scarps cut across recently formed alluvial fans, evidence of recent fault movement. Light-colored rocks along the line of the fault are ash-flow tuff from Tertiary volcanoes; the ash marks the beginning of activity in the Marysvale Volcanic Field. Older buildings in many towns in the region were made from this easily worked tuff.

Large gravel quarries mark the valley floor near Elsinore and Joseph. The water table is close to the surface in the valley, so quarries must constantly be pumped out.

South of Joseph the valley narrows abruptly due to the Antelope Range, which is composed of volcanic rock. At Sevier, I-70 turns west up Clear Creek, with the Tushar Mountains to the south and the Pahvant Range to the north. The highest range between the Sierras and the Rockies, the Tushars are a large pile of Tertiary volcanic rocks, part of the Marysvale Volcanic Field. Volcanic rocks dominate the scenery along the interstate between Joseph and Cove Fort.

The Tertiary Sevier River Formation preserves fossil mammals, including abundant camels. After the layers were tilted, erosion beveled them into a sloping pediment and then cut through the pediment to make these hills. —Felicie Williams photo

Tertiary time saw plenty of fireworks here: the Marysvale Volcanic Field was a range of immense composite volcanoes that were active from 34 to 18 million years ago. Their magma may have risen from the melting Farallon Plate, which was then moving close beneath Earth's crust as it was subducted. At the time, volcanism was widespread west of the Colorado Plateau. A number of calderas—large depressions ringed by faults—formed where the land collapsed into partly emptied magma chambers.

Later, regional uplifting and stretching led to normal faulting and added tilt. The normal faulting was accompanied by basalt and rhyolite volcanism. Finally, the Pleistocene brought a wetter climate and fast erosion.

Some of the massive layers seen along Clear Creek are ashflow tuff, pink rock that forms when falling ash is so hot that it welds together in layers when it comes to rest on the ground. Tuff often weathers into vertical cliffs or cone-shaped teepee rocks. Other layers are volcanic breccia, which can be rock composed of shattered fragments blown out of the volcanoes or mudflow deposits that formed when or soon after the volcanoes were active. Included too are greenish-white lake deposits, which in places were faulted against the pink tuff.

Cove Fort, the namesake of the town, was built in 1867. It is constructed of local volcanic rocks: basalt for the walls and pink tuff for the fireplaces and chimneys. The basalt came from a lava flow near the junction with I-15. The basalts, along with faulting near Cove Fort, frequent earthquakes, and nearby hot springs, prove that this area is still tectonically active.

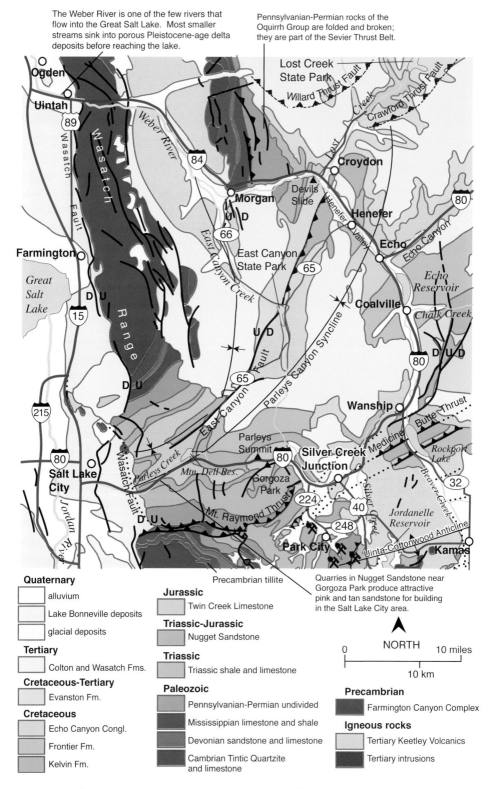

The Weber River is one of the few rivers that flow into the Great Salt Lake. Most smaller streams sink into porous Pleistocene-age delta deposits before reaching the lake.

Pennsylvanian-Permian rocks of the Oquirrh Group are folded and broken; they are part of the Sevier Thrust Belt.

Precambrian tillite

Quarries in Nugget Sandstone near Gorgoza Park produce attractive pink and tan sandstone for building in the Salt Lake City area.

Quaternary
alluvium
Lake Bonneville deposits
glacial deposits

Tertiary
Colton and Wasatch Fms.

Cretaceous-Tertiary
Evanston Fm.

Cretaceous
Echo Canyon Congl.
Frontier Fm.
Kelvin Fm.

Jurassic
Twin Creek Limestone

Triassic-Jurassic
Nugget Sandstone

Triassic
Triassic shale and limestone

Paleozoic
Pennsylvanian-Permian undivided
Mississippian limestone and shale
Devonian sandstone and limestone
Cambrian Tintic Quartzite and limestone

Precambrian
Farmington Canyon Complex

Igneous rocks
Tertiary Keetley Volcanics
Tertiary intrusions

NORTH

0   10 miles

10 km

Geology along the I-80 and I-84 loop between Salt Lake City and Uintah.

I-80 and I-84
Salt Lake City—Uintah
78 miles (126 km)

Starting at the I-15 junction, I-80 crosses delta deposits built out along the edge of Pleistocene Lake Bonneville. It then plunges past the Wasatch Fault into the Wasatch Range via Parleys Canyon. Red beds at the canyon's mouth are the Ankareh Formation; the beds were deposited when Earth's continental plates had united into the supercontinent Pangaea. Straddling the equator, Pangaea's interior was hot and dry with only seasonal rains. The climate caused the widespread oxidation of iron, which gives these and other Triassic red beds their color.

Within 1 mile (1.6 km) of entering the canyon, I-80 passes between high roadcuts of Triassic-Jurassic Nugget Sandstone, the age equivalent to the Navajo Sandstone of the Colorado Plateau. Its tall crossbeds and well-sorted sand tell of its origins in a dune field.

About 3 miles (5 km) up the canyon, the Twin Creek Limestone is quarried for Portland cement. This limestone was deposited in a shallow continental seaway that extended from Canada to southern Utah during Jurassic time. It is the same age as the Carmel Formation to the south.

Rocks in Parleys Canyon were folded into a syncline with a northeasterly axis during the Sevier Orogeny. Older layers on the outer margins of the fold are out of sight of the highway. The fold is cut off on its west end by the Wasatch Fault. Eastward, near Mountain Dell Reservoir, the Cretaceous Kelvin Formation (equivalent to the Cedar Mountain Formation elsewhere in the state), a red and tan sandstone interbedded with conglomerate and shale, and the Frontier Formation, dark sandstone and shale, are the youngest rocks in the syncline's center.

I-80 parallels the contact between the Twin Creek Limestone on the south and Cretaceous beds on the north until Parleys Summit. Then the highway drops through the Nugget Sandstone and underlying Triassic beds and crosses

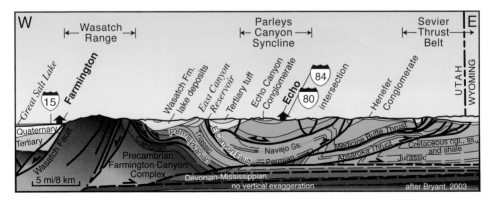

West-east cross section across the Wasatch Range from Farmington. Thrust faults developed during the Sevier Orogeny. The region west of the Wasatch Fault, a still-active normal fault, has dropped down since Pliocene time.

the Mount Raymond Thrust Fault. Buried here by Quaternary sediment, the fault is visible in mountains to the south where the Triassic and Permian beds overlie the younger Twin Creek Limestone. Soon after crossing the fault we lose the older rocks under a wild jumble of Tertiary Keetley Volcanics.

The Keetley Volcanics, between Silver Creek Junction and Wanship, are mostly volcanic breccia that erodes into bizarre shapes. The breccia is fairly uniform, lacking flow surfaces to distinguish individual flows. Only now and then are there reddish soil zones, baked by hot lava flows, to break the uniformity. The flows were very thick, formed of particularly stiff magma. As the magma cooled, crusts formed, which were then broken up and incorporated into the still-moving portions of the flow. These broken crusts became the chunks of rock giving the breccia its coarse texture.

In these steep mountains, landslides play a major role in erosion. Watch for hummocky slide surfaces; all but the youngest are well vegetated. Some have even damaged the pavement and roadcuts.

Peaks to the south mark the highest part of the Wasatch Range. They are the eroded tops of Tertiary granitic batholiths that intruded at about the same time the Keetley Volcanics were erupting. After the thrust-faulted mass of mountains of the Sevier Orogeny had formed, they began to settle downward due to their weight. As they did so, the mountains' bottommost layers melted. The resulting magma worked its way upward along the fault edging the north flank of the Uinta-Cottonwood Anticline and other faults.

Rockport Lake, on the Weber River south of Wanship, is edged with tilted, thrust-faulted Cretaceous rocks, including marine sandstone and shale. They are well exposed in roadcuts. The faults roughly parallel the rocks' bedding.

From Rockport Lake, I-80 and then I-84 descend along the Weber River, which flows west from the high peaks of the Uinta Mountains. A few minutes north of Wanship, lake deposits appear west of the highway. In Tertiary time,

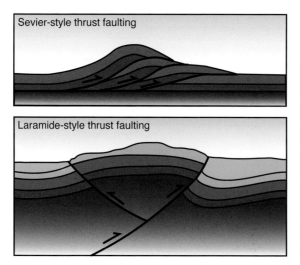

The area in and around Wanship is near the junction of two very different mountain ranges. To the west, the Sevier Orogeny built the Wasatch Range by thrusting sheets of rock eastward great distances. To the east, the Uinta Mountains of the Laramide Orogeny slid only a short distance along a few almost vertical faults.

after the region's mountains had risen, large lakes spread across the valleys and, for millions of years, filled with sediment eroded from the mountains.

At Coalville, Cretaceous rocks dip northwest off the flanks of the Uinta Anticline. Their hogbacks contain the coal that is mined here. Oil is produced in the Tertiary rock just east of Coalville. Echo Reservoir, to the north, is filling rapidly with sediment eroded from these soft Cretaceous and Tertiary rocks.

Near the junction with I-84 at Echo, cliffs of Echo Canyon Conglomerate come into view. In places more than 3,000 feet (900 m) thick, this rough, bouldery rock was alluvial fan gravel washed east from land elevated by the Crawford Thrust Fault 89 to 84 million years ago. The Crawford is one of the thrust faults of the Late Cretaceous Sevier Orogeny.

From Echo I-80 climbs east to Wyoming through Echo Canyon, which is walled by Echo Canyon Conglomerate and then younger conglomerates. The Cretaceous sequence of sedimentary rock is tremendously thick here—3 to 4 miles (5 to 6 km). The sediment accumulated in the large basin that formed east of the mountains of the Sevier Orogeny.

The tectonic thrusting of the orogeny worked its way eastward. Younger thrust sheets developed below and in front (east) of earlier ones, carrying piggyback-style the older ones, so the conglomerates become younger and younger to the east as well. The lower thrust fault planes surface in Wyoming in long arcuate ridges.

The Echo Canyon Conglomerate was deposited in alluvial fans at the base of an ancient mountain range. The lower layers here were warped upward by continued mountain building before being covered by the nearly horizontal layers that cap the hill. Where the angled layers and nearly flat layers meet is an angular unconformity. —Felicie Williams photo.

Sedimentary beds are almost vertical at Devils Slide, seen here from the viewpoint along I-84.
—Felicie Williams photo

From Echo, I-84 continues north past high bluffs of very coarse Echo Canyon Conglomerate. The older Henefer Conglomerate is exposed in the hills surrounding Henefer Valley.

I-84 crosses the East Canyon Fault near the end of Henefer Valley. Slightly dipping Tertiary rocks give way first to Cretaceous layers then abruptly to nearly vertical Mesozoic and then Paleozoic rocks. The Jurassic Twin Creek Limestone, quarried for cement at Croydon, also forms the two ridges of Devils Slide, created by differential erosion of parallel beds of limestone. These rocks are on the west limb of the same regional syncline we crossed on I-80 in Parleys Canyon. Here, though, the road also crosses the older formations of the syncline.

A single, massive fossil-rich bed of gray Mississippian limestone appears near milepost 105. West of it are still-older marine strata right down to Cambrian limestone north of Morgan. Between mileposts 109 and 108, all the Paleozoic layers are folded and resemble roller coasters; father west, they dip less steeply. They were bent, folded, and faulted by movement along the great thrust faults of the Sevier Orogeny. Near milepost 108, horizontal Tertiary lake deposits reappear south of us.

Morgan sits in a half graben, a faulted valley that only dropped down along one side—here, along the east side. Volcanic rocks mantle older rocks in the hills to the west. I-84 passes through them between Morgan and about milepost 92, riding on Pleistocene terraces and beds deposited in an arm of Lake Bonneville.

Dark rocks near and west of the milepost 92 rest stop are Precambrian Farmington Canyon Complex schist and gneiss. Metamorphosed—nearly melted—by mountain building between 1.8 and 1.65 billion years ago, their original sediment was deposited offshore. The timing of their initial deposition is unknown because the radiometric age of their mineral content was reset during metamorphosis. These ancient rocks were thrust up during the Sevier Orogeny.

As the road emerges from the mountains, it crosses the Wasatch Fault. Steep slopes and rushing streams give way to the sediment and old shorelines left by Pleistocene Lake Bonneville.

US 6
Price—Spanish Fork
68 miles (109 km)

US 6 parallels the Price River out of Price, heading quickly into the Book Cliffs. The low gray hills and slopes in Price Canyon are Cretaceous Mancos Shale; the cliffs above are four other Cretaceous formations that belong to the Mesaverde Group—sandstones separated by darker shales and seams of coal. The top two, the Castlegate Sandstone and the Price River Formation, are named for exposures here. Carefully studied and described, these cliffs make up a type section, a yardstick against which other exposures can be compared.

The Mesaverde Group was deposited along a coast of the Cretaceous Interior Seaway, with offshore barrier bars, beaches, lagoons, and swamps where plant debris accumulated, later to be compressed into coal. Some sandstones were deposited in stream channels that drained east from the newly rising mountains of the Sevier Orogeny; coarser material corresponds to times of mountain building, when invigorated rivers and streams eroded deeper and carried coarser sediment. Sandstone tends to form cliffs and ledges, while shale weathers into slopes, and coal seams are often hidden by talus.

Near the long-gone town of Castle Gate, a monument records the death of 172 coal miners in 1924, caused by three explosions of ignited coal dust in the mine. Both this and a similar 1900 disaster in Scofield fueled attempts to unionize and force owners to provide safer working conditions.

The mineable coal is in the Blackhawk Formation, the lowermost coal-rich layer in the Mesaverde Group. Its thick seams are mined underground. Rock south of milepost 229 was baked brick red by an underground coal fire. Lightning or forest fires start most coal fires. The large power plant at Castle Gate is schedule to be removed soon.

Small mine dumps east of Soldier Summit mark abandoned ozocerite mines. Ozocerite is a brown, black, or dark-green natural wax used in insulation, lubricants, and inks.

Displays at the Utah State University Eastern Prehistoric Museum in Price and the Western Mining and Railroad Museum in Helper zero in on dinosaurs, ice age animals, archaeology, and coal mining.

Thistle is the site of a large landslide that blocked the river and flooded the town in 1983.

Scofield Reservoir lies in a low spot at the intersection of two sets of faults.

Scofield was the site of a mining disaster in 1900; 206 miners were killed by a coal dust explosion.

NORTH

0   10 miles

0   10 km

Geology along US 6 between Price and Spanish Fork.

Quaternary
alluvium
landslides
Lake Bonneville deposits
Pleistocene gravel

Tertiary
Uinta Fm.
Green River Fm.
Flagstaff and Wasatch Frms.

Cretaceous–Tertiary
North Horn Fm.

Cretaceous
Mesaverde Gp.
Mancos Shale

Jurassic
undivided

Triassic
undivided

Paleozoic
Permian
Pennsylvanian–Permian
older Paleozoic
Cambrian

Precambrian

Igneous rocks
Tertiary volcanics

Castle Gate, where Utah 191 heads north, was named for its stone promontories in the Castlegate Sandstone. One promontory has since been felled to widen the highway.
—George L. Beam photo, circa 1929

Past Castle Gate, US 6 climbs through the Price River Formation and the Cretaceous-Tertiary North Horn Formation, both deposited in delta and river environments near the receding Cretaceous Interior Seaway.

Paleogene lake deposits appear near milepost 223. The fossiliferous Flagstaff Formation forms a sharp-edged ridge above the bridge at milepost 221. Above it, the Colton Formation's reddish mudstone and siltstone accumulated on a delta that spread into one of the lakes.

Near milepost 213, US 6 crosses the White River, named for the clayey Paleogene lake sediment it carries. The meanders that wind across the valley floor are typical of sluggish low-gradient rivers.

Soldier Summit lies close to the meeting of the northern Wasatch Plateau, the Uinta Basin, and the Wasatch Range. This geologically complex region of anticlines, synclines, and faults formed during the Sevier and Laramide Orogenies when rocks were thrust eastward. However, at Soldier Summit only a simpler blanket of Tertiary formations is visible. At and south of town are Paleogene lake deposits that include Flagstaff Formation limestone and delta and floodplain beds of the Colton Formation. Slopes north of the summit are younger lake deposits of the Green River Formation.

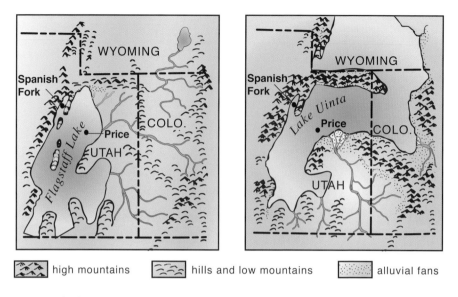

high mountains   hills and low mountains   alluvial fans

During Paleocene and Eocene time, large lakes occupied parts of Utah,
Colorado, and Wyoming. Flagstaff Lake formed before Lake Uinta.

West of Soldier Summit, strata dip north. They slide easily down the hill-
slope south of Soldier Creek. Slides dam the valley at mileposts 208 and 209.
Hillsides are scooped where slides have broken loose. Trees with twisted trunks
prove some sliding is quite recent; the trees were bent during a slide and then
resumed skyward growth.

Unusual Eocene bird tracks have been found in lakebed shale near milepost
204. West of the rest stop the Green River Formation appears: sandstone, grayish siltstone, white limestone, and seams of coal.

Coarse Cretaceous conglomerate between mileposts 192 and 191 was deposited in alluvial fans along the east side of the mountains of the Sevier Orogeny.
The conglomerate's pebbles include Precambrian quartzite and Paleozoic limestone, proving that these mountains, which began to rise early in Cretaceous
time, had deep canyons that reached all the way down to their cores by Late
Cretaceous time. Notice the weathered pinnacles and the occasional window
up high. Sediments of the same age back in Price Canyon were much finer, having been deposited in marine and floodplain environments. This short drive
has taken us from the low-energy environment near the edge of the Cretaceous
Interior Seaway to the high-energy river environment close to the mountains.

Near Thistle, steeply dipping beds of Jurassic Navajo Sandstone and Twin
Creek Limestone line the highway. The highway overlooks the destruction
caused by the Thistle Landslide, which dammed Thistle and Soldier Creeks in
the spring of 1983.

The Green River Formation, well exposed in roadcuts, contains fossil fish, snails, clams, turtle fragments, and algae-coated logs. —Felicie Williams photo

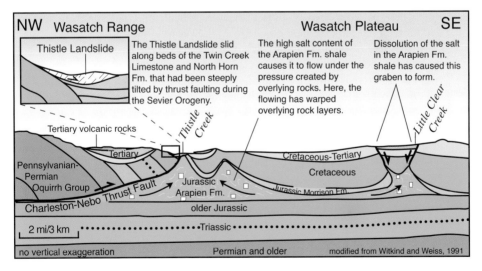

NW Wasatch Range   Wasatch Plateau   SE

Thistle Landslide

The Thistle Landslide slid along beds of the Twin Creek Limestone and North Horn Fm. that had been steeply tilted by thrust faulting during the Sevier Orogeny.

The high salt content of the Arapien Fm. shale causes it to flow under the pressure created by overlying rocks. Here, the flowing has warped overlying rock layers.

Dissolution of the salt in the Arapien Fm. shale has caused this graben to form.

Tertiary volcanic rocks

Thistle Creek

Little Clear Creek

Tertiary

Cretaceous-Tertiary

Cretaceous

Pennsylvanian-Permian Oquirrh Group

Charleston-Nebo Thrust Fault

Jurassic Arapien Fm.

Jurassic Morrison Fm.

older Jurassic

2 mi/3 km ·····Triassic·····

no vertical exaggeration   Permian and older   modified from Witkind and Weiss, 1991

Cross section showing the Charleston-Nebo Thrust Fault, which formed during the Sevier Orogeny, and the Thistle Landslide near Thistle. The Arapien Formation, soft and easily deformed, provided a weak layer for faults to move along. The formation flowed and developed into thicker masses in areas with the fewest overlying beds or along the broken-up areas adjacent to faults.

Between Thistle and the mouth of Spanish Fork Canyon, US 6 passes many fault-sliced outcrops of gray, cliff-forming Pennsylvanian and Permian limestone and a few remnants of softer Triassic sandstone and siltstone. These rocks were pushed east above the Charleston-Nebo Thrust Fault, which extends from Charleston, near Heber City, clear to Mt. Nebo near Nephi. Later, thrust faults developed under this one, and movement along them folded both the earlier fault surface as well as the overlying sedimentary layers, carrying the rock layers even farther east. Altogether, the rocks moved 60 miles (100 km) east over younger strata.

Near the mouth of Spanish Fork Canyon, steep alluvial fans merge with Pleistocene stream deltas that extended into former Lake Bonneville; the deltas now form a wide plain stretching toward Utah Lake. The deltas are part of the Provo Shoreline, a level the lake stayed at for thousands of years. Sediments deposited along higher shorelines of Lake Bonneville line the mountain face.

The Wasatch Fault has close to 3 miles (5 km) of displacement at Spanish Fork. It holds an ever-present threat of earthquakes. The soft sediments in Utah Valley are very worrisome; in a large quake, they would liquefy, causing great damage.

US 89
Kanab—Panguitch
68 miles (109 km)

Backed by Triassic and Jurassic rocks of the Vermilion Cliffs, Kanab is the gateway to Utah's Grand Staircase, the succession of Vermilion, White, Gray, and Pink Cliffs that edge the Colorado Plateau. Kanab sits on Quaternary stream deposits colored pink by rocks eroded from the Vermilion Cliffs.

At the base of the Vermilion Cliffs, the Triassic Chinle Formation is fine-grained red and purple shale that was deposited in a complex and ever-changing river and delta environment. It contains lots of volcanic ash derived from volcanoes to the west. Its banded coloring is due to minute amounts of iron that oxidized to varying degrees in the rapidly changing depositional settings. Above the Chinle Formation, the top part of the Vermilion Cliffs is composed of the brilliant Triassic-Jurassic Moenave Formation; its fine sediment was deposited on a lake-dotted floodplain. The floodplain edged the dune field of the Wingate Sandstone, so the two formations interfinger. Above the Moenave and Wingate are the Kayenta Formation and Navajo Sandstone. The Kayenta is composed of sand and silt that was deposited in rivers; its top layers form a bench. The Navajo is a thick, cliff-forming dune deposit.

The deep gullies near milepost 67 formed in the fifty years following 1890. For a long time the introduction of livestock was blamed for their development. Recent research, with the ability to date different soil levels and erosional events, has suggested that gullying has occurred repeatedly over geologic time, and it is almost certainly tied in some way to climate cycles.

From milepost 69 you can glance upstream at the long, straight canyon of Kanab Creek. Its two sides don't quite match because the canyon lies along a fault.

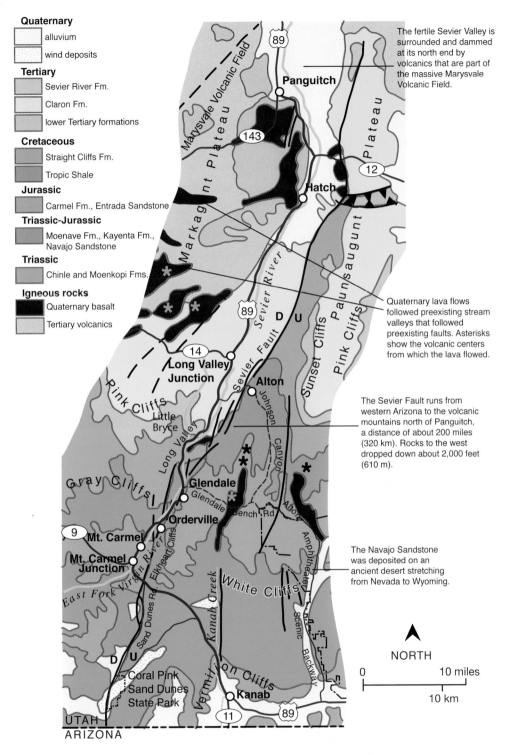

Quaternary
- alluvium
- wind deposits

Tertiary
- Sevier River Fm.
- Claron Fm.
- lower Tertiary formations

Cretaceous
- Straight Cliffs Fm.
- Tropic Shale

Jurassic
- Carmel Fm., Entrada Sandstone

Triassic-Jurassic
- Moenave Fm., Kayenta Fm., Navajo Sandstone

Triassic
- Chinle and Moenkopi Fms.

Igneous rocks
- Quaternary basalt
- Tertiary volcanics

The fertile Sevier Valley is surrounded and dammed at its north end by volcanics that are part of the massive Marysvale Volcanic Field.

Quaternary lava flows followed preexisting stream valleys that followed preexisting faults. Asterisks show the volcanic centers from which the lava flowed.

The Sevier Fault runs from western Arizona to the volcanic mountains north of Panguitch, a distance of about 200 miles (320 km). Rocks to the west dropped down about 2,000 feet (610 m).

The Navajo Sandstone was deposited on an ancient desert stretching from Nevada to Wyoming.

Marysvale Volcanic Field

Markagunt Plateau

Sevier Plateau

Paunsaugunt Plateau

Panguitch

Hatch

Sevier River

Sevier Fault

Sunset Cliffs

Pink Cliffs

Pink Cliffs

Little Bryce

Long Valley Junction

Alton

Johnson Canyon

Gray Cliffs

Long Valley

Glendale

Glendale Bench Rd.

Orderville

Mt. Carmel

Mt. Carmel Junction

Elkheart Cliffs

East Fork Virgin River

Kanab Creek

White Cliffs

Amphitheater

Scenic Backway

Sand Dunes Rd.

Coral Pink Sand Dunes State Park

Vermilion Cliffs

Kanab

UTAH
ARIZONA

NORTH

0 .... 10 miles
0 .... 10 km

Geology along US 89 between Kanab and Panguitch.

The soft Chinle Formation weathers and erodes easily. —Lucy Chronic photo

In several places between mileposts 71 and 73, the Navajo Sandstone preserves slumps (the swirly layers in this photo) that formed on the steep downwind slopes of sand dunes.
—Felicie Williams photo

North of the bridge, the Navajo Sandstone shows its signature buff-white to pink color and sweeping, large-scale crossbedding. Because it is strong and thick, with uniform grains held firmly by calcium carbonate cement, it forms another rank of prominent cliffs: the White Cliffs of the Grand Staircase.

North of milepost 73, US 89 passes through coral-colored sand hills composed of secondhand sand weathered from surrounding sandstones. From the divide at milepost 77, the turrets and cliffs of Zion National Park, carved in Navajo Sandstone, rise to the west.

US 89 crosses the Sevier Fault near milepost 79. To the east, the Navajo Sandstone is high above the highway; to the west, it appears in the canyon below highway level. Do not confuse the Sevier Fault with the thrust faults of the Sevier Orogeny. The Sevier Fault is much younger, one of several outlying normal faults of the Basin and Range. These outlying faults are part of the transition zone between the Colorado Plateau and the Basin and Range. Rocks have moved downward along the west side of each. Here, displacement is more than 3,000 feet (900 m).

To the northwest of Mount Carmel Junction and Mount Carmel are low hills of soft, easily eroded Jurassic Carmel Formation siltstone, named for this area in which the formation's gypsum-bearing beds are so well exposed. There is a large mass of white gypsum near the highway just west of Mount Carmel. Seams of low-grade coal in gray Cretaceous layers mark some of the hills north of town.

Farther up Long Valley, which follows the Sevier Fault, is Orderville; it boasts several gem and mineral stores. Here, the Navajo Sandstone and overlying Carmel Formation dip north more and more steeply. Just north of Orderville, US 89 passes the Cretaceous Cedar Mountain Formation, Dakota Sandstone, and Straight Cliffs Formation on the west side of the valley and fault. These rocks don't show up on the east side of the valley and fault until Glendale. In Glendale they stand tall, forming the Gray Cliffs of the Grand Staircase.

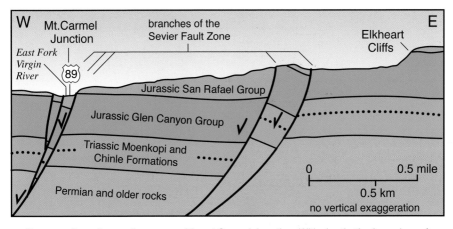

Cross section of a small area near Mount Carmel Junction. With depth, the branches of the Sevier Fault Zone will curve westward and merge as the main Sevier Fault.

Now surrounded by farms and orchards, Glendale began as a mining town, exploiting coal in the Dakota Sandstone. North and east of town, gypsum gives the hills of older Jurassic rocks a strange gray cast.

One mile (1.6 km) past the rest stop near milepost 95, a line of pink cliffs appears to the west. These pink cliffs are Tertiary Claron Formation that has been eroded into breaks along the edge of the Markagunt Plateau. US 89 climbs through a lava flow between mileposts 96 and 97. To the east, Bryce Canyon National Park, also eroded in the Claron Formation, forms the rim on the east side of the Paunsaugunt Plateau. To the west, on the west side of the Sevier Fault, the Claron Formation is at highway level.

The Claron Formation's soft pink beds are less spectacular in this flat landscape than they are on Bryce's cliffs. The beds were deposited on river floodplains and in lakes and contain unusually good evidence of soil-forming processes: bioturbation, or mixing by animals and roots; traces or fossils of roots; thin limy deposits of caliche left from evaporation; and the oxidation of iron in the soil, which caused the pink color. The thin, white limestone beds in higher layers formed in broad, shallow lakes and are more abundant at Cedar Breaks National Monument.

Utah 14 branches west at Long Valley Junction and passes by Cedar Breaks National Monument. North of Long Valley Junction, US 89 continues down the valley of the Sevier River, which is still following the Sevier Fault. Here, the river's gradient is low; it meanders across a wide floodplain.

East of milepost 121 the Sevier Fault shows up well: pink Claron Formation beds east of the fault abut Quaternary lava to its west. —Lucy Chronic photo

To the west are Tertiary volcanics, part of the huge Marysvale Volcanic Field, a massive, explosive outspilling of volcanic rock and ash. A much younger dark basalt flow near Hatch, visible to the north from milepost 118, originally followed the valley of a tributary to the Sevier River and forced the stream aside. The stream then made new cuts, removing softer rock next to the basalt and isolating it as a resistant ridge.

North of the Utah 12 intersection, gravel terraces lie about 25 and 50 feet (8 and 15 m) above the river. The gravel, which also shows up in roadcuts, includes rocks washed from volcanic highlands to the west. Its cobbles and pebbles are rounded and sorted—sure signs of transportation and deposition by water.

After milepost 129, the Sevier River's floodplain is marked by abandoned channels, which are lower and contain different vegetation than the surrounding floodplain. Some of the river has been artificially channeled to protect surrounding fields from changes in its course.

US 89
Panguitch—Sevier
60 miles (97 km)

The dark volcanic uplands of the Markagunt and Sevier Plateaus surround Panguitch. The Sevier Fault runs along the east edge of the Sevier River valley, which Panguitch sits in, and separates the two plateaus. The fault dips westward steeply at the surface and then more gently as it gets deeper, so the great block of rock west of it rotated as it slid down and to the west. The block's eastern edge dropped dramatically, forming the Sevier River valley, while the western edge remained high, becoming the Markagunt Plateau.

Northeast of Panguitch US 89 passes jagged cliffs eroded in great piles of volcanic rock. River terraces and alluvial fans scallop the sides of the valley. Because this part of the valley was plugged at its north end by Pliocene volcanic rocks, the river has a very gentle gradient, wandering across a wide marsh-dotted floodplain underlain by lake deposits. Oxbow lakes mark abandoned river channels.

Visible to the north from milepost 139 is Circleville Mountain, a partly eroded Miocene volcano in the Tushar Mountains. Another Miocene volcano, Mt. Dutton, rises to the northeast. Geoscientists believe that, between about 40 and 20 million years ago, as the Farallon Plate foundered during its subduction, hot asthenosphere moved in above it, providing a source for great volumes of magma that rose in large bodies, often erupting violently once near the surface. In this part of Utah, the magma formed the Marysvale Volcanic Field, which dominates the scenery for the rest of this route.

Both Mt. Dutton and Circleville Mountain mark where composite volcanoes once stood. These great mountains were akin to Japan's Fujiyama and America's Mount St. Helens, having been composed of volcanic ash layers and lava flows with plenty of mudflow breccias mixed in.

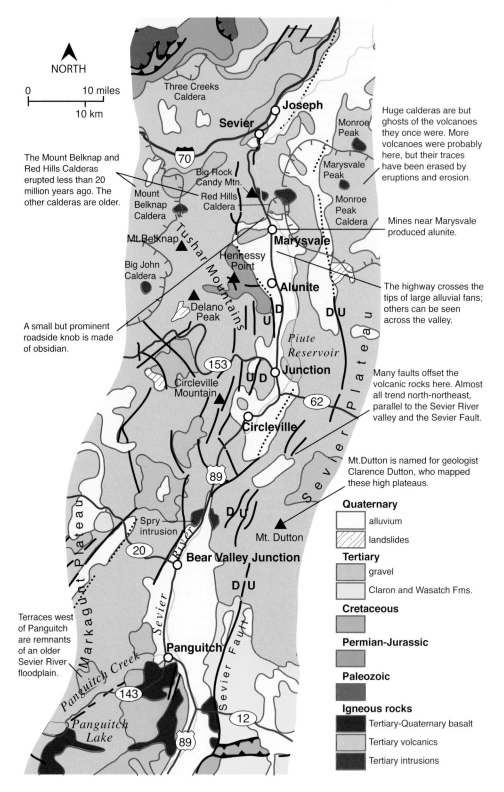

NORTH

0   10 miles

10 km

Three Creeks Caldera

Sevier

Joseph

Monroe Peak

Marysvale Peak

Huge calderas are but ghosts of the volcanoes they once were. More volcanoes were probably here, but their traces have been erased by eruptions and erosion.

The Mount Belknap and Red Hills Calderas erupted less than 20 million years ago. The other calderas are older.

70

Big Rock Candy Mtn.

Mount Belknap Caldera

Red Hills Caldera

Monroe Peak Caldera

Mt Belknap

Hennessy Point

Marysvale

Mines near Marysvale produced alunite.

Big John Caldera

Tushar Mountains

Alunite

D

U

D U

The highway crosses the tips of large alluvial fans; others can be seen across the valley.

A small but prominent roadside knob is made of obsidian.

Delano Peak

Piute Reservoir

153

U D

Junction

Circleville Mountain

62

Many faults offset the volcanic rocks here. Almost all trend north-northeast, parallel to the Sevier River valley and the Sevier Fault.

Circleville

Sevier Plateau

Mt.Dutton is named for geologist Clarence Dutton, who mapped these high plateaus.

89

Quaternary

alluvium

landslides

Spry intrusion

D U

Tertiary

gravel

Claron and Wasatch Fms.

Mt. Dutton

20

Sevier River

Bear Valley Junction

D U

Cretaceous

Permian-Jurassic

Paleozoic

Markagunt Plateau

Terraces west of Panguitch are remnants of an older Sevier River floodplain.

Sevier River

Sevier Fault

Panguitch

Panguitch Creek

143

Panguitch Lake

89

12

Igneous rocks

Tertiary-Quaternary basalt

Tertiary volcanics

Tertiary intrusions

Geology along US 89 between Panguitch and Sevier.

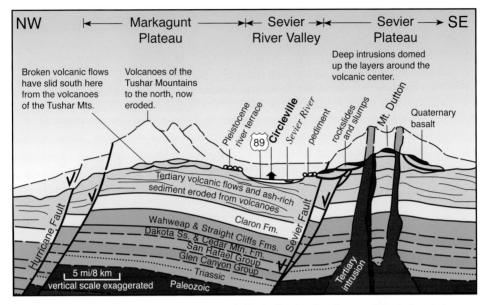

NW ←—— Markagunt ——→|←— Sevier —→|←—— Sevier ——→ SE
         Plateau   River Valley   Plateau

Broken volcanic flows   Volcanoes of the   Deep intrusions domed
have slid south here   Tushar Mountains   up the layers around the
from the volcanoes   to the north, now   volcanic center.
of the Tushar Mts.   eroded.

Pleistocene river terrace   Circleville   Sevier River   rockslides and slumps   Mt. Dutton   Quaternary basalt

89

Sevier River pediment

Tertiary volcanic flows and ash-rich sediment eroded from volcanoes

Claron Fm.

Wahweap & Straight Cliffs Fms.
Dakota Ss. & Cedar Mtn. Fm.
San Rafael Group
Glen Canyon Group
Triassic

Hurricane Fault

Sevier Fault

Tertiary intrusion

5 mi/8 km
vertical scale exaggerated   Paleozoic

Northwest-southeast cross section through Circleville showing the possible skyline during Miocene time. The vertical scale is exaggerated about 3x to illustrate underlying formations.

Made of broken and ground-up volcanic rock and ash, this volcanic breccia has eroded into tents. This type of erosional form in volcanic breccia is common. After the hot flow stopped moving, steam escaped through it, rising as if funneling upward. Minerals developed where the steam rose, cementing portions of the flow. The uncemented part eroded away, leaving the tents. Field of view 45 feet (14 m) wide. —Lucy Chronic photo

A second phase of volcanism began in the Colorado Plateau and Great Basin about 17 million years ago, during late Miocene time, and continues today. In this phase, deep faults of the Basin and Range reach through the brittle crust and tap dark basalt magma in the Earth's mantle. Rising, the magma melts a little of the crust it passes through, generating light-colored rhyolite magma. The two magmas do not mix, so the volcanism is termed *bimodal*. There is much more basalt than rhyolite.

Eroded pediments and alluvial fans edge the Sevier River valley. Though they may look like alluvial fans, pediments are carved into the solid rock of the mountains and covered with just a thin veneer of gravel, whereas alluvial fans are composed of loose gravel. The pediments here are as much as 100 feet (30 m) above the present level of the river, cut off at their toes by the Sevier River as it swings back and forth on its valley floor.

North of milepost 146, we encounter the Spry intrusion. Its magma intruded the Claron Formation 25 million years ago, possibly as a laccolith. This intrusion and all the volcanic rock on Mt. Dutton belong to the Oligocene Mount Dutton Formation, a composite of many separate eruptions from several closely spaced volcanoes.

Road and river crowd into a narrow canyon north of milepost 152. The great volcanoes of Mt. Dutton, now to the east, and the Tushar Mountains, to

The granitic Spry intrusion cooled slowly below the surface, allowing small but visible crystals of minerals to form. —Felicie Williams photo

The dip of the volcanic rock here is nearly parallel to the original angle of the sides of the volcano it erupted from. Layering like this is what built the steep slopes of Utah's stratovolcanoes. —Felicie Williams photo

the north, are very close. At first we mostly see volcanic breccias of the Mount Dutton Formation. Then there is ashflow tuff, a rock that formed when incandescent ash shot down a volcano's flanks, welding itself together upon settling on the sides of the volcano and surrounding landscape. The tuff erodes into steep and often pinnacled slopes; it is light buff in color or weathered to dark brown. Faulted, jointed, and irregularly cemented, both tuff and volcanic breccias crumble into rockslides that spill down the canyon walls.

An interesting red layer is exposed in the roadcut near milepost 155. It was a thick soil that developed on top of a cooled lava flow, only to be covered by another lava flow and baked red in the process.

North of Circleville the valley widens out again, and the Sevier River meanders calmly across another floodplain, once the floor of a small lake. To the west, high peaks crown the Tushar Mountains. Its immense piles of volcanic rock are the tallest mountains between the Rockies and the Sierras. A small area of sedimentary rock, some of it altered by volcanism, shows beneath the pile of volcanic rock west of Alunite.

Picturesque Marysvale is a mining town centered between volcanic calderas. Several kinds of mineral deposits occur in the area, though none are large, and currently most mines are closed. In the 1990s most of the old mine openings were sealed. The mineral deposits are found where the rocks were altered by hydrothermal water—water heated by igneous intrusions. The circulating water apparently reached Jurassic evaporites of the Arapien Formation beneath the volcanic rocks; the hot water heated gypsum in the formation and formed sulfuric acid. The acid dissolved trace elements from all the rocks it flowed through and redeposited them in surrounding rock where conditions were right. These concentrated ore minerals are what miners have exploited in the region.

Miners began prospecting for gold in the Marysvale area in 1868. In the Kimberly District, 8 miles (13 km) northwest of Marysvale, gold and silver occur with pyrite in veins in altered rocks. In other areas, lead, zinc, gold, and silver minerals occur in veins along fault surfaces and joints and in limestone adjacent to the volcanic rocks.

Other altered rock has produced phosphates, used for fertilizer and explosives, and alunite, used for alumina and sulfuric acid and to make explosives during World War I. Between 1949 and 1966, an area north of Marysvale produced uranium; its ore, which also contains molybdenum, gold, lead, and zinc, occurs where an intrusion broke up older volcanic rocks, allowing hot mineral-bearing water into the older rock.

About 1.5 miles (2.4 km) north of Marysvale, a cliff of obsidian just west of the highway marks the edge of a caldera. Obsidian is black volcanic glass that formed from lava that cooled so quickly it didn't have time to crystallize. Tall pinnacles near the milepost-185 rest area formed in breccia from Mt. Belknap, another former volcano, to the west. To the north, similar rocks were altered to yellowish clay by hydrothermal water.

About 4.5 miles north of Marysvale, Big Rock Candy Mountain—named for a song made famous by Burl Ives—is visible to the west. It is made of layers of volcanic rock that were altered about 21 million years ago. The type of alteration that occurred was determined by several factors: distance from the underlying intrusion, variations in heat, and variations in the acidity of circulating hydrothermal water. Close to the intrusion, alunite formed from aluminum and potassium-rich feldspar. Farthest out, iron oxide developed, which gives clay and springwater their rusty yellow color. As you can see, the altered rocks erode readily on the mountain's slopes.

About 5 miles (8 km) north of Big Rock Candy Mountain, both river and highway veer eastward around a mass of volcanic breccia erupted by Mt. Belknap. Just north of the I-70 junction are tilted greenish and pinkish ash-rich layers of the Tertiary Sevier River Formation; their top surface has been eroded into a gently sloping pediment. (The I-70: Fremont Junction—Cove Fort road guide covers the stretch of US 89 between Sevier and Salina.)

US 89
Salina—Thistle
84 miles (135 km)

North of Salina along US 89, the slopes of the Pahvant Range to the west are mostly Cretaceous and lower Tertiary sedimentary rocks. Pink and white bands mark the Flagstaff Formation. The Flagstaff was deposited in and near a large lake—Flagstaff Lake—that spread between the newly risen Wasatch and Uinta ranges. The bands represent variations in the oxidization of minute amounts of iron, which was dependent upon the original amount of iron present and the environmental conditions in which it oxidized.

Geological features of different ages parallel one another along the Utah Hingeline, which trends north to south through this area: East of the highway the Ephraim Fault may be the east edge of a late Precambrian rift. The Gunnison Thrust Fault, far beneath Tertiary layers that top the Wasatch Plateau east of US 89, is the easternmost thrust fault related to the Sevier Orogeny. The Sevier and Sanpete Valleys are large Tertiary grabens that formed as a result of Basin and Range extension. East of the highway the long Basin and Range monocline that edges the Wasatch Plateau is along the fault that edges the grabens. These valleys are much studied because of their potential for oil.

The Arapien Formation of Jurassic age, deposited in a narrow ocean basin that extended like an arm down from a sea in Canada, contains salt and gypsum interbedded with mudstone. Between Salina and Gunnison, this formation makes low white and pink hills. The formation often contains distorted bedding that developed for two major reasons: the salt- and gypsum-rich sediments were easily contorted under pressure from overlying beds and by nearby

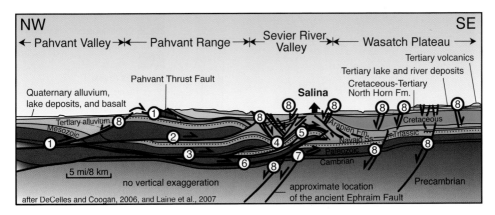

In this cross section, faults are numbered in the order in which they occurred: 1 through 7 are thrust faults that formed during the Sevier Orogeny, when subduction along the west coast was pushing land to the east. When one large thrust fault locked up, another formed beneath it, often bending the overlying thrust faults and rock layers in the process. Faults numbered 8 are normal faults formed by Basin and Range extension, or stretching, which followed the end of the west coast subduction and the startup of the lateral movement along the San Andreas Fault.

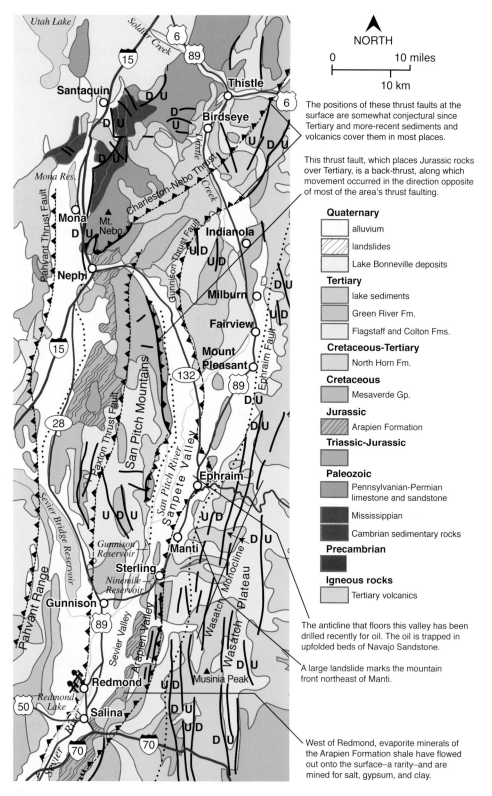

Utah Lake
Soldier Creek
6
15
89

Santaquin
D U
D U
Thistle
6
Birdseye
D U
D U
U D
Mona Res.
Thistle Creek
Charleston-Nebo Thrust
Pahvant Thrust Fault
Mona
Mt. Nebo
D U
Indianola
U D
U D
Gunnison Thrust Fault
Nephi
D U
Milburn
U D
Fairview
15
Mount Pleasant
San Pitch Mountains
132
89
D U
Paxton Thrust Fault
28
D U
Ephraim
Sevier Bridge Reservoir
San Pitch River
Sanpete Valley
U D U
U D
D U
Gunnison Reservoir
Manti
Sterling
Ninemile Reservoir
Wasatch Monocline
Pahvant Range
Gunnison
89
Sevier Valley
Arapien Valley
Wasatch Plateau
D U
Redmond
U D
Musinia Peak
50
Redmond Lake
Salina
D U
U D
Sevier River
70
70
D U

NORTH
0   10 miles
10 km

The positions of these thrust faults at the surface are somewhat conjectural since Tertiary and more-recent sediments and volcanics cover them in most places.

This thrust fault, which places Jurassic rocks over Tertiary, is a back-thrust, along which movement occurred in the direction opposite of most of the area's thrust faulting.

Quaternary
alluvium
landslides
Lake Bonneville deposits

Tertiary
lake sediments
Green River Fm.
Flagstaff and Colton Fms.

Cretaceous-Tertiary
North Horn Fm.

Cretaceous
Mesaverde Gp.

Jurassic
Arapien Formation

Triassic-Jurassic

Paleozoic
Pennsylvanian-Permian limestone and sandstone
Mississippian
Cambrian sedimentary rocks

Precambrian

Igneous rocks
Tertiary volcanics

The anticline that floors this valley has been drilled recently for oil. The oil is trapped in upfolded beds of Navajo Sandstone.

A large landslide marks the mountain front northeast of Manti.

West of Redmond, evaporite minerals of the Arapien Formation shale have flowed out onto the surface—a rarity—and are mined for salt, gypsum, and clay.

Geology along US 89 between Salina and Thistle.

faulting; and the gysum and salt dissolved readily, leaving the other sediments to fold into spaces left by their dissolution.

East of these hills, the Wasatch Plateau's west-dipping monocline exposes Cretaceous and Tertiary rocks. They are sliced into narrow slivers by north-south-trending normal faults, part of the Basin and Range extension. High, white-topped Musinia Peak (also known as Marys Nipple) is a sliver of Paleocene and Eocene Flagstaff Formation limestone.

The Sevier River valley near Salina is a geothermal area. Its warm springwater (less than 122°F, or 50°C) is probably rain and snowmelt from the Wasatch Plateau. Flowing down through the rock layers and along faults, it is warmed deep beneath the surface before returning to the surface in hot springs and wells.

In Pleistocene time, an arm of Lake Bonneville reached south to just north of Salina; its shorelines etch the slopes near Gunnison. The fertile valley floor, so flat that the Sevier River seems at a loss to know which way to go, is a legacy of that lake, though it is covered with younger sediment.

At Gunnison, US 89 leaves the Sevier River and follows the valley of the San Pitch River, which is called Sanpete Valley. To the southeast, beds of the Flagstaff Formation can be seen behind hogbacks of the slightly younger, gray-colored Green River Formation. After milepost 245, the road crosses through a ridge of faulted, vertical Cretaceous beds. Some are composed of thick conglomerate deposited during early mountain building of the Sevier Orogeny; red hogbacks of the vertical Cretaceous beds are plain to see west of Gunnison Reservoir, though they are too narrow to show on the geologic map.

US 89 crosses a small ridge of the Arapien Formation just south of Ninemile Reservoir. To the south, the Arapien Valley is the type locality for this rock, where it forms gray, yellow, and pink badlands.

Cuestas of the Green River Formation, varying from sand and silt to lake limestone, help contain the water in Gunnison Reservoir. Some cuestas are half buried beneath the flat floor of the Sanpete Valley. One forms the hill on which the Manti Utah Temple stands in Manti. Another was quarried for limestone used in buildings. The limestone is oolitic; its small spheres, called ooids, look like fish eggs but formed as grains rolled around on the shallow bottoms of chemically supersaturated lakes—lakes much like the Great Salt Lake, where ooids form today. The grains collected layer upon layer of calcium carbonate on their surfaces, ending up smooth and round. The formation contains fossils of fish, alligators, turtles, and other inhabitants of the Eocene lakes.

Landslides, rockfalls, and earthflows are common in the Sanpete Valley. Mountains on either side of the valley are still young and steep. Tertiary beds are only poorly cemented, and rocks slide easily down the slope of the Wasatch Monocline.

The valley narrows north of Mount Pleasant. Mid-Tertiary volcanics, and sedimentary layers eroded from them, form the tops of many of the hills to the west.

Between Milburn and Indianola US 89 crosses into the drainage of Thistle Creek, a tributary of Soldier Creek. Soldier Creek flows into Utah Lake near the city of Spanish Fork. North of milepost 272, Thistle Creek has made a shortcut for itself. Its former course is now an oxbow lake.

In Manti Canyon, east of Manti, Flagstaff Formation limestone makes a dramatic escarpment above less-resistant Tertiary and Cretaceous layers. —Lucy Chronic photo

Partly reddish volcanic breccia borders the highway between Indianola and Birdseye. The breccia came from a stratovolcano that was west of here. Higher dark summits of Pennsylvanian and Permian limestone appear ahead and to the west. These are part of the Sevier Thrust Belt of Cretaceous time. High on the peaks are cup-shaped cirques carved by Pleistocene glaciers.

A short distance downstream from Birdseye, you can see dead trees that drowned in a lake that formed in the spring of 1983 behind a landslide down-river from Thistle. After an unusually snowy winter followed by incessant rain, a large slide developed in a body of loose sediment that had accumulated in a steep side-valley west of the Spanish Fork River. The side-valley had formed over a layer of steeply dipping sedimentary rock. The slide moved slowly, at most 3.5 feet (1.1 m) per hour, inexorably blocking the Spanish Fork River and flooding the town of Thistle.

By pumping water out of the newly forming lake, the Army Corps of Engineers averted a much greater disaster. Left alone, the lake may have overtopped the slide, cut through it, and released its water all at once. Later the corps tunneled through bedrock around the toe of the slide to make a permanent channel for the river. The highway and railway were soon moved up off the canyon floor. Many smaller slides occurred as the new highway was being built, and parts of the big slide have moved again in wet years.

The Thistle Landslide. The town of Thistle, flooded by the lake backed up behind the slide, is out of view to the left of the slide. Parts of the slide continue to move in wet years. —Felicie Williams photo

Abandoned houses in Thistle. The whiter areas at the bottoms of the slopes define the high-water mark of the lake that formed behind the slide. —Felicie Williams photo

US 89
Brigham City—Garden City
64 miles (103 km)

The Wellsville Mountains, north and east of Brigham City, are composed of faulted east-dipping Precambrian metamorphic rocks and Paleozoic sedimentary rocks. A fairly complete Cambrian to Permian sequence appears as US 89 traverses the range. The range is edged on the west by the Wasatch Fault. Rocks west of the fault have dropped more than 14,000 feet (4,300 m) relative to those east of it.

North of Brigham City, most of the mountains' escarpment is Geertsen Canyon Quartzite. This hard cliff-forming rock was sand along the edge of a late Proterozoic to early Cambrian sea. Known by different names in different places, the quartzite stretches halfway across the continent. Its layers become younger eastward because the sea crept slowly in that direction, depositing a sandy shoreline as it went.

The delta at the front of the Wellsville Mountains in Brigham City was built out into Lake Bonneville during Pleistocene time. This ancient lake's shorelines are now terraces on both sides of the delta. Quarries expose the delta's westward-sloping gravel layers. It would have been quite a sight to see mammoths and other Pleistocene animals roaming the shores of this freshwater lake.

US 89 turns north near Mantua, soon crossing into Ordovician and then Silurian rocks deposited in marine and nearshore environments. Some contain fossils: corals, trilobites, bryozoans, sponges, echinoderms, brachiopods, and graptolites. The sandstone, limestone, and dolomite layers form cliffs and ledges, whereas shale layers form slopes. The highest cliffs are thick Mississippian limestone. Like the Cambrian quartzite, this limestone reaches halfway across the continent and has many names, but unlike the quartzite, it was deposited in open sea with little or no sediment from land.

The gentler east slope of the Wellsville Mountains is composed of Pennsylvanian and Permian limestone and sandstone. These layers show a change over time: they were deposited in shallower water as the sea retreated from the region. Occasionally, there are beds of salt, phosphate, and other evaporite minerals. They precipitated from isolated basins of evaporating seawater.

All of the Paleozoic-age rocks we have seen so far were caught up in a thrust sheet of the Sevier Thrust Belt and moved about 40 miles (64 km) east relative to underlying rocks. After the Sevier Orogeny, younger underlying thrust sheets transported them an additional 75 miles (120 km) eastward.

East of the Wellsville Mountains, US 89 descends into Cache Valley, a graben dropped between the West Cache Fault Zone at the base of the Wellsville Mountains and the East Cache Fault Zone at the edge of the Bear River Range. During Pleistocene time, Cache Valley was an arm of Lake Bonneville, Great Salt Lake's ice age ancestor. Old shorelines from multiple lake levels mark the surrounding slopes. At its highest, the lake overflowed northward through Cache Valley into the Snake River in Idaho.

At Logan the highway climbs onto a Pleistocene delta that was built by the Logan River as it emptied into Lake Bonneville. Gravels deposited in this delta

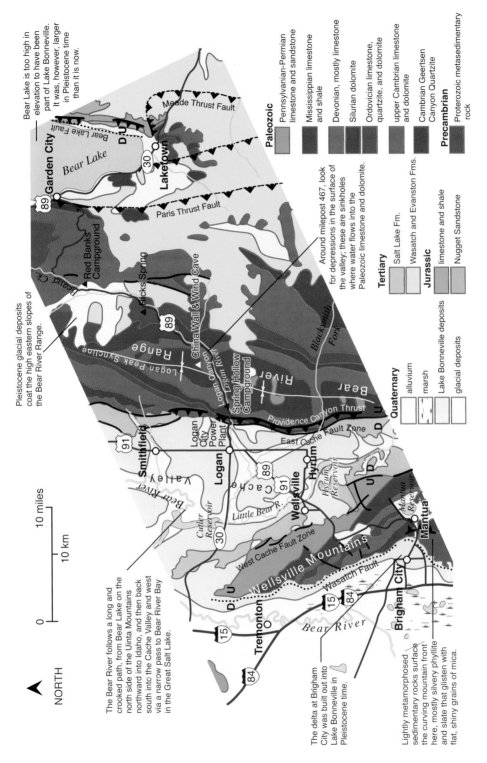

NORTH

10 miles

10 km

0

Bear Lake is too high in elevation to have been part of Lake Bonneville. It was, however, larger in Pleistocene time than it is now.

Meade Thrust Fault

Bear Lake Fault

D U

Garden City

Bear Lake

30

Laketown

89

Paris Thrust Fault

Bear Ck.

Red Banks Campground

Ricks Spring

China Wall & Wind Cave

89

Logan Peak Syncline

Bear River Range

Logan Canyon

Spring Hollow Campground

Logan River

Blacksmith Fork

Bear River

Around milepost 467, look for depressions in the surface of the valley; these are sinkholes where water flows into the Paleozoic limestone and dolomite.

Pleistocene glacial deposits coat the high eastern slopes of the Bear River Range.

Smithfield

91

Bear River Valley

Logan City Power Plant

Logan

89

East Cache Fault Zone

Providence Canyon Thrust

D U

Cutler Reservoir

30

Cache Valley

89

91

Little Bear R.

Wellsville

Hyrum

Hyrum Reservoir

Mantua Reservoir

Mantua

U D

West Cache Fault Zone

Wellsville Mountains

D U

Tremonton

15

Wasatch Fault

Brigham City

15

84

Bear River

84

The Bear River follows a long and crooked path, from Bear Lake on the north side of the Uinta Mountains northward into Idaho, and then back south into the Cache Valley and west via a narrow pass to Bear River Bay in the Great Salt Lake.

The delta at Brigham City was built out into Lake Bonneville in Pleistocene time.

Lightly metamorphosed sedimentary rocks surface the curving mountain front here; mostly silvery phyllite and slate that glisten with flat, shiny grains of mica.

Paleozoic

Pennsylvanian-Permian limestone and sandstone

Mississippian limestone and shale

Devonian, mostly limestone

Silurian dolomite

Ordovician limestone, quartzite, and dolomite

upper Cambrian limestone and dolomite

Cambrian Geertsen Canyon Quartzite

Precambrian

Proterozoic metasedimentary rock

Tertiary

Salt Lake Fm.

Wasatch and Evanston Fms.

Jurassic

limestone and shale

Nugget Sandstone

Quaternary

alluvium

marsh

Lake Bonneville deposits

glacial deposits

Geology along US 89 between Brigham City and Garden City.

Lake Bonneville delta deposits in western Cache Valley provide material for construction.
—Lucy Chronic photo

are visible near the mouth of Logan Canyon. The delta developed at the ancient lake's Provo Shoreline level; wave-cut shorelines of the Bonneville Shoreline level are higher.

Triangular facets along the face of the Bear River Range and a clear scarp that cuts the youngest Pleistocene sediments near the mouth of the canyon mark the East Cache Fault Zone—and prove it is still active. Both the facets and scarp formed relatively recently as the valley dropped downward along the fault; had they formed long ago, their angularity would have been subdued by erosion. Rocks west of the fault zone may be as much as 12,000 feet (3,660 m) lower. The fault is nearly vertical at the surface, but seismic studies, in which small earthquake waves from man-made shocks are reflected off deep rock layers, show it gradually flattens out westward, deep beneath the surface.

The crest of the Bear River Range is one long syncline. Many of its layers are the same as those of the Wellsville Mountains. Starting near the Logan City Power Plant, Ordovician limestone and then quartzite are followed by Silurian dolomite, which is black at first and then lighter between mileposts 463 and 464. Highly fractured Devonian dolomite and limestone follow. The Devonian layers flatten for the next several miles as US 89 crosses the syncline's axis, though there are some low-angle thrust faults and folds. A roadside sign near Spring Hollow Campground, at milepost 465, points out the gravelly beach terrace that marks the farthest reach of Lake Bonneville up Logan Canyon. A little beyond it, the massive much-jointed dolomite has eroded into picturesque pinnacles. The syncline axis is just before milepost 466.

Above the highway, a distinctive cliff of Mississippian limestone is informally known as the China Wall. A trail at milepost 467 offers a pleasant 2-mile (3 km) hike to Wind Cave, which was dissolved in the wall. The limestone contains abundant fossils of creatures such as crinoids and brachiopods that lived in a warm equatorial sea.

Strange marks on a rock surface near milepost 471 may be burrows that small conical cephalopods left while searching for food. —Lucy Chronic photo

The faulted rocks near Ricks Spring are riddled with caverns and undergound channels (solution features) dissolved by groundwater. The water becomes acidic by absorbing carbon dioxide from the air and soil. As is true with much Paleozoic limestone and dolomite, solution features are very common here. Most formed in the wetter past, but they still provide pathways for lots of water; Ricks Spring and others like it contribute much of the water that flows in the Logan River.

Ricks Spring flows from a cave in Ordovician limestone. Its maximum flow, more than 30,000 gallons (113,600 liters) per minute, comes in late spring and early summer from melting snow in the surrounding mountains. The spring dries up completely in winter.

Above Ricks Spring the Logan River has cut a deep canyon into Ordovician quartzite. Before excavating this cliff-walled trench, the river for some time meandered across the surface of the hard quartzite; its meanders are preserved as tight loops in the present canyon.

The highest part of the Bear River Range was glaciated in Pleistocene time. As you drive, look for cirques on distant peaks, rounded uplands, and till.

Red slopes near Red Banks Campground formed in much softer mudstones and limestones of the Wasatch Group, which is widespread east and southeast of here. They were deposited in and near Tertiary lakes that formed between the newly risen Wasatch Range and Rocky Mountains. They are much younger than the rocks we have been driving through in Logan Canyon.

US 89 returns to older rocks, this time Cambrian limestone and shale, in the upper reaches of Beaver Creek. Between mileposts 488 and 490, roadcuts reveal intricate faulting in the layers.

Bear Lake comes into view west of the summit of the Bear River Range, its brilliant turquoise water contrasting with sage-covered hills. Like Cache Valley, this valley is a graben. The fault along the east shoreline is still active, and the lake bottom is gradually dropping downward along it.

Rocks east of the lake are mostly Mesozoic in age, with Triassic-Jurassic Nugget Sandstone forming the resistant ridge that confines the east side of the lake. More reddish Tertiary deposits—the Wasatch Group again—lap up onto these Mesozoic rocks.

Very fine grains of calcium carbonate minerals suspended in Bear Lake are the cause of the brilliant turquoise color of its water. Rain and snowmelt become saturated with calcium as they make their way through crevices and caves in limestone before surfacing in springs near the lake. Tiny molecules of calcium carbonate form right in the water. Individual springs along the lake have their own microenvironments, within which various species of fish and invertebrates such as ostracods evolved.

For many years springs were the only source of water for the lake, but in 1912 part of the Bear River was diverted here to be stored for irrigation. This changed the lake's delicate chemical balance and added silt and clay. The color has been changing ever since; the lovely blue may eventually disappear. Many of the lake's endemic species have become extinct or are close to it.

An ostracod is a very small crustacean with a bivalve shell. Field of view is 0.12 inch (3 mm).
—Gerard Visser photo

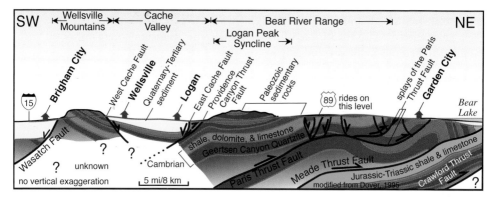

Southwest-northeast cross section between Brigham City and Bear Lake. This is the heart of the Sevier Thrust Belt. Fault after fault surfaces here; each carried huge sheets of sedimentary rock eastward into this region during the Sevier Orogeny. The Paris Thrust Fault, which comes to the surface on the west side of Bear Lake, was active 120 to 113 million years ago; the rock it carried here broke along smaller faults, or splays, where the main fault neared the surface. The Meade Thrust Fault was active 113 to 98 million years ago; it surfaces southeast of Bear Lake.

US 189 and US 40
Provo—Silver Creek Junction
50 miles (80 km)

The high levels of Lake Bonneville formed pronounced terraces, like bathtub rings, which extend north and south along the mountain front near Provo. The most pronounced shoreline here is the Provo Shoreline, and subdivisions of Provo have expanded onto this terrace.

Where the Provo River flowed into Lake Bonneville during Pleistocene time, it built a delta, now also covered with houses. The ancient delta and its houses obscure the Wasatch Fault, which runs along the base of the mountain front. Truncated mountain ridges reveal that there has been fairly recent movement along the fault, though not in historic time, and the mountains north of Provo show the scars of several large landslides. Earthquakes and the landslides they engender threaten Utah's densely populated urban corridor; ultimately, disastrous quakes are almost certain to come.

The front of the Wasatch Range here is composed entirely of the Oquirrh Group: older limestone and shale beneath a thick sequence of younger limestone, shale, and sandstone—all of Pennsylvanian and Permian age. All the rocks you see were thrust eastward up to 25 miles (40 km) above the great Charleston-Nebo Thrust Fault during Cretaceous time. They are part of the Sevier Thrust Belt. The journey bent and broke them and, in places, stood them on end.

The cliffs near the entrance to Provo Canyon are Mississippian Great Blue Limestone, a member of the Oquirrh Group. The thin beds indicate this rock was deposited far offshore, out of the reach of, or inhospitable to, animals that would have burrowed and destroyed the layering. —Lucy Chronic photo

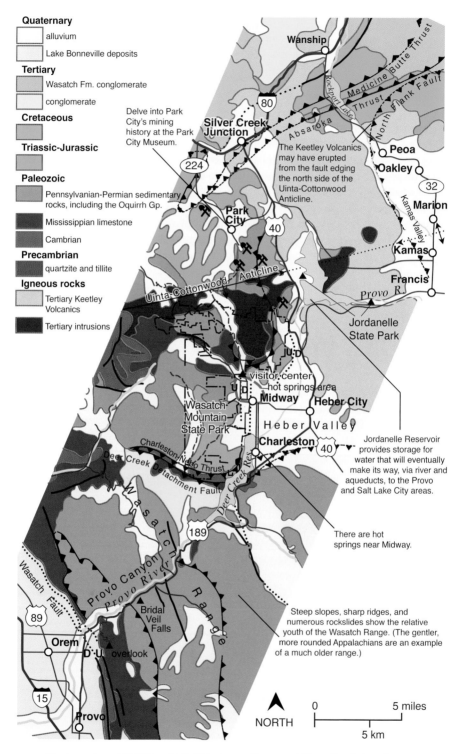

Quaternary
- alluvium
- Lake Bonneville deposits

Tertiary
- Wasatch Fm. conglomerate
- conglomerate

Cretaceous

Triassic-Jurassic

Paleozoic
- Pennsylvanian-Permian sedimentary rocks, including the Oquirrh Gp.
- Mississippian limestone
- Cambrian

Precambrian
- quartzite and tillite

Igneous rocks
- Tertiary Keetley Volcanics
- Tertiary intrusions

Delve into Park City's mining history at the Park City Museum.

The Keetley Volcanics may have erupted from the fault edging the north side of the Uinta-Cottonwood Anticline.

Jordanelle Reservoir provides storage for water that will eventually make its way, via river and aqueducts, to the Provo and Salt Lake City areas.

There are hot springs near Midway.

Steep slopes, sharp ridges, and numerous rockslides show the relative youth of the Wasatch Range. (The gentler, more rounded Appalachians are an example of a much older range.)

Wanship

Silver Creek Junction

Peoa

Oakley

Marion

Park City

Kamas

Francis

Provo R.

Jordanelle State Park

Midway

Heber City

Heber Valley

Wasatch Mountain State Park

Charleston

visitor center — hot springs area

Uinta-Cottonwood Anticline

Rockport Lake Thrust

Medicine Butte Thrust

North Flank Fault

Absaroka Thrust

Kamas Valley

Charleston-Nebo Thrust

Deer Creek Detachment Fault

Deer Creek Res.

Wasatch Range

Provo Canyon

Provo River

Bridal Veil Falls

overlook

Wasatch Fault

Orem

Provo

0   5 miles
NORTH
5 km

Geology along US 189 and US 40 between Provo and Silver Creek Junction.

Across from Bridal Veil Falls a long valley opens up, revealing the vast thickness of the Oquirrh Group. Almost 5 vertical miles (8 km) of sea-bottom mud accumulated on the rapidly sinking floor of the Oquirrh Basin, a large basin that extended from south-central Utah well into Idaho during late Paleozoic time. —Lucy Chronic photo

Folding during the Laramide Orogeny affected this area as well. Rocks in Provo Canyon dip east-southeast off the southern flank of the Cottonwood Anticline, a gentle fold that crosses the Wasatch Range. It lines up with and is often grouped with the Uinta Anticline of the Uinta Mountains.

Rocks of the Oquirrh Group extend to the eroded edge of the Charleston-Nebo Thrust Sheet, near Charleston. US 189 crosses the thrust fault near Deer Creek Reservoir. Later faulting during Basin and Range extension caused the Heber Valley to drop down relative to adjacent hills and mountains.

Snowmelt and rain in the Wasatch Range feed several dozen hot springs near Midway. The water percolates down through caverns and and fractures to depths where the rocks are hot and then travels upward through faults in limestone on the west side of the Heber Valley. At the surface, the hot water forms mounds of tufa—you can actually swim inside a mound.

Heber City lies on the floodplain of the Provo River. Here, US 189 joins US 40. North of town the highway follows the river upstream.

The Wasatch Range, west of Heber City, is at the intersection of two ancient Precambrian structures; both have influenced the geology of much of the state. The first structure is the east-west suture between two ancient continents. The suture has been a zone of weakness ever since it formed, and because of this it was the line along which late Precambrian continental extension took place, forming an extensional basin. The extensional basin filled with a glut of sediment, here called the Big Cottonwood Formation, that during the Laramide Orogeny was uplifted as the core of rock in the Uinta-Cottonwood Anticline.

The second structure is the Utah Hingeline, the north-south line that roughly parallels I-15. It divides thinner crust to the west from thicker crust to the east. Thicker layers of sediment accumulated over the thinner crust to the west as it slowly flexed downward under the sediment's weight. In contrast,

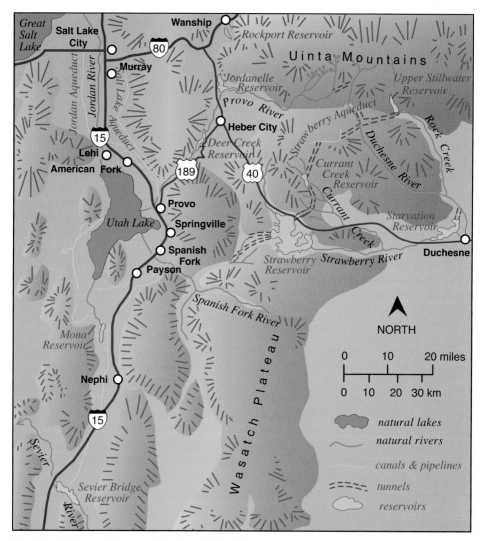

Deer Creek Reservoir is part of the Central Utah Project. Its water flows through an intricate system of reservoirs, canals, and tunnels—starting on the southern side of the Uinta Mountains—and is passed on as needed through a pipeline down Provo Canyon to the cities west of the Wasatch Front. Without these diversions, the water would flow to the Green River.

only thin layers were deposited on the thick, inflexible crust to the east. During mountain building, the thin crust west of the hingeline faulted and folded easily to accommodate compression, while the overlying thick sedimentary layers were carried as cohesive masses great distances by thrust faults. To the east, the thicker crust broke instead into almost vertical fractures, and the thin sedimentary layers folded over the faults.

Wasatch Mountain State Park lies on the eastern flanks of the Wasatch Range. The park's high, glaciated valleys rise dramatically west of Heber Valley. Each cirque (the curving cliff at the head of a glaciated valley) was filled with a glacier during the Pleistocene. Because the park is at the intersection of the Uinta-Cottonwood Anticline and the Utah Hingeline, driving the roads that wind their way through the park is like driving through a microcosm of Utah's geology.

North of Heber Valley, US 40/189 climbs onto low hills composed of the Tertiary Keetley Volcanics and crosses the Cottonwood Anticline. On the anticline's north side, beds in the mountains dip to the north. The hills east of Jordanelle Reservoir, in Jordanelle State Park, are composed of the Keetley Volcanics: massive light-colored breccias and tuffs produced from what must have been truly enormous eruptions from multiple volcanoes. Many of the eruptions were lahars: rapidly moving clouds of glowing volcanic ash that, upon stopping, welded into breccia and tuff. During some of the eruptions there were huge mudflows that lithified within the volcanics as mudflow breccias. In places the rocks are streaked with whitish and yellowish clay that formed as acidic steam rose through the rock. Steam probably continued to rise through the ash long after the eruptions had ended.

The ridge northeast of Jordanelle Reservoir is volcanic breccia. —Lucy Chronic photo

It seems strange that the Provo River should have cut a canyon through this ridge of volcanic rocks since there is a much easier route to follow. As Pleistocene time came to a close, the upper portion of the Provo River, coming from the west end of the Uinta Mountains, flowed north across the Kamas Valley to join the Weber River near Oakley. The lower Provo River, the part we saw in Provo Canyon, gradually eroded headward through the volcanic rocks until it reached the upper Provo and captured, or pirated, its water.

North of the reservoir, Utah 248 leads west to Park City. Park City became a mining town after ores bearing silver, lead, zinc, and copper were discovered in 1869; the city has since become a ski and tourist destination.

The ore in Park City formed during Tertiary time. Diorite magma intruded the region's quartzite and limestone, doming, fracturing, and crushing it in the process. Hot mineral-rich water escaping the cooling magma precipitated ore-bearing minerals on the walls of fractures; in the limestone, the water followed the bedding, dissolving the rock and depositing ore minerals, which sometimes preserved the original bedding.

Hazards caused by mining at Park City are ever present: Open shafts and adits (horizontal mining tunnels) occasionally collapse. Old mine tailings may contain toxic metals. Tailings and dumps on the hillsides can be unstable. Fortunately, toxic elements are now removed from the water before it reaches Jordanelle Reservoir. Nonetheless, there are several Superfund sites in the area that the Environmental Protection Agency is responsible for cleaning up; they are in varied stages of completion.

US 191 and Utah 44
Vernal—Wyoming State Line
52 miles (84 km)

North of Vernal, US 191 crosses rows of cuestas and hogbacks of Mesozoic and Paleozoic sedimentary rocks that were tilted by the rise of the Uinta Mountains. The erosion-resistant ridges are separated by valleys of softer rock. Both circumvent the range, like racetracks (valleys) and lines of bleachers (hogbacks and cuestas). The core of the range is Precambrian metamorphic and sedimentary rock. The Uintas are one vast anticline with an east-west axis. The range is faulted along its northern and southern margins.

Driving northward, the rocks change from youngest to oldest, starting with yellow hills of Cretaceous Mancos Shale around Vernal. US 191 parallels hogbacks of Cretaceous sandstone for a time while following a valley that developed in soft shales of the Cretaceous Cedar Mountain Formation and Jurassic Morrison Formation. Adjacent hogbacks and the fine, impervious clay-rich shale between them contains the water of Steinaker Reservoir. The clay—weathered volcanic ash—swells when it gets wet and then shrinks when it dries, causing the highway to buckle near the reservoir. Colorful rocks in Red Fleet State Park record waves of mountain building during Mesozoic time.

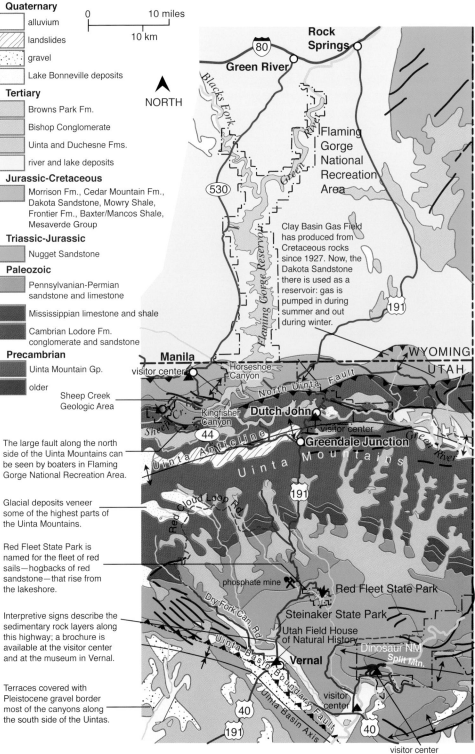

Quaternary
- alluvium
- landslides
- gravel
- Lake Bonneville deposits

Tertiary
- Browns Park Fm.
- Bishop Conglomerate
- Uinta and Duchesne Fms.
- river and lake deposits

Jurassic-Cretaceous
- Morrison Fm., Cedar Mountain Fm., Dakota Sandstone, Mowry Shale, Frontier Fm., Baxter/Mancos Shale, Mesaverde Group

Triassic-Jurassic
- Nugget Sandstone

Paleozoic
- Pennsylvanian-Permian sandstone and limestone
- Mississippian limestone and shale
- Cambrian Lodore Fm. conglomerate and sandstone

Precambrian
- Uinta Mountain Gp.
- older

0   10 miles

10 km

NORTH

Rock Springs

Green River

80

530

Blacks Fork

Green River

Flaming Gorge National Recreation Area

Flaming Gorge Reservoir

191

Clay Basin Gas Field has produced from Cretaceous rocks since 1927. Now, the Dakota Sandstone there is used as a reservoir: gas is pumped in during summer and out during winter.

Manila
visitor center

Horseshoe Canyon

WYOMING
UTAH

North Uinta Fault

Sheep Creek Geologic Area

Sheep Cr.

Kingfisher Canyon

44

Dutch John
visitor center

Uinta Anticline

Greendale Junction

Green River

The large fault along the north side of the Uinta Mountains can be seen by boaters in Flaming Gorge National Recreation Area.

U i n t a   M o u n t a i n s

Glacial deposits veneer some of the highest parts of the Uinta Mountains.

Red Cloud Loop Rd.

191

Red Fleet State Park is named for the fleet of red sails—hogbacks of red sandstone—that rise from the lakeshore.

phosphate mine

Red Fleet State Park

Dry Fork Can. Rd.

Steinaker State Park

Interpretive signs describe the sedimentary rock layers along this highway; a brochure is available at the visitor center and at the museum in Vernal.

Utah Field House of Natural History

Uinta Basin

Vernal

Dinosaur NM
Split Mtn.

Uinta Basin Boundary Fault

Uinta Basin Axis

Terraces covered with Pleistocene gravel border most of the canyons along the south side of the Uintas.

40

191

visitor center

40

visitor center and Carnegie Quarry

Geology along US 191 and Utah 44 between Vernal and the Wyoming state line.

Signs after milepost 215 explain how phosphate mining, processing, and reclamation occur. The rock containing concentrated phosphate is ground and, with added water, piped as a slurry to Rock Springs, Wyoming. It is used in fertilizers, pharmaceuticals, and other products.

Flat-topped mesas that stretch southward from the Uintas are surfaced with Tertiary Bishop Conglomerate, a poorly cemented, well-rounded gravel that was eroded from the mountains and deposited in alluvial fans at the foot of the mountains. Later, mountain streams, much enhanced by glacial meltwater and a steep gradient, carved the fans into separate mesas.

Excellent rock exposures in Red Fleet State Park include this dinosaur trackway, part of a site with more than 350 individual tracks in the Nugget Sandstone. The reservoir's waves helped expose this site, which is partly below the high-water mark.
—Felicie Williams photo

The Precambrian rocks that make up the main mass of the Uinta Mountains belong to the 770-to-740-million-year-old Uinta Mountain Group (not to be confused with the Tertiary Uinta Formation of the Uinta Basin). The group, deposited in an ancient extensional basin, is mostly composed of purplish-red sandstone with interbedded shale and conglomerate. The formation is up to 28,000 feet (8,500 m) thick.

US 191 through Flaming Gorge National Recreation Area

The Green River flows south from the low hills of Wyoming into Flaming Gorge, snakes through Horseshoe and Kingfisher Canyons, abruptly crosses the North Uinta Fault, and enters Red Canyon. Flaming Gorge Reservoir occupies not only Flaming Gorge, but the other three canyons as well. The reservoir's dam was constructed in Red Canyon.

North of milepost 235, Utah 191 continues to descend through the Uinta Mountain Group on its way to Flaming Gorge Reservoir. Red Canyon was named for the unit's dark red rock; interlayered red sandstone, shale, and siltstone are well exposed near Flaming Gorge Dam in the town of Dutch John. These sedimentary rocks reflect the varied environments of the extensional basin they were deposited in: deltas, river floodplains, and nearshore marine environments.

In Red Canyon, the Uinta Mountain Group drops to the river in a succession of terraces. John Wesley Powell explored and named these canyons in 1869. —Lucy Chronic photo

Flaming Gorge Dam, built between 1958 and 1964, is part of the Colorado River Storage Project. The project's aims were many: regulating flow in the Colorado River Basin so each state's apportioned water would be available to it, making Green River water available for irrigation, improving conditions for wildlife, enhancing recreation, providing flood control, and generating power.

The reservoir swings west away from the road, winding west through Red Canyon. The road crosses the dam and then climbs through more of the Uinta Mountain Group. At milepost 247, the sedimentary layers become almost vertical. At the base of the long, east-west-trending cliff to the north is the North Uinta (or Flank) Fault, the major east-west thrust fault that edges the north side of the Uintas. The rocks south of the fault were thrust up relative to the ones north of it, so the formations north of the fault are much younger (Triassic-Jurassic) than the Uinta Mountain Group (Precambrian) to the south. (For a close-up diagram of the fault, see Sheep Creek Geological Area, below.)

North of the fault, the cliff consists of the eolian Nugget Sandstone. The road winds through it and then drops into a succession of younger and younger units, starting with a valley of less-resistant Morrison and Cedar Mountain Formations.

These are followed by a series of formations that tell a story of multiple advances and retreats of the Cretaceous Interior Seaway. The yellow-brown Dakota Sandstone hogback is a beach deposit signaling the first advance of the seaway. A valley of marine Mowry Shale, famous for the many fish scales it contains, is followed by the Frontier Formation sandstone, another beach-deposited formation. A wider valley of Baxter Shale (the age-equivalent to the Mancos Shale), also deposited in the sea, is followed by two more hogbacks separated by shale north of milepost 252, near the Wyoming state line. These are the Mesaverde Group sandstones with their seams of coal; the last sandstone signals the final retreat of the seaway.

As US 191 leaves Utah, the layers of rock level out, no longer influenced by the North Uinta Fault. After the Wyoming line, the road first crosses the Fort Union and Uinta Formations, both derived from the rising Uinta Mountains during Tertiary time, and then heads toward Interstate 80 on thick beds of the Green River Formation, which is famous for its oil shale and fossil fish. It was deposited in a large lake.

Utah 44 through Flaming Gorge National Recreation Area

Utah 44 is an alternate route to Wyoming that goes around the reservoir's south and west sides. From the US 191 junction, Utah 44 crosses a hilly upland of red rocks of the Uinta Mountain Group. For nearly 10 miles (16 km) the highway parallels the crest of the Uinta Anticline before turning north. Higher mountains farther west were glaciated during Pleistocene time. (The road that loops through the Sheep Creek Geological Area takes off to the west, returning to Utah 44 about 6 miles, 10 km, farther north; see the next section.)

Utah 44 continues north, crossing the North Uinta Fault, passing onto Mississippian shale and limestone, and then dropping into the drainage of Sheep Creek. The brilliant red rocks exposed along the creek are the Triassic Chinle and Moenkopi Formations, for which Flaming Gorge was named.

The red Triassic Moenkopi and Chinle Formations, visible north of Utah 44, were laid down by rivers and streams and in tidal flats. They contain tracks and fossils of amphibians and dinosaurs. Above them the yellow Nugget Sandstone forms a resistant cliff. —Lucy Chronic photo

Upper Cretaceous layers, seen here across Flaming Gorge Reservoir from Utah 44, define a perfect racetrack valley on the east side of the reservoir. The two resistant hogbacks are sandstone members of the Mesaverde Group separated by less-resistant shale. —Lucy Chronic photo

The road crosses another fault—a smaller one—separating the softer Triassic deposits from the ridge-forming Nugget Sandstone. The massive Triassic-Jurassic Nugget Sandstone, known as the Navajo Sandstone farther south, was deposited in a field of dunes larger than the state of Utah. At the fault, the sandstone is almost vertical. The road continues through younger rocks, starting with a valley of less-resistant Morrison Formation, followed by Cedar Mountain Formation shale and limestone, and then by the hogback-valley-hogback sequence of Dakota Sandstone, Mowry Shale, and Mesaverde Group sandstone, described above along Utah 191. At Manila, Utah 44 crosses into almost flat-lying Green River Formation shale and continues on it into Wyoming and to the city of Green River.

Sheep Creek Geological Area

The Sheep Creek Geological Area provides a close-up look at the North Uinta Fault, which flanks the north side of the Uinta Mountains. These mountains were raised along the faults of this zone, and the smaller faults on the south side of the range, during the Laramide Orogeny. The range was also folded into one large east-west-oriented anticline.

The turnoff to the geological area is off Utah 44 about 15 miles (24 km) west of US 191. The road starts in the Precambrian Uinta Mountain Group, dark-red gritty layers that were deposited in a gradually subsiding extensional basin.

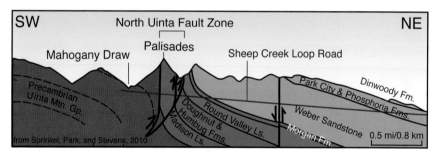

Southwest-northeast cross section of the Sheep Creek Geological Area. Signs along the road label both formations and faults. Vertical exaggeration is 1.7x.

Soon after zigzagging down into Sheep Creek Canyon, the road passes Palisades Memorial Park, the location of a disastrous debris flow that demolished a campground and took seven lives in 1965. Landslides are common when bedding surfaces are parallel to a slope, as they are here. In this canyon, rock slides downward along soft shale layers and accumulates in stream bottoms until a big enough flood carries it roaring down the canyon.

Just past the memorial park is the main thrust fault line, running east-west. The canyon narrows at the fault. Formations to the south were uplifted relative to the formations to the north, so that Precambrian rocks in Sheep Creek are now adjacent to Mississippian ones. Beds closest to the fault were deformed by

The light-colored swath running across the hillside is the North Uinta Fault Zone, separating Uinta Mountain Group sandstones on the left from Mississippian Madison Limestone on the right. The limestone cliff is called the Palisades. —Lucy Chronic photo

motion along the fault, pulled up or down along it like taffy, until they were much thinned and almost vertical or even overturned. Two other parallel faults cut through the Mississippian rocks north of the main fault; these are probably splays of it.

Passing by a few softer layers of Mississippian and Pennsylvanian rocks, which have eroded into steep hillsides rather than cliffs, the road enters an area of the canyon edged by cliffs of the erosion-resistant, light-colored Weber Sandstone of Permian age. As the road cuts northward through the sandstone, you can see that its layers are less and less influenced by movement that occurred along the fault. By the end of the canyon, the layers are almost horizontal, indicating that the fault barely disturbed them.

Younger red-colored Permian and Triassic formations appear above the road after the Weber Sandstone cliffs; they drop gradually to road level where the road turns southeast. Maroon shale with white streaks of evaporites right at the turn is the Phosphoria Formation—the same rock mined for phosphate adjacent to US 191 on the south side of the Uinta Mountains.

Utah 12
Torrey—US 89 near Panguitch
124 miles (200 km)

Utah 12 spans most of the distance between Capitol Reef and Bryce Canyon National Parks, and for much of its length it follows the northern edge of Grand Staircase–Escalante National Monument.

Utah 12 heads south from Torrey through red-brown Triassic mudstone and siltstone of the Moenkopi Formation. Where the road rides across the Miners Mountain Anticline, the layers are gently rippled, and then they dip steeply at the Cocks Comb. The Cocks Comb, the fold that defines the south side of the Miners Mountain Anticline, is ruptured by the Teasdale Fault, one of the easternmost faults resulting from Basin and Range stretching. Rocks have dropped on the southwest side of the fault, revealing Jurassic Navajo Sandstone close to road level on the southwest side of the fault. Sculpted hills of Navajo Sandstone have checkerboard surfaces—evenly spaced joints that cut across the sandstone's bedding; both the joints and bedding are weathered and thus somewhat recessed.

For a stretch, the highway follows a contact between the Navajo Sandstone and an apron of landslide debris that covers most of the eastern shoulder of Boulder Mountain; then the highway climbs up onto the landslide debris. Boulders of volcanic rock are part of the debris. Large quantities of lava flows and volcanic ash built Boulder Mountain 29.5 to 20 million years ago, in Tertiary time. The immense pile of volcanic rock, possibly sagging into a partly emptied magma chamber, appears to have caused the sedimentary rock layers below and around it to sag too, so that they dip toward Boulder Mountain.

Many flat-topped mesas—characteristic of the Colorado Plateau—can be seen from vantage points along this highway. In the wild and beautiful terrain between here and the Henry Mountains, far to the east, flat-lying rocks abruptly swoop downward in the long ridges of Waterpocket Fold of Capitol Reef National Park. The prominent light-colored rock along the summit of this fold is the Navajo Sandstone.

Utah 12 travels for many miles on the landslide aprons that surround Boulder Mountain. Their characteristic hummocky slopes are spotted with marshy depressions. The slides were probably most active in Pleistocene time, when ice topped the mountain and the climate was wetter, but there is plenty of evidence of recent movement, including bent tree trunks and damage to the pavement.

The viewpoint at milepost 99 overlooks the surreal landscape of the Escalante country. The Straight Cliffs, in the distance to the south, are carved in gray Cretaceous rocks and edge the Kaiparowits Plateau. The round dome of Navajo Mountain, a single laccolith still covered with Jurassic rocks, is 70 miles (113 km) to the south. The rapid, deep down-cutting of the Colorado River and its tributaries has exposed this incredible country to view—most likely within the last 6 million years.

The town of Boulder is an oasis. Boulder Mountain receives quite a bit of precipitation, and its melting snow and rainfall recharge many springs and fill the region's creeks. Boulder is named for the large boulders of volcanic rock

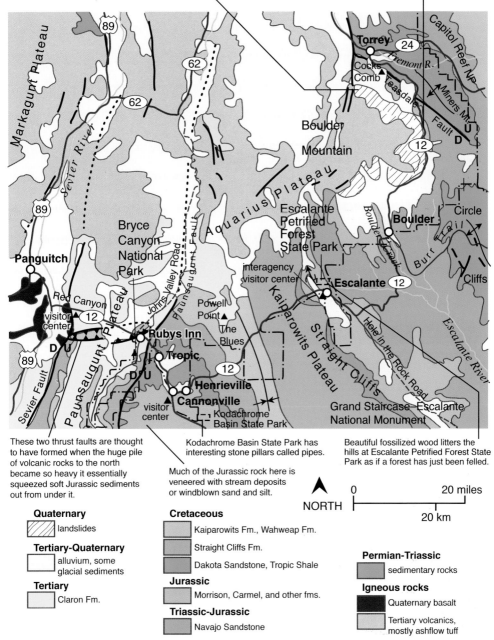

Volcanic rocks of Boulder Mountain, cut by joints caused by cooling and shrinking, break up easily to form a landslide apron around the mountain.

The ashflow tuff near milepost 113 is hard, firmly welded rock that looks almost like porcelain. Ashfall tuff, which cools before reaching the ground, is generally softer.

These two thrust faults are thought to have formed when the huge pile of volcanic rocks to the north became so heavy it essentially squeezed soft Jurassic sediments out from under it.

Kodachrome Basin State Park has interesting stone pillars called pipes.

Much of the Jurassic rock here is veneered with stream deposits or windblown sand and silt.

Beautiful fossilized wood litters the hills at Escalante Petrified Forest State Park as if a forest has just been felled.

NORTH

0   20 miles

20 km

Quaternary
landslides

Tertiary-Quaternary
alluvium, some glacial sediments

Tertiary
Claron Fm.

Cretaceous
Kaiparowits Fm., Wahweap Fm.

Straight Cliffs Fm.

Dakota Sandstone, Tropic Shale

Jurassic
Morrison, Carmel, and other fms.

Triassic-Jurassic
Navajo Sandstone

Permian-Triassic
sedimentary rocks

Igneous rocks
Quaternary basalt

Tertiary volcanics, mostly ashflow tuff

Geology along Utah 12 between Torrey and US 89.

that pepper the surface in the area. Rivers and streams probably moved many of them here from Boulder Mountain during the wetter, icier Pleistocene.

Beyond Boulder the road twists in and out of Grand Staircase–Escalante National Monument, named for the staircase-like landscape formed in Mesozoic and Cenozoic rocks. Erosion-resistant formations are the cliffs and steps, and softer, more easily eroded ones are the slopes.

In this vast ocean of Navajo Sandstone, you can see large-scale crossbedding that formed in dunes interrupted by thin, flat, pale-pink layers of siltstone that accumulated between dunes. Occasional gray limestone bands show that there were shallow ponds that came and went as the dunes shifted. The thick sandstone often breaks into arching recesses.

The Navajo Sandstone extends from Wyoming across Utah and into Arizona. In Jurassic time this area lay at the same latitude as today's Sahara—a latitude with dry winds that suck up moisture but bring little or no rain. And to the west at that time, new mountains also blocked moisture.

Utah 12 descends through the Navajo to cross the Escalante River, which is quite small at the crossing despite its deep canyon, and then climbs up through the sandstone again. Here, the rock is partly pink, as it is farther west at Zion National Park. The color, from tiny amounts of iron oxide, cuts across bedding and crossbedding. Originally present throughout the sediment, the iron that is

This part of Grand Staircase–Escalante is surfaced with Navajo Sandstone that has been carved into a labyrinth of slot canyons by the Escalante River and its tributaries. Part of Utah 12 travels along a knife-edged ridge worthy of a good thriller, with cliffs that drop nearly 1,000 feet (300 m) on either side.
—Felicie Williams photo

responsible for the color has been moved around by groundwater—dissolved and carried away in some areas, deposited in greater amounts in others.

Near milepost 69, the lava-capped Aquarius Plateau is visible to the north; its pink cliffs, composed of soil and lake bed deposits of Tertiary age, are similar to those at Bryce Canyon National Park. The north end of the Straight Cliffs and the Kaiparowits Plateau rise to the southwest.

At the town of Escalante, Utah 12 descends into the Escalante River drainage once more. Whereas we just crossed *over* the top of a plateau, the river has cut right through it. East of town the river enters the steep-walled canyon it has carved through the plateau.

Since the viewpoint at milepost 99, Utah 12 has descended the gently sloping western side of the Circle Cliffs Uplift. In the broad valley near the town of Escalante, the road crosses onto younger, softer layers of the Morrison Formation, a soft, colorful, easily eroded shale that contains abundant volcanic ash.

Escalante Petrified Forest State Park, near Escalante, preserves a fallen forest, mostly in the upper part of the Morrison Formation. This formation is famous for its dinosaur bones in other parts of Utah. The beautiful colors of the logs come from small amounts of impurities, such as iron and manganese, that precipitated with the silica.

The Morrison is rich in volcanic ash. Composed of tiny shards of silica glass, ash breaks down readily to clay, leaving excess silica dissolved in water. Ashy mud buried the trees of this state park, cutting off oxygen and delaying their

Cracks that formed as the original woody layers dried are responsible for the checked pattern in this sample of petrified wood from Escalante Petrified Forest State Park. The tree is thought to have been a conifer. —Lucy Chronic photo

decay. Silica-rich groundwater saturated the buried wood, replacing the woody material with silica—one molecule at a time. Often much of the original detail of the wood was preserved.

Heading west on Utah 12 from Escalante, Cretaceous formations of the Straight Cliffs are visible: the Tropic Shale and the Straight Cliffs and Kaiparowits Formations, all named for nearby sites. These rocks contain none of the red tints seen in the Triassic, Jurassic, and Tertiary rocks of the region. Reds and pinks come from tiny amounts of oxidized iron. The Cretaceous rocks contain iron as well, but the high quantity of plant material they contain used up the available oxygen as it decayed, so that little or none was left to oxidize the iron. The formations of the Straight Cliffs make up one of the most complete Cretaceous terrestrial sequences in North America. They have yielded fossils of numerous dinosaurs in recent years.

From milepost 39 we first glimpse Bryce Canyon National Park: its ornately sculpted pink stream and lake deposits rim the Paunsaugunt Plateau. The Paunsaugunt Fault lies just this side of the park, along the eastern edge of the plateau. It is a normal fault that formed in response to Basin and Range extension. North of Utah 12, Powell Point, composed of the same Tertiary layers found in Bryce, is higher than those in the national park because the land west of the fault, including the national park, dropped downward relative to that east of the fault.

Below Powell Point, the subtle gray badlands of the Blues are composed of upper Cretaceous Kaiparowits Formation, thick layers of highly fossiliferous, muddy sandstone that were deposited rapidly in rivers and freshwater lakes northeast of highlands of the Sevier Thrust Belt. Its many fossils include small mammals, sharks, crocodiles, lizards, turtles, insects, snails, and many varieties of dinosaurs. This formation was deposited in an abundantly rich and diverse ecosystem, and the Blues is one of the most spectacular outcrops of dinosaur-rich Cretaceous rocks exposed along a paved road in the United States.

East of Henrieville, prominent pink and white Jurassic Entrada Sandstone erodes into cliffs and hoodoos, fantastic columns or pinnacles of rock. Watch for small faults that offset the layers. The oft-photographed Grosvenor Arch and Kodachrome Basin State Park, both south of Cannonville, are in the same colorful rocks.

Kodachrome Basin State Park has numerous sentinels, called sedimentary pipes, that eroded out of the surrounding rock. Geologists believe the sediment in the pipes jetted up from underlying water-saturated layers when the pressure of thick overlying sand became too great. Similar smaller pipes are fairly common in sedimentary rocks, but these at Kodachrome are large specimens; an earthquake may have contributed to the creation and size of its pipes by shaking the sediment and causing it to react more fluidly.

Utah 12 crosses the Paria River at milepost 26, near Cannonville, and then follows it up through mustard-colored Cretaceous Tropic Shale marked with thin seams of coal. West of the town of Tropic, these rocks erode into gentle slopes and steep, irregular ridges.

West of Tropic, Utah 12 crosses over the Paunsaugunt Fault into the sculpted pink beds of the Claron Formation. Erosion of these Tertiary lake, soil, and

The subtle gray badlands of the Blues below Powell Point. —Lucy Chronic photo

The pipes in Kodachrome Basin State Park have outer rims composed of fine-grained sand, well sorted and densely packed by the flushing action that occurred within the pipes. The rims make them more resistant to erosion than surrounding rock. —Lucy Chronic photo

The Paunsaugunt Fault is visible north of the highway just beyond the sign at the Bryce Canyon National Park boundary. Here, the fault separates gray Cretaceous rocks from pink layers of the younger Tertiary Claron Formation. —Lucy Chronic photo

stream deposits is particularly rapid here because they are soft and the elevational relief is great.

The rest area east of milepost 10 looks westward across the Sevier River valley to the high volcanic mountains of the Markagunt Plateau. Soon after, Utah 12 drops into Red Canyon, where the Claron Formation is more red than pink and includes conglomerates that filled in river channels. The Sevier Fault, another Basin and Range fault and the western boundary of the Paunsaugunt Plateau, is just west of milepost 3. It is easy to spot north of the highway where dark-gray Quaternary lava flows abut the Claron Formation. The lava probably ascended along the fault. Crossing the fault, Utah 12 descends on the slope of an alluvial fan to the Sevier River and US 89.

Utah 14
Cedar City—Long Valley Junction
41 miles (66 km)

Utah 14 winds east up Cedar Canyon through the great escarpment of the Hurricane Cliffs, the western edge of the transition zone between the Colorado Plateau and the Basin and Range. The cliffs formed in a wide Basin and Range fault zone, a complex of many faults and small folds across which the rock on the west side of the zone has dropped several thousand feet (more than 1 km).

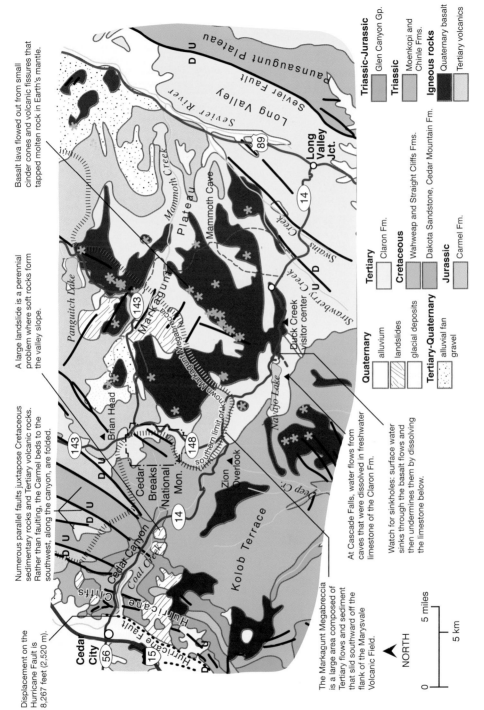

Displacement on the Hurricane Fault is 8,267 feet (2,520 m).

Numerous parallel faults juxtapose Cretaceous sedimentary rocks and Tertiary volcanic rocks. Rather than faulting, the Carmel beds to the southwest, along the canyon, are folded.

A large landslide is a perennial problem where soft rocks form the valley slope.

Basalt lava flowed out from small cinder cones and volcanic fissures that tapped molten rock in Earth's mantle.

The Markagunt Megabreccia is a large area composed of Tertiary flows and sediment that slid southward off the flank of the Marysvale Volcanic Field.

At Cascade Falls, water flows from caves that were dissolved in freshwater limestone of the Claron Fm.

Watch for sinkholes: surface water sinks through the basalt flows and then undermines them by dissolving the limestone below.

Quaternary
- alluvium
- landslides
- glacial deposits

Tertiary-Quaternary
- alluvial fan gravel

Tertiary
- Claron Fm.

Cretaceous
- Wahweap and Straight Cliffs Fms.
- Dakota Sandstone, Cedar Mountain Fm.

Jurassic
- Carmel Fm.

Triassic-Jurassic
- Glen Canyon Gp.

Triassic
- Moenkopi and Chinle Fms.

Igneous rocks
- Quaternary basalt
- Tertiary volcanics

NORTH

0   5 miles
0   5 km

Geology along Utah 14 between Cedar City and Long Valley Junction.

The road is periodically closed because of landslides; watch for evidence of these as you drive.

Triassic rocks, the deep red-brown and purple siltstone, sandstone, and mudstone of the Moenkopi and Chinle Formations, were deposited on deltas and floodplains near an ancient sea. They contrast with the pale Jurassic Navajo Sandstone, which has the expansive crossbedding and well-rounded grains of desert dunes, though here the sandstone has been crushed and broken by faulting. The Kayenta and Moenave Formations below the Navajo pass by very quickly since their layers are comparatively thin in Cedar Canyon. Pink and cream-colored, candy-striped sandstone, siltstone, gypsum, and limestone farther up the canyon are nearshore deposits of the Jurassic Carmel Formation.

Cretaceous rocks appear farther up the canyon, near milepost 8. By and large they are gray and yellowish gray, much less colorful than the older rocks. And they dip less steeply than the Triassic and Jurassic rocks, leveling out at the top of the Hurricane Cliffs.

Throughout southeastern Utah, there are four Cretaceous rock groups: conglomerate, sandstone, and shale of the Cedar Mountain Formation is at the base of the column, followed by the thin but resistant Dakota Sandstone. Soft, easily eroded dark-gray marine Mancos or Tropic Shale is next, and above it is a thick section of interbedded sandstone, shale, and coal of the Straight Cliffs, Wahweap, and Kaiparowits Formations, which are known farther east as the Mesaverde Group.

Together, these rocks tell of an advance of the Cretaceous Interior Seaway that happened when the area was bowed down in front of the mountains of the Sevier Orogeny, and then the retreat of the seaway. The conglomerates, sandstones, and coals are land and shoreline deposits; shelly limestone formed in

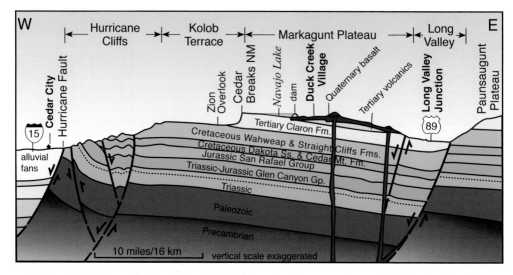

Westward stretching of the crust in the last 17 million years has caused normal faults to develop, along which most fault blocks dropped down to the west. The faults curve, so the blocks tilt gently eastward.

shallow, brackish lagoons; and shale built up when the water was deep and calm. When the seaway was at its greatest extent, its western edge was where Cedar Canyon is today.

Above the highway, particularly near the campground at milepost 12, glimpses of the "breaks," ornately eroded patches of soft, brilliantly colored Tertiary siltstone, sandstone, and limestone of the Claron Formation, appear along the west rim of the Markagunt Plateau. The Claron and overlying Tertiary and Quaternary volcanic rocks surface the plateau.

East of Zion Overlook, Utah 148 turns off to Cedar Breaks National Monument, where the Claron Formation is brilliantly exposed. Its beds are the soils and occasional lake limestones of a wide vegetated plain that spread between the rising mountains of the Sevier (to the west) and Laramide Orogenies (to the east).

Continuing eastward, Utah 14 skirts the young volcanic rocks that cap much of the Markagunt Plateau. The rocks are gray ash flows and breccias, erupted between 30 and 20 million years ago from volcanoes that are exposed west of here along Utah 143, and thin basalt flows erupted between 5.3 million and a few thousand years ago. Navajo Lake, south of the highway, was dammed by a basalt flow, though the basalt is now topped by a man-made earthen dam. The

The Claron Formation is exposed in Cedar Breaks National Monument. The brightly colored pink layers are silty stream deposits that, shortly after they were deposited, were reworked by roots and burrowing animals. The white layers are muddy lake limestone.
—Felicie Williams photo

basalt flows erupted from many vents. Some flows formed lava tubes, which are common features in basalt because its lava is very liquid. Tubes form as a basalt flow's surface chills into a crust while the lava inside continues flowing until it flows out, leaving behind a tube-shaped empty space. Mammoth Cave, a lava tube about 5 miles north of Duck Creek Village on US Forest Service Road 067 (Mammoth Creek Road), is open to the public.

Few streams flow over the basalt flows. Water moves easily down through cracks in the basalt and uses gravel-filled river and stream channels that existed before the lava flows erupted. Meadows, small ponds, and bogs in low spots may have formed where the basalt flows crossed and dammed creek channels.

The gentle eastward slope of the Markagunt Plateau leads to Long Valley Junction and US 89, crossing small faults near Strawberry and Swains Creeks; the creeks follow weakened rocks along the faults. Long Valley Junction is on the divide between the East Fork of the Virgin River, flowing south, and the Sevier River, flowing north. East of the junction the Paunsaugunt Plateau was lifted upward along the Sevier Fault relative to land west of the fault, bringing Cretaceous rocks abruptly to the surface. In the distance, back from the plateau's rim-like edge, more breaks of the Claron Formation cap the Paunsaugunt Plateau.

Utah 24
Sigurd—Fruita
71 miles (114 km)

Sigurd is the site of two large gypsum plants. Look for chunks of almost pure, white gypsum along Utah 24 near the mines. The high-quality gypsum here is called alabaster and, like all gypsum, is soft enough to carve with a pocketknife. The gypsum occurs in Arapien Formation shale that precipitated in a long ocean inlet during Jurassic time. The shale was mined for salt and gypsum, the latter used for sheetrock. The formation's fine clays were used in the oil-refining process and in cosmetics. Most of the shale has been mined out.

South of the outcrops of Arapien Formation shale, Utah 24 enters a valley between hills of lava flows and light-colored volcanic ash and breccia at the north end of the Sevier Plateau. The breccia, broken fragments of volcanic rock mixed with volcanic ash, forms much of the plateau's northern end. This rock formed from slides that traveled down the volcanoes' slopes.

Unusually deep gullying in occurred in the fine alluvium of the valley near milepost 18; gullying like this is common throughout the West. Current research suggests it is caused by long-term climate variations.

Leaving the valley, the road climbs to a summit on the Sevier Plateau. The volcanic Fish Lake Hightop Plateau is visible to the east and the Awapa Plateau to the southeast. Utah 24 winds down into the flat-floored Plateau Valley, a long north-south-trending graben that subsided between parallel faults. East of the junction with Utah 62, the highway climbs the east side of this valley, here called Grass Valley. To the south you can look down the length of the graben

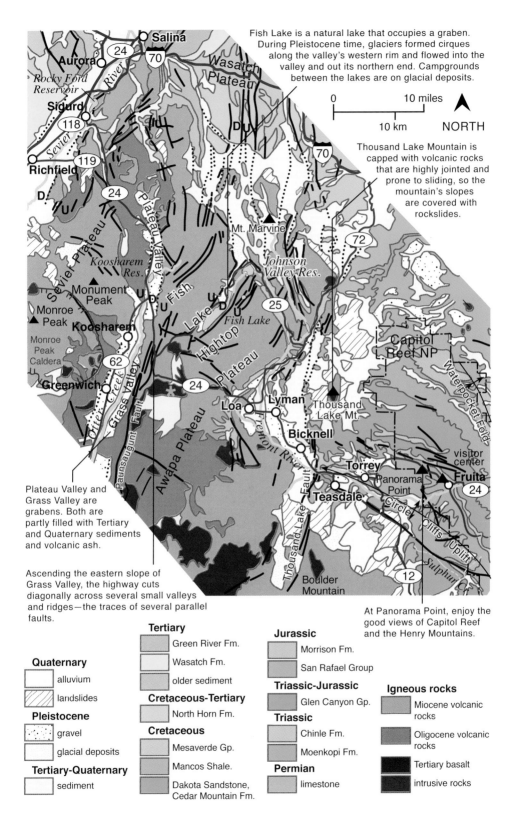

Fish Lake is a natural lake that occupies a graben. During Pleistocene time, glaciers formed cirques along the valley's western rim and flowed into the valley and out its northern end. Campgrounds between the lakes are on glacial deposits.

Thousand Lake Mountain is capped with volcanic rocks that are highly jointed and prone to sliding, so the mountain's slopes are covered with rockslides.

Plateau Valley and Grass Valley are grabens. Both are partly filled with Tertiary and Quaternary sediments and volcanic ash.

Ascending the eastern slope of Grass Valley, the highway cuts diagonally across several small valleys and ridges—the traces of several parallel faults.

At Panorama Point, enjoy the good views of Capitol Reef and the Henry Mountains.

Quaternary
alluvium
landslides

Pleistocene
gravel
glacial deposits

Tertiary-Quaternary
sediment

Tertiary
Green River Fm.
Wasatch Fm.
older sediment

Cretaceous-Tertiary
North Horn Fm.

Cretaceous
Mesaverde Gp.
Mancos Shale.
Dakota Sandstone, Cedar Mountain Fm.

Jurassic
Morrison Fm.
San Rafael Group

Triassic-Jurassic
Glen Canyon Gp.

Triassic
Chinle Fm.
Moenkopi Fm.

Permian
limestone

Igneous rocks
Miocene volcanic rocks
Oligocene volcanic rocks
Tertiary basalt
intrusive rocks

Geology along Utah 24 between Sigurd and Fruita.

and see the town of Koosharem. Broad alluvial fans edge the valley, their sediment whittled from the plateaus.

The volcanic rocks that blanket this area erupted from several stratovolcanoes west of here in the Marysvale Volcanic Field—before there was a valley in between. Erosion and faulting have removed the summits of the volcanoes and revealed the calderas beneath them. The highest peaks visible to the west are remnants of the Monroe Peak Caldera.

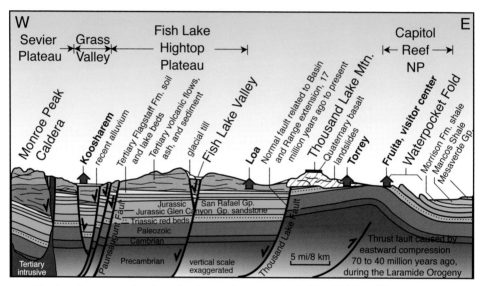

West-east cross section from the Sevier Plateau to Capitol Reef National Park. Between Sigurd and Fruita, Utah 24 crosses the faulted western edge of the Colorado Plateau, crossing several high plateaus separated by faulted valleys. The thickness of the beds is exaggerated roughly 2x in order to provide detail.

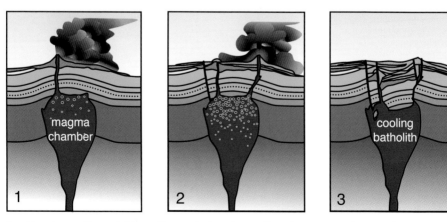

A caldera is usually much bigger than a single volcano. Over time, many volcanoes form above one magma chamber. Eventually, eruptions partly empty the chamber. When its roof drops down to fill the empty space, a caldera forms, an event that may trigger another, potentially massive, eruption.

The great volumes of Tertiary volcanic rocks along Utah 24 were deposited by different kinds of flows that traveled down the sides of volcanoes. Here, two rock-studded mudflows look different from each other because of the amount of iron in them. Most of the flows look gnarled due to the fragments embedded in them. —Felicie Williams photo

The highway turns east across the south edge of Fish Lake Hightop Plateau, where rocks are mostly hidden by soil and sagebrush. Thousand Lake Mountain, to the east, is capped with basalt flows. Beneath the flows, middle Eocene lake beds show their pink color here and there. They preserve fossils of titanotheres, early rhinoceros-like mammals.

The pointed summits of the Henry Mountains, a cluster of small Oligocene intrusions, are visible to the southeast in the far distance. To the south is Boulder Mountain, a Miocene shield volcano. An ice cap covered the mountain during the Pleistocene ice ages. In the valley below, red sedimentary rocks of the Colorado Plateau contrast with the gray volcanic rocks of the mountains the road has been in.

Tributary streams have carved deep canyons into the side of the Awapa Plateau, south of the highway, and the Fish Lake Hightop Plateau; the canyons are visible as the highway approaches Loa. These and other tributaries join the Fremont River, which drains the Fish Lake Hightop Plateau and Thousand Lake Mountain.

Loa, Lyman, and Bicknell lie on broad gravel-covered slopes below these highlands. Most of the gravel was deposited in Pleistocene time, when streams were laden with coarse glacial debris. Between Lyman and Bicknell, gray lava boulders litter the surface, some coated with white caliche deposited by carbonate-rich groundwater.

At Bicknell, watch for a visible and sudden change from volcanic to sedimentary rocks. Just north of Bicknell is the familiar Arapien Formation shale, which we saw near Sigurd. Here it also has distorted bedding and patches of pale gypsum. Just east of Bicknell, Utah 24 crosses the Thousand Lake Fault. Mesozoic sedimentary strata are at the surface here because the fault block west of the fault dropped downward. North of Thousand Lake Mountain and the Fish Lake Hightop Plateau the fault continues into the Wasatch Plateau along Lower Joes Valley. To the south, it separates Boulder Mountain from the Awapa Plateau.

The red cliffs east of Bicknell are steeply tilted Wingate Sandstone, a Triassic-Jurassic dune deposit. The massive, crossbedded, knobby white and pink rock is the Navajo Sandstone, a younger Jurassic dune deposit. East of the fault, the highway follows the southern flank of the Circle Cliffs Uplift.

Utah 24 enters a canyon of the Fremont River, in which Triassic rocks are exposed: down low is dark brick-red mudstone and siltstone of the Moenkopi Formation, its thin layers deposited on a tidal flat; the resistant Shinarump Conglomerate forms a caprock above the Moenkopi, and above the conglomerate is the paler Chinle Formation, a floodplain deposit.

Nestled between Boulder Mountain, to the south, and the uplands around Capitol Reef National Park, to the northeast, Torrey lies on a pediment cut in

Ornately sculpted cliffs of red siltstone and mudstone of the Moenkopi Formation border the Fremont River. The slopes below the cliffs are not cones of talus, but in-place shale, weathered low once the resistant caprock of Shinarump Conglomerate was lost. The Shinarump Conglomerate, here mostly sandstone, is the brown resistant cliff at the top. —Felicie Williams photo

the Moenkopi Formation. Lava boulders and other coarse debris cover parts of the surface; powerful streams brought the sediment here during the Pleistocene ice ages.

Utah 12 branches off to the south beyond Torrey. It is a particularly scenic and geologically interesting route through Grand Saircase–Escalante National Monument. Near the junction, more of the Shinarump Conglomerate and softer Moenkopi Formation red beds below it are exposed.

The Shinarump, a thin layer spread over much of northern Arizona and southern Utah, is mostly sandstone in this region of Utah. To the south it often contains pebbles that were eroded from the mountains of central Arizona during Triassic time. It was deposited in wide, low valleys by two large river systems in southwest and southeast Utah. About 30 feet (9 m) thick here, the conglomerate is much thinner near the Capitol Reef National Park visitor center, near the divide separating the two river systems.

The overlying remainder of the Chinle Formation is weak mudstone that erodes into colorful badlands; it was formerly soil layers that were full of volcanic ash. Above it is the thick, cliff-forming Wingate Sandstone. Just east of Torrey, south of the highway in the canyons of Sulphur Creek and some of its tributaries, white ledges of Permian limestone appear below the Chinle Formation.

Fruita is in Capitol Reef National Park. Capitol Reef is part of a 70-mile-long (113 km) monocline known as Waterpocket Fold. This sharp, one-way fold, bent down to the east, is the eastern limb of the Circle Ciffs Uplift, which is an anticline. Here, the word *reef* is used in its older sense, meaning "barrier." It is certainly that. The park visitor center is built of Moenkopi Formation sandstone that was quarried nearby.

Utah 30 and Utah 16
Garden City—Woodruff
41 miles (66 km)

The brilliant blue color of Bear Lake is caused by light reflected off fine grains of calcite, dolomite, and aragonite suspended in the water. Groundwater flowing through Paleozoic limestone dissolves the carbonates and then enters the lake through springs. The lakewater is supersaturated with the minerals, so they precipitate out continually. Besides springwater, Bear Lake is fed by the Bear River; this water originates on the quartzite-clad slopes of the Uinta Mountains and has been diverted to the lake since the early twentieth century to be stored for irrigation. The river water is diluting the carbonates, causing the water's color to fade and leading to the extinction of some ostracods, gastropods, and other endemic species that use the carbonate minerals for their shells.

Irrigation water is released from the north end of the lake back into the Bear River's channel. The river flows north to agricultural areas in southern Idaho before turning south and draining at last into the Great Salt Lake. This is the largest river feeding into the great lake.

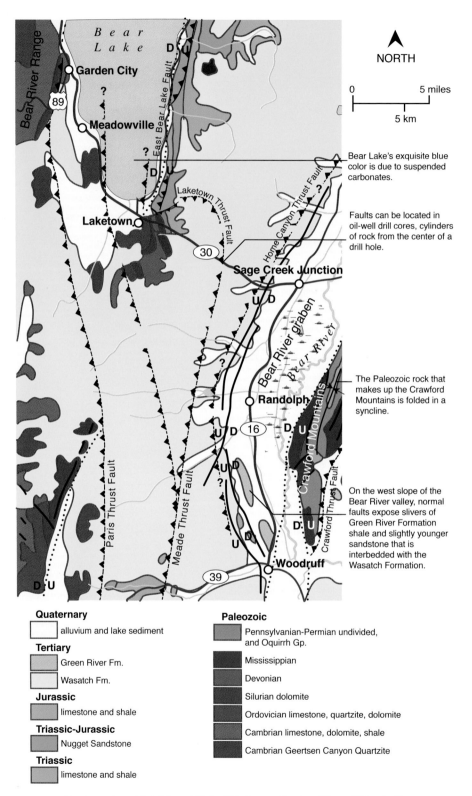

Geology along Utah 30 and Utah 16 between Garden City and Woodruff.

The brilliant blue of Bear Lake is light reflected off fine carbonate minerals suspended in the water. —Lucy Chronic photo

Bear Lake lies in a graben between the Bear River Range, to the west, and the East Bear Lake Fault, a normal fault related to Basin and Range extension. The graben started dropping down along the fault at the beginning of Holocene time and continues to drop today.

Utah 30 crosses several thrust faults between Garden City and Sage Creek Junction. In Cretaceous time, the thrust-faulted mountains of the Sevier Thrust Belt must have been impressive, rivaling the Canadian Rockies. But erosion, burial, and faulting related to Basin and Range extension have taken their toll.

Curving south around the lake south of Meadowville, Utah 30 parallels a ridge of Cambrian-age Geertsen Canyon Quartzite. Here, close to a thrust fault, the quartzite is very fractured and almost vertical.

East of the lake the highway swings up through a short canyon and passes an amazingly complete sequence of Paleozoic beds. Intensely folded and faulted because they are close to the Laketown Thrust Fault, they start abruptly with vertical ridges of Silurian-age Laketown Dolomite and end just a few miles later with Triassic-age red beds. Corals and crinoids are abundant in some of the limestone and dolomite ridges. At the top of the canyon, the relatively flat-lying Tertiary Wasatch Formation lies above a surface eroded into the Paleozoic and Mesozoic strata.

The surfacing Crawford Thrust Fault breaks into several fault splays in the Crawford Mountains, east of Sage Creek Junction. The range is a long syncline with a north-south-trending axis. The western side of the syncline is more steeply bent than the eastern side; note the east-dipping beds in the mountain face on the northern end of the mountains.

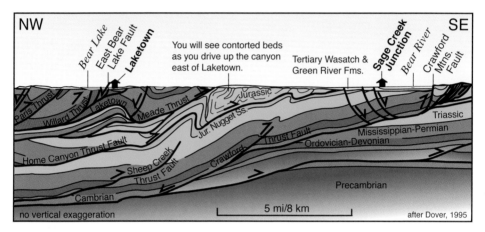

Northwest-southeast cross section across Bear Lake. The Sevier Thrust Belt surrounds the lake. The rocks here were pushed many miles east above huge thrust faults. The East Bear Lake Fault, a normal fault along which the lake's basin has dropped, parallels and may merge with an older thrust fault deep beneath the surface.

The Tertiary Wasatch Formation is mostly coarse red river gravel. Its source is the Uinta Mountains to the south. —Lucy Chronic photo

At Sage Creek Junction, turn south onto Utah 16. Quaternary alluvium covers the valley floor. Alluvial fans emerge from several valleys along the front of the Crawford Mountains. Although not visible from this distance, the fans are cut in places by normal faults.

The road follows the Bear River graben south to Woodruff. The graben parallels the mountains, and the Bear River flows through its valley. After the thrust faulting of Cretaceous time ended, rocks of the graben moved down along a splay of the Crawford Thrust Fault. And then during Basin and Range extension, the rocks dropped down even further. Movement like this, up and down already established fault planes, has happened repeatedly in the Sevier Thrust Belt. The Bear River was much grander during the Pleistocene, its larger flow carrying vast quantities of sediment from fields of glaciers in the Uinta Mountains and depositing it in the gradually subsiding graben valley.

Utah 39
Woodruff—Ogden
63 miles (101 km)

After leaving Woodruff, Utah 39 climbs westward into Walton Canyon, and gently sloping beds of poorly cemented sediment get older—from Tertiary to Cretaceous—and coarser. These conglomerates blanket the underlying rocks, and they record, with unconformities and coarsening layers, the waves of mountain building that occurred here during the Sevier Orogeny.

Around milepost 61, a few beds of highly fractured, gray Jurassic Twin Creek Limestone are visible. The limestone formed in a large shallow bay that opened northward onto the Jurassic sea; it is an important source of oil near Pineview, 37 miles (60 km) to the south.

During the Sevier Orogeny, the Willard Thrust Sheet, the large body of rock above the Willard Thrust Fault, was pushed over this limestone, contorting it until it was tightly folded and almost vertical; in some places the limestone beds were even overturned.

Almost immediately after a snow gate at milepost 68 the canyon narrows between walls of quartzite that are Precambrian to Cambrian in age. The older quartzite overlies the limestone; we have just crossed the Willard Thrust Fault.

The Willard Thrust Fault affected much of northeastern Utah and is the highest and oldest thrust fault in this region. It moved eastward between 125 and 95 million years ago. Under it, younger thrust faults, also part of the Sevier Orogeny, developed and carried it even farther east. One of these younger thrust faults—the Crawford—moved 90 to 84 million years ago; it surfaces in the Crawford Mountains to the northeast.

The broad surface above the canyon is veneered with loosely consolidated conglomerate of the Tertiary Wasatch Formation and pockets of Quaternary sand and gravel. Occasionally the road passes outcrops of Ordovician, Silurian, and Devonian limestone. They are part of a broad anticline that underlies the Tertiary and Quaternary deposits, and their layers help define the overall shape of the rise.

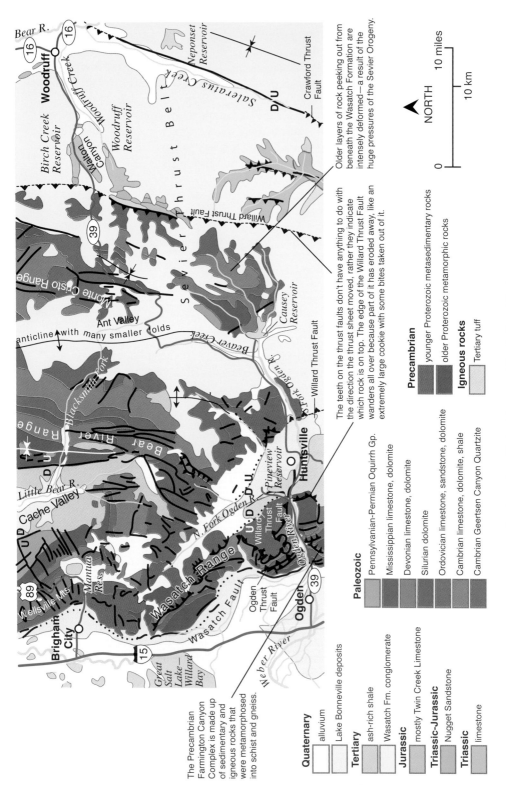

Geology along Utah 39 between Woodruff and Ogden.

Older layers of rock peeking out from beneath the Wasatch Formation are intensely deformed—a result of the huge pressures of the Sevier Orogeny.

The teeth on the thrust faults don't have anything to do with the direction the thrust sheet moved, rather they indicate which rock is on top. The edge of the Willard Thrust Fault wanders all over because part of it has eroded away, like an extremely large cookie with some bites taken out of it.

The Precambrian Farmington Canyon Complex is made up of sedimentary and igneous rocks that were metamorphosed into schist and gneiss.

Quaternary
alluvium
Lake Bonneville deposits

Tertiary
ash-rich shale
Wasatch Fm. conglomerate

Jurassic
mostly Twin Creek Limestone

Triassic-Jurassic
Nugget Sandstone

Triassic
limestone

Paleozoic
Pennsylvanian-Permian Oquirrh Gp.
Mississippian limestone, dolomite
Devonian limestone, dolomite
Silurian dolomite
Ordovician limestone, sandstone, dolomite
Cambrian limestone, dolomite, shale
Cambrian Geertsen Canyon Quartzite

Precambrian
younger Proterozoic metasedimentary rocks
older Proterozoic metamorphic rocks

Igneous rocks
Tertiary tuff

NORTH

0   10 miles

0   10 km

Utah 39 climbs onto the upland of the Monte Cristo Range, with westward views of the distant Bear River Range and Cache Valley. Ant Valley, just to the west, has dropped down along several parallel normal faults, a result of Basin and Range extension. The road gradually descends, passing mostly over the blanket of the Wasatch Formation. Then it turns south and drops into the drainage of Beaver Creek, where the few exposures of bedrock are limestones and dolomites of Pennsylvanian through Mississippian age.

Soon after the turnoff to Causey Dam and Memorial Park, Utah 39 descends through a sequence of older and older rocks, starting with Cambrian quartzites and ending with Precambrian quartzites. All of the rocks dip to the east here in the canyon of the South Fork of the Ogden River. Ripple marks in some of the beds trace their heritage back to shallow water; they were deposited between 720 and 580 million years ago. They are on the hanging wall, or upper sheet, above the Willard Thrust Fault. The route crosses into younger Cambrian rocks at the canyon mouth, which are in the fault's footwall.

Huntsville lies in another graben. The still-active main fault of the graben runs northwestward along the west edge of the valley. The valley is filled with up to 750 feet (230 m) of sediment. This area was a bay in Lake Bonneville 25,000 years ago.

The low hills south of Huntsville are composed of Tertiary volcanic tuff that is close to 5,000 feet (1,500 meters) thick. The narrowest part of Pineview Reservoir—its western end—lies between rises of metamorphosed Precambrian rock.

Quartzite cobbles and boulders near milepost 31 have eroded out of the Wasatch Formation. The largest boulders are 3 feet (1 m) across. They give a sense of the energy of the Tertiary river systems that eroded the rocks above the Willard Thrust Fault. —Lucy Chronic photo

A Z-fold exposed by erosion on the north side of Ogden Canyon developed in Mississippian limestone when rocks of the Willard Thrust Sheet were pushed eastward over the limestone, dragging and folding it. —Lucy Chronic photo

West of Huntsville, in Ogden Canyon, the highway rapidly crosses older and older beds, but thrust faults, folding, and normal faults that branch from the Wasatch Fault add complexity. Right after the dam, dramatically folded Mississippian beds that lie under the Willard Thrust Fault are plain to see on the north side of the canyon. Within 1 mile (1.6 km) of this Z-fold, the road passes through thin layers of folded and fractured Devonian through Ordovician limestone and shale.

Grayish-yellow Cambrian dolomite comes next, exposed on the steep hillsides above the road, and then the less-resistant Ophir Shale, and finally the resistant, pale-tan Tintic Quartzite. The Tintic Quartzite and overlying Ophir Shale and Cambrian dolomite record the transgression of the sea after the continent of Rodinia had rifted apart. The quartzite was the beach, the shale was deposited in shallow water with some input of fine sediment, and the dolomite was deposited in a shallow sea with very little sediment input from land.

The quartzite and shale sequence is repeated by three thrust faults in the canyon; look for repeats, especially, of the Tintic Quartzite's prominent yellow cliff. All of the formations are tilted back toward the east; they are on the east side of an anticlinorium—a large anticline that was folded up by underlying, later thrust faulting.

Precambrian Farmington Canyon Complex gneiss and granite, metamorphosed during an orogeny about 1.8 billion years ago, are exposed on the steep slopes near the mouth of the canyon. The road crosses the Wasatch Fault at the mouth of the canyon; rocks west of here have dropped several thousand feet (more than 1 km) along the fault. The fault is active, with the potential to generate large earthquakes.

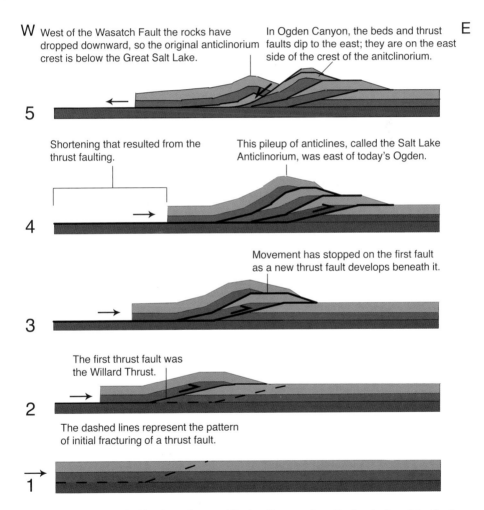

W — West of the Wasatch Fault the rocks have dropped downward, so the original anticlinorium crest is below the Great Salt Lake.

In Ogden Canyon, the beds and thrust faults dip to the east; they are on the east side of the crest of the anitclinorium.

E

5

Shortening that resulted from the thrust faulting.

This pileup of anticlines, called the Salt Lake Anticlinorium, was east of today's Ogden.

4

Movement has stopped on the first fault as a new thrust fault develops beneath it.

3

The first thrust fault was the Willard Thrust.

2

The dashed lines represent the pattern of initial fracturing of a thrust fault.

1

The history of thrust faulting in and around Ogden Canyon since the beginning of the Sevier Orogeny. Drawings 1 through 4 show how tectonic compression during the Sevier Orogeny shortened the crust, building a stack of thrust sheets called an anticlinorium, the remnants of which can be seen in Ogden Canyon. After the tectonic pressure was released, thrust sheets slid back along the faults a small distance. Drawing 5 shows the beginning of Basin and Range extension and the development of normal faults. As the crust was pulled westward, the crest of the anticlinorium dropped down along the Wasatch Front.

Landslides are frequent in the loose Lake Bonneville sediment along the Wasatch Front; look for one to the south in Bonneville sediment exposed by erosion along the Ogden River. Much larger slides could occur if an earthquake were to strike in spring, when the ground is wet. Liquefaction—when ground shaking causes loose sediment to act like a liquid—could also be extremely destructive.

Utah 143
Panguitch—Parowan
48 miles (77 km)

Utah 143 climbs from Panguitch and heads westward across the Markagunt Plateau, the westernmost of Utah's high plateaus that make up the transition zone between the Colorado Plateau and the Basin and Range. The transition zone experienced thrust faulting during the Sevier Orogeny, though the faulting was far less extreme than it was west of here, and the zone is offset by Basin and Range normal faults as well, though land in the transition zone did not drop as low nor stretch apart as much as country to the west.

Less than 2 miles (3 km) south of Panguitch, Utah 143 climbs onto a basalt flow, one of at least fifty on the slopes of the Markagunt Plateau. About 5.3 million to only a few thousand years old, many of these flows rose up along deep Basin and Range faults. The White Member of the Tertiary Claron Formation rests beneath the flows and shows in roadcuts; the member is marl, a soft, muddy freshwater limestone. Farther up the slope, older Tertiary ashfall and ashflow tuffs appear between the Claron and the basalt.

Older volcanic rocks in this region came from large volcanoes to the west between 30 and 20 million years ago. Their magmas contained more silica and were thicker and more viscous and gassy than basalt magma. This made the eruptions explosive. Here, they produced thick layers of tuff and volcanic breccia. Remnants of these volcanoes lie on the Markagunt Plateau, just to the west, or have been dropped down by the faults edging the west side of the plateau and obscured by sediment.

Utah 143 winds its way to Panguitch Lake through outcrops of broken volcanic rock. They are part of the southern end of an extremely large landslide that moved between 22.8 and 20 million years ago, during Tertiary volcanism and well before Basin and Range faulting and the region's basalt flows occurred. The slide, a jumble of lava, tuff, ash, and breccia, was at least 25 miles (40 km) wide and possibly 45 miles (72 km) long. It moved down a gentle slope on top of the muddy Claron Formation and above layers of fine sediment that had accumulated between the flows. The movement broke up the volcanic rocks and mixed in the mud, and some of the slide's pieces are 5 miles (8 km) across! The slide appears to have mostly moved southward, away from the volcanoes. The rock of the resulting slide body is aptly named the Markagunt Megabreccia.

Panguitch Lake is a natural lake, but an earth dam has deepened it. Utah 143 passes several basalt flows south and west of the lake, even turning to avoid the

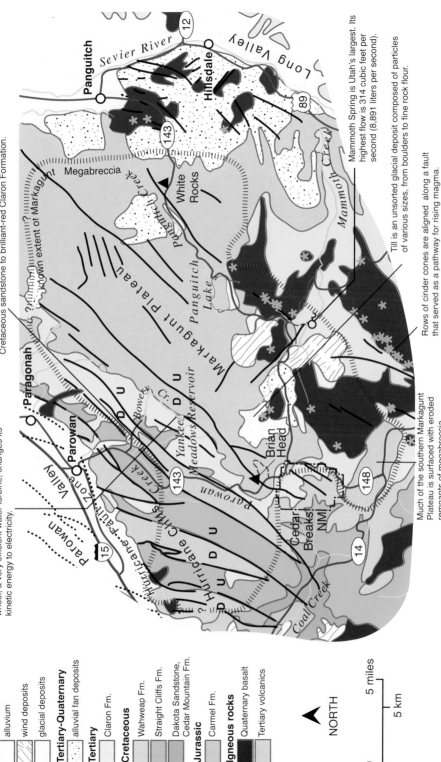

One of the Hurricane Fault Zone faults can be seen very clearly 0.4 mile (0.6 km) up Yankee Meadows Road. When the road crosses the fault, the canyon walls change from buff-colored Cretaceous sandstone to brilliant-red Claron Formation.

Water from Yankee Meadows Reservoir flows through a pipeline to Parowan, where a Pelton wheel, a very efficient water turbine, changes its kinetic energy to electricity.

Mammoth Spring is Utah's largest. Its highest flow is 314 cubic feet per second (8,891 liters per second).

Till is an unsorted glacial deposit composed of particles of various sizes, from boulders to fine rock flour.

Rows of cinder cones are aligned along a fault that served as a pathway for rising magma.

Much of the southern Markagunt Plateau is surfaced with eroded remnants of megabreccia.

Panguitch

Sevier River

Hillsdale

Long Valley

12

143

89

Mammoth Creek

Megabreccia

known extent of Markagunt

White Rocks

Panguitch Creek

Panguitch Lake

Markagunt Plateau

Paragonah

Parowan

Bowery Cr.

Yankee Meadows Reservoir

143

Parowan

Brian Head

D U

D U

Hurricane Fault Zone

Parowan Valley

Hurricane Cliffs

Cedar Breaks NM

148

15

14

Coal Creek

D U

Geology along Utah 143 between Panguitch and Parowan.

Quaternary

alluvium

wind deposits

glacial deposits

Tertiary-Quaternary

alluvial fan deposits

Tertiary

Claron Fm.

Cretaceous

Wahweap Fm.

Straight Cliffs Fm.

Dakota Sandstone, Cedar Mountain Fm.

Jurassic

Carmel Fm.

Igneous rocks

Quaternary basalt

Tertiary volcanics

NORTH

0   5 miles

    5 km

toe of one. Hundreds of thousands of years can go by between flows, so this basalt field can still be considered active. We may see another basalt flow some time in the future.

North of the Utah 148 junction, the road swings around Brian Head, a welded tuff headland. The roughly horizontal fractures on it are weak planes that developed as the tuff flowed after it had landed. Basin and Range country is visible to the west; eastward, layers of the Colorado Plateau show their

If you walk along a streambed you can see the range of light-colored to dark igneous rocks on the Markagunt Plateau. As their colors vary, so did the way their lava behaved. Dark basalt (right), with a great deal of iron and manganese, erupted as the most liquid of the lavas and flowed the most quietly. The lava that became pale rhyolite (left) was so thick and slow that it formed small, high, domelike flows or, if it contained a lot of gas, foamed and exploded as it reached the surface. Porphyry (center) has large mineral grains (here, feldspar) that grew while its magma stewed in an underground chamber and fine grains that crystallized quickly once the magma erupted at the surface. —Felicie Williams photo

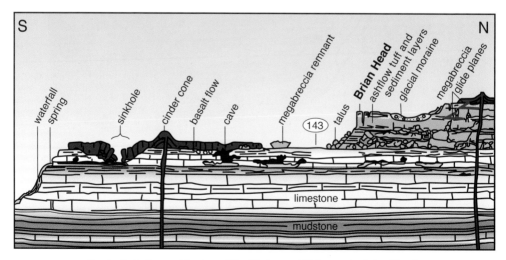

Geologic features of the top of the Markagunt Plateau near Brian Head.

horizontal lines. Pink cliffs of Claron Formation rim the Paunsaugunt Plateau. Below it are gray Cretaceous cliffs, and then white cliffs of Jurassic Navajo Sandstone far to the south in Zion National Park. These are three of the great steps of the Grand Staircase.

Starting down the steep grade to Parowan, Utah 143 passes hummocky landslides of Tertiary sediment and then descends quickly through the soft pink Claron Formation. Below the Claron are cliffs of the Grand Castle Formation, white Paleocene conglomerate and sandstone that are 590 feet (180 m) thick. Dinosaur tracks have been found in its middle sandstone member. Below it, the canyon is floored with Cretaceous sandstone and mudstone that were deposited on a floodplain. They are the oldest rocks seen along this canyon.

Around milepost 10, Utah 143 starts to cross the Hurricane Fault Zone. Rocks seen earlier along Utah 143 will be seen again where wedges of them dropped down along faults: look for light-colored tuff, darker basalt flows, and outcrops of the familiar Claron Formation.

Parowan sits in the Parowan Valley. In the dry climate of this region, crops must be irrigated. Snowmelt and rainfall are captured in reservoirs in the high elevations and distributed to agricultural areas via ditches. More water is pulled

Brian Head is made of welded tuff. The cliff recedes as large chunks break away along vertical joints that formed as the tuff cooled and shrank. —Felicie Williams photo

from wells drilled into the thick alluvial fans that descend into the region's valleys. The "mining" of this groundwater, with not enough recharge, is causing the ground of Parowan Valley to subside. The same is true for Cedar Valley, southwest of Parowan Valley.

Utah 150
Kamas—Wyoming State Line
55 miles (89 km)

East of Kamas, Utah 150 follows Beaver Creek into the heart of the Uinta Mountains. A horseshoe of Paleozoic rocks edges this western end of the Uinta Anticline. As in all anticlines, the oldest rocks are nearest the center. The layers are bent upward along the anticline's flanks but have eroded from its heights. Gray Pennsylvanian limestone is exposed here and there on slopes above the highway.

Farther up Beaver Creek, Mississippian strata appear between mileposts 2 and 3. This massive, fossil-bearing limestone is part of a very extensive layer, proof that a sea spread over much of the continent during Mississippian time. In this immediate area the gray limestone lies directly on top of pinkish Precambrian rocks; their contact is visible between mileposts 4 and 5. The only older Paleozoic rocks in the area are small remnants of Cambrian quartzite. Other older layers may have been here at one time, but if so, they eroded off the Precambrian rocks before the Mississippian limestone was deposited.

The Precambrian rocks—the Uinta Mountain Group—were deposited between 770 and 740 million years ago in an east-west extensional basin, the extent of which is closely mirrored by the modern Uinta Mountains. Much of

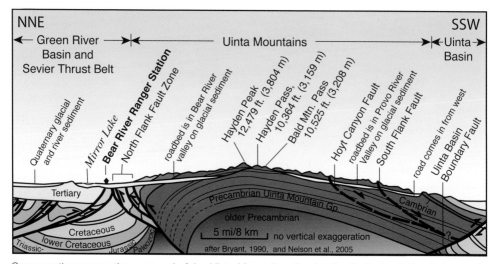

Cross section across the west end of the Uinta Mountains showing the faulted anticline that extends the full length of the range.

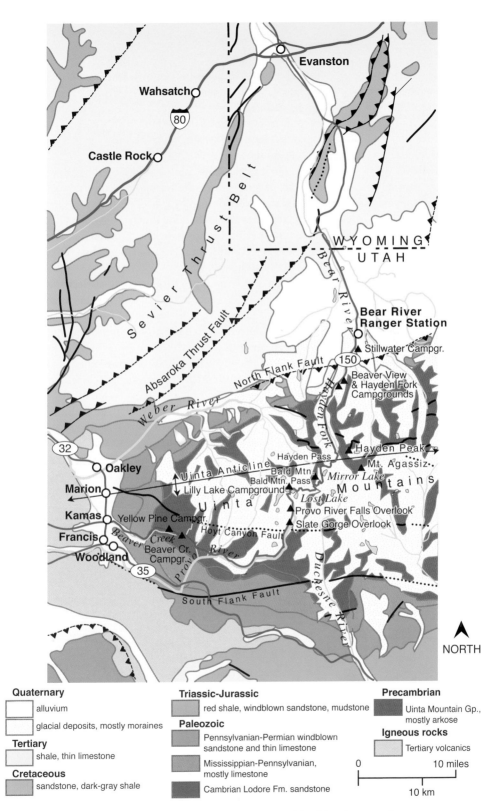

Geology along Utah 150 between Kamas and the Wyoming state line.

Quaternary
- alluvium
- glacial deposits, mostly moraines

Tertiary
- shale, thin limestone

Cretaceous
- sandstone, dark-gray shale

Triassic-Jurassic
- red shale, windblown sandstone, mudstone

Paleozoic
- Pennsylvanian-Permian windblown sandstone and thin limestone
- Mississippian-Pennsylvanian, mostly limestone
- Cambrian Lodore Fm. sandstone

Precambrian
- Uinta Mountain Gp., mostly arkose

Igneous rocks
- Tertiary volcanics

0   10 miles

10 km

NORTH

Evanston

Wahsatch

80

Castle Rock

S e v i e r   T h r u s t   B e l t

Absaroka Thrust Fault

WYOMING
UTAH

Bear River

Bear River Ranger Station

Stillwater Campgr.

150

Beaver View & Hayden Fork Campgrounds

North Flank Fault

Weber River

Hayden Fork

Hayden Pass

Hayden Peak

Mt. Agassiz

32

Oakley

Uinta Anticline

Bald Mtn.

Bald Mtn. Pass

Mirror Lake

M o u n t a i n s

Marion

Lilly Lake Campground

Lost Lake

Kamas

Yellow Pine Campgr.

U i n t a

Provo River Falls Overlook

Slate Gorge Overlook

Francis

Beaver Creek

Hoyt Canyon Fault

Woodland

Beaver Cr. Campgr.

Provo River

Duchesne River

35

South Flank Fault

the sediment of the group came from highlands to the north and east and was deposited initially in rivers and deltas and then reworked by a rising sea. Only microfossils are in these rocks, including some of Utah's earliest fossils. During the Laramide Orogeny, tectonic stresses arched the basin into an anticline and broke the crust along old faults. The faults became reverse faults, which today bound the present-day Uinta Mountains.

Beaver Creek valley widens out into glaciated country, and soon the highway cuts through a glacial moraine, its loose rocks often dropping onto the road. Moraines are made of unsorted material dumped by glaciers as they melt. Lateral moraines, long, mounded deposits, were left along the sides of valleys, and terminal moraines cross valleys in curving, bumpy hills. The terminal moraines near the Yellow Pine and Beaver Creek Campgrounds tell of the gradual retreat (melting) of the glacier that occupied the valley.

More moraines, in some places alternating with sorted river-deposited gravel, are visible near where Utah 150 approaches the Provo River. The highway follows the river from here to Bald Mountain Pass. Rocks of the Uinta Mountain Group and the moraines derived from them become more and more visible as we go farther into the mountains.

Rockslides are common in the cliffs and steep slopes in this region. Many shift perhaps a few inches or feet (tens of cm) a year, preventing vegetation from establishing itself.

The Provo River's headwaters are the small lakes south of Bald Mountain Pass, not far from the headwaters of the Weber, Bear, and Duchesne Rivers; this area is a major water source for Utah.

Above timberline, which ranges between 10,000 and 11,000 feet (3,050 and 3,350 m) at this latitude, the Uinta Mountain Group is clearly exposed. Bald Mountain is on the crest of the Uinta Anticline.

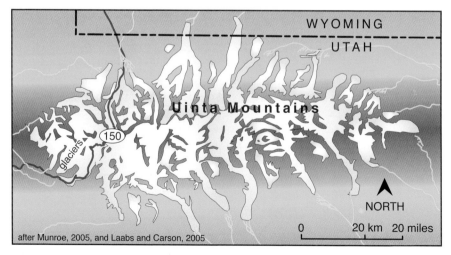

Glaciers extended well beyond the edge of the Uinta Mountains during Pleistocene time.

Provo River Falls cascades over thin beds of sandstone. —Lucy Chronic photo

At the viewpoint near milepost 30, Hayden Peak and Mt. Agassiz, both over 12,000 feet (3,660 m) high, rise to the northeast and east. Louis Agassiz was a Swiss biologist, paleontologist, and geologist who, between 1836 and 1840, studied the movements and effects of glaciers in the Alps, demonstrating decisively that much of Europe had once been buried in glacial ice. Later, at Harvard University, he revolutionized the teaching of natural history, encouraging his students to observe things for themselves, to "strive to interpret what really exists." Hayden Peak was named for Ferdinand V. Hayden, a geologist who, between 1867 and 1879, led government geologic and geographic surveys of the western territories. Saddles north and south of Hayden Peak formed in weak rock along some of the faults that parallel the Uinta Anticline.

This view also reveals the plateau of the western Uintas. Here, as the Pleistocene closed, the glacial ice cap that completely covered the top part of the western Uintas stagnated and melted, leaving a blanket of rocky ground moraine over almost everything except for scattered peaks. Melting blocks of ice covered with sediment left depressions that have become hundreds of small, scenic lakes known as kettles. Other lakes occupy cirques, the curved heads of valleys where glaciers once originated. The saddle between Bald Mountain Pass and Hayden Pass shows these features well.

North of Hayden Pass, Utah 150 descends along the Hayden Fork of the Bear River, following a gentle S-curve northward to the Wyoming border. For much of its length the valley has the broad, U-shaped profile typical of a valley that has been gouged out by a glacier. Recessional moraines, left at the successive tips of a retreating glacier, curve across the valley. The river's route probably predated the Pleistocene ice ages, but glaciers smoothed and straightened it.

Near the Beaver View and Hayden Fork Campgrounds, the highway once more crosses the horseshoe-shaped band of Paleozoic layers, though here it is concealed by glacial debris. Between the Hayden Fork and Stillwater Campgrounds, the highway crosses the North Flank Fault, a reverse fault that bounds the Uinta Mountain Anticline. North of the fault the Green River Basin contains a vast deposit of oil shale—the largest such deposit in the world. The voluminous outwash plain extending north past the Wyoming border was deposited by meltwater from the Uinta Mountain's glaciers.

THE GREAT BASIN AND THE WASATCH FRONT

GAUNT MOUNTAIN RANGES AND EMPTY BASINS

Utah's western desert is part of the immense Great Basin, a region characterized by rank upon rank of long, narrow mountain ranges that alternate with basins glutted with gravel and sand chiseled from the mountains. The pattern of basins and ranges has formed due to the tilting and dropping of fault blocks.

Storm-fed streams reach most of these basins only occasionally, bringing dissolved calcium carbonate, salt, and other minerals picked up from rocks and soil. Sometimes playa lakes form. As these temporary bodies of water evaporate in the hot desert sun, the minerals remain, each time adding a new layer of brilliant white to the basin floor. The shimmering playas of closed intermountain basins are among the flattest surfaces on Earth. Utah's largest playa, covering 4,000 square miles (10,360 square km), is the Great Salt Lake Desert. A small part of it has attained fame as the Bonneville Speedway; the desert's immensely large, flat expanse makes it a great locale for setting land-speed records.

During wet years, as in this photo, the Great Salt Lake Desert becomes a shallow sea-sized lake.
—Lucy Chronic photo

197

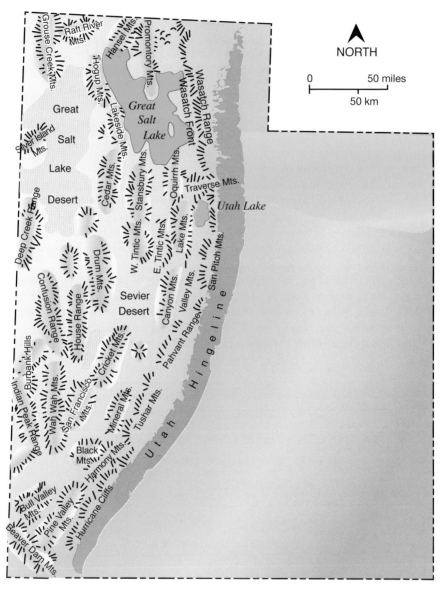

Utah's share of the Great Basin contains many individual mountain ranges, most of them trending north to south. The ranges are separated by arid desert basins, some of which drain out of the Great Basin; others do not, and instead have flat, salty playas marking their lowest points.

Erosion and Weathering in a Landlocked Basin

In the desert mountains of the Great Basin, weathering proceeds even with little rain: overnight frost and plant roots pry rocks apart, gravity tugs at steep cliffs, wind hammers at the rocks with grains of sand. Heavy rains instantly convert dry desert washes into roiling torrents with heavy loads of rock debris—the fabled walls of water of western lore. Overburdened streams deposit their loads of rounded pebbles, boulders, and sand in alluvial fans where their courses become less steep—at the mouths of canyons along mountain fronts.

Dust storms and dust devils that spiral against the summer sky remove fine material from the Great Basin, leaving behind desert pavement—an armor of pebbles that slows further erosion. Any rain or snow that doesn't evaporate in the region's dry air eventually sinks into the gravel and sand of valley floors. Only in the far southwest corner of Utah do a few of the valleys drain south, into the Colorado River and ultimately the Gulf of California. As in playa lakes, water evaporating from the desert surface leaves its minerals behind. This gradually forms caliche (ca-LEE-chee), a light-colored calcareous crust that coats sand grains and pebbles and often cements them together into a rocklike layer up to a few feet below the surface of the soil. Caliche reduces the permeability of soil, slows erosion, and limits the growth of plant roots.

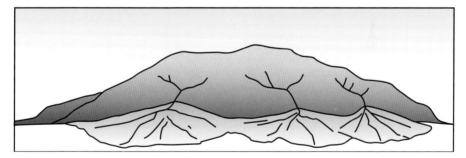

Alluvial fans merge to form alluvial aprons (tan color), or bajadas, around desert ranges. Fans may even reach a valley's center, meeting with others from neighboring ranges.

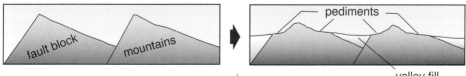

As mountain fronts erode in arid regions, they slowly recede, leaving lightly graveled pediments of solid rock that merge smoothly with surrounding alluvial fans and bajadas.

THICK SEDIMENTS, HUGE VOLCANOES, AND MOUNTAIN BUILDING

Both tectonic and climatic events, overlapping in time and space, helped form the gaunt ranges and arid, sunbaked basins of Utah's western deserts. Rifting long ago caused the crust to grow thinner west of the Utah Hingeline, allowing it to subside beneath vast accumulations of oceanic sediment during latest Precambrian and Paleozoic time. Pennsylvanian and Permian layers are very thick where they filled the Oquirrh Basin in northwest Utah. Fossils are abundant in the Paleozoic sedimentary rocks, some illustrating evolution on a grand scale.

Tectonic pressure from the west during Mesozoic and early Cenozoic time, caused by the North American Plate overriding the Farallon Plate in a subduction zone, sent a wave of mountain building eastward across the West. In western and central Utah, one thrust sheet after another piled up as mountains 125 to 50 million years ago, during the Sevier Orogeny of Cretaceous time. As pressure from the subduction ended, the mountains relaxed, sliding back a small amount along the thrust faults they had risen on.

By 35 million years ago, a wave of igneous activity was heading southeastward across Utah. The crust had grown thicker due to the piling up of thrust sheets during the Sevier Orogeny. Deeply buried crust was heated and started to melt. The magma rose up faults and blasted through to the surface to build huge stratovolcanoes. The heat of the intrusions and volcanics altered surrounding rocks, and in places valuable minerals were deposited.

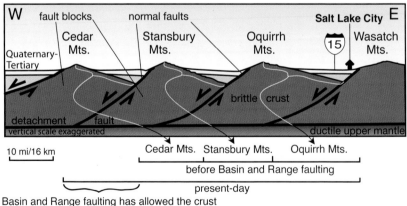

Basin and Range faulting has allowed the crust to stretch 14 miles (23 km) to the west!

In the Basin and Range, Earth's crust is much thinner and hotter than on other parts of the continent. Its lower portion acts in a plastic manner: it stretches out horizontally like taffy. The brittle upper crust that sits above breaks into fault blocks that are bound by normal faults. This faulting allows the crust to extend, or stretch. The normal faults that edge the mountain ranges merge at depth with detachment faults. The fault blocks slide along these gently westward-sloping surfaces, which lie about 10 miles (16 km) beneath the surface and extend under almost all of the Great Basin. Detachment and normal faulting have nearly doubled the Great Basin's width.

Beginning about 17 million years ago, the Great Basin began to stretch due to the start of movement along the San Andreas Fault. Fault blocks in the Great Basin moved along old and new faults. The faults are listric: they flatten out with depth, so as the fault blocks slide westward they tip, and their western edges remain high, forming mountain ranges, while their eastern edges drop, forming basins. Movement on the faults continues today; land west of active faults drops an average 0.016 to 0.025 inch (0.4 to 0.6 mm) every year relative to the land east of them.

As the ranges formed, rock and sand eroded from them filled the basins. During the wetter Pleistocene time, many of the basins contained lakes, including Lake Bonneville, which covered most of the expanse of the Great Salt Lake Desert. Sediment deposited in them settled out and was reworked into perfectly flat surfaces. Increasing aridity in the last 10,000 years has dried up most streams and lakes in the Great Basin and reduced Lake Bonneville to today's Great Salt Lake.

In the Shadow of Faults

The Wasatch Fault is among the easternmost normal faults of the Basin and Range in Utah. Displacement on the Wasatch Fault is pronounced: the land west of it has dropped significantly, leaving a high, prominent scarp known as the Wasatch Front. Very strong earthquakes must have accompanied the great degree of movement along this fault. Currently, however, it is fairly quiet compared to other active faults in the Basin and Range; some geologists suspect that it may be locked, or stuck in place, and that stress is building within it that will eventually lead to a big earthquake.

Much of Utah's population lives along the Wasatch Front, including the urban centers of Salt Lake City, Provo, and Odgen. These urban areas are continuously monitored with an array of seismographs, and people here are aware that an earthquake may happen; extensive safety plans are in place should a large quake occur. If one does, it might shake the valley's delta gravels and clayey lake deposits until they liquefy, and it would shake landslides loose from the steep mountain front. It could also drop parts of the city below the level of the region's lakes and destroy buildings that lie near branches of the fault.

Rarely do detachment faults appear at the surface, but some do in the Basin and Range as rounded summits called metamorphic core complexes. A thick blanket of thrust-faulted sedimentary rock once buried these mountains, but it slid off them during detachment faulting. With the weight removed, the mountain cores rose upward, revealing the detachment fault surfaces: broken, crushed, partly melted metamorphic rock that had been smeared out by the friction and heat of fault movement. Precambrian metamorphic rocks from the middle and lower continental crust lie below the fault surfaces.

People and the Vast Desert

A few of western Utah's ranges contain radioactive minerals. Many contain ores of gold, silver, iron, copper, and other valuable minerals. Almost without exception, these minerals occur where Tertiary igneous intrusions penetrated

Paleozoic or Mesozoic limestone. Heated fluids circulated within the limestone and cooling intrusions, concentrating mineral solutions, and with the limestone acting as a catalyst, the fluids precipitated the ore minerals, enriching both the limestone and the igneous rocks.

Mineral hunting is good in most of the mining areas, especially in dumps near the mines. But be very careful. Old mines are extremely dangerous. Stay out of them. Use considerable care in choosing collecting sites, and, of course, obtain permission before searching on private property.

In the huge volcanic fields of southwestern Utah, a few geothermal generators hint at promising geothermal resources. And there are hot springs for people to soak in. Many alluvial fans have become irrigated cropland; the salty playas, though, remain pristine, flooding in wet years and always imparting a particular vastness to Basin and Range scenery.

THE GREAT SALT LAKE AND LAKE BONNEVILLE

One of Utah's largest geologic attractions, the Great Salt Lake spreads wide west and north of Salt Lake City, covering land that has dropped downward west of the Wasatch Fault. The lake doesn't have an outlet, so evaporation has made it salty. In addition to common table salt (sodium chloride), the lake contains significant amounts of sulfate, magnesium, calcium, and potassium. Its salinity varies quite a bit, but in historic times it has always been far saltier than seawater. In 1963, at its lowest historic level, the entire lake was saturated, meaning it could not dissolve any more salt. It was eight times saltier than seawater with nearly 27 percent dissolved salt by weight. In 1988, at its historic high after four years of above-normal precipitation, the salt content was between 6 percent and 13 percent.

The lake is divided in two by a stone railway causeway built between Promontory Point and Lakeside in the late 1950s. With just two small openings, there is limited circulation between the two sides of the lake. The southern part, which receives almost all of the freshwater inflow, contains about 11 to 12 percent salt, whereas the slightly lower-elevation northern part is almost saturated.

The lake is shallow: its deepest spot, near the center of the causeway, has historically varied between 28 and 49 feet (8.5 and 15 m). The lake's bottom is fairly smooth and only slightly down-curved, aside from steep drop-offs along the faulted western edges of the islands and nearby ranges. Deltas and alluvial fans have buried the drop-off one would expect along the Wasatch Front.

The earliest lakes here, of Miocene and Pliocene age, were probably small. Their beds form the lowest layers on the floor of the Great Salt Lake. By about 25,700 years ago, the lakes had merged into one: Lake Bonneville. An abundance of fossils found in Lake Bonneville sediments indicate that animals of the Pleistocene ice ages frequented its shore, including mammoths, musk oxen, sabertoothed cats, horses, bighorn sheep, bison, camels, deer, short-faced bears, giant ground sloths, mastodons, pocket gophers, and even an extinct species of trout.

For about 1,700 years, evaporation was balanced with inflow, and the landlocked lake cut a series of shorelines called the Stansbury Shoreline. Then the lake began to fill higher, probably because of water added during the ice

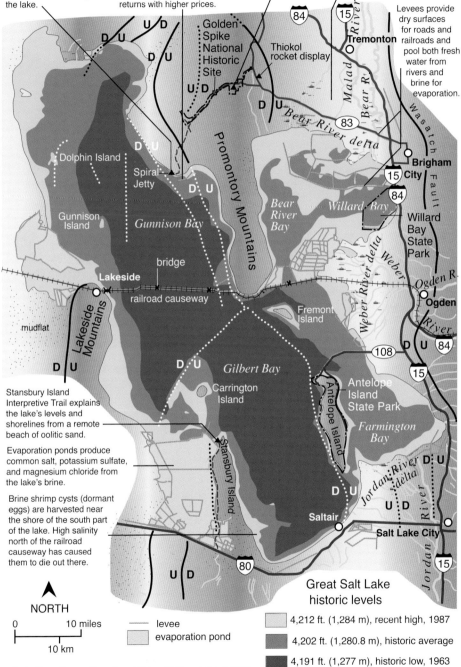

Spiral Jetty, a work of landscape art constructed by Robert Smithson in 1970, serves as a rallying point for preservation of the wildlands and scenery of the lake.

Natural seeps of low-quality, thick, tarry oil near Rozel Point come from Tertiary rocks. Attempts to produce the oil in the 1930s were abandoned, but interest returns with higher prices.

Chinese Arch, named for the Chinese workers who helped build the cross-continent railroad in 1869, is a small natural arch in Paleozoic rock.

Now camouflaged by farmland, ditches, towns, and cities, deltas of the Bear, Weber, and Jordan Rivers make up much of the eastern shore of the lake.

Levees provide dry surfaces for roads and railroads and pool both fresh water from rivers and brine for evaporation.

U D
D U
D U
U D
D U

Golden Spike National Historic Site

Thiokol rocket display

84 15 River

Malad River Bear River

Tremonton

83 Bear River delta

D U

Dolphin Island

Spiral Jetty

D U

D U

Brigham City

15

84

Promontory Mountains

Bear River Bay

Willard Bay

Willard Bay State Park

Wasatch Fault

Gunnison Island

Gunnison Bay

bridge

Lakeside

railroad causeway

Weber River delta Weber River

Ogden R.

Ogden

River

mudflat

Lakeside Mountains

Fremont Island

108 D U 84

15

D U

Gilbert Bay

Carrington Island

Antelope Island

Antelope Island State Park

Farmington Bay

Stansbury Island Interpretive Trail explains the lake's levels and shorelines from a remote beach of oolitic sand.

Evaporation ponds produce common salt, potassium sulfate, and magnesium chloride from the lake's brine.

Brine shrimp cysts (dormant eggs) are harvested near the shore of the south part of the lake. High salinity north of the railroad causeway has caused them to die out there.

Stansbury Island

Jordan River delta

D U

U D

Jordan River

D U

U D

Saltair

Salt Lake City

15

80

U D

Jordan

NORTH

0 10 miles

10 km

—— levee

evaporation pond

Great Salt Lake historic levels

4,212 ft. (1,284 m), recent high, 1987

4,202 ft. (1,280.8 m), historic average

4,191 ft. (1,277 m), historic low, 1963

adapted from Currey, Atwood, and Mabey, 1984, and Oviatt, 2012

Map of the Great Salt Lake.

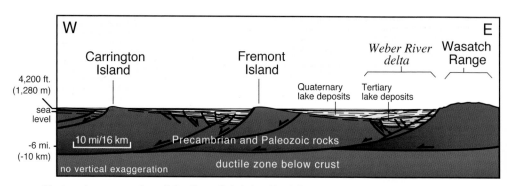

West-east cross section of the Great Salt Lake. The lake occupies several valleys that have dropped down along listric faults. As the fault blocks slid westward they tipped, and their western edges remained high, forming the irregular rows of islands and promontories. Both faults and valleys trend about 20 degrees west of north, parallel to the front of the Wasatch Range.

ages. Around 18,000 years ago, it overflowed northward into the Snake River across a weak barrier of alluvial fans and Tertiary rocks near Downey, Idaho. Lake Bonneville remained close to this, its highest level, for about 500 years. It carved several benches, together known as the Bonneville Shoreline, into the surrounding mountains.

Gradually, the lake's water seeped through layers of loose sediment near Downey, Idaho, and dissolved the underlying limestone. About 17,500 years ago part of the land holding back the water collapsed. Water raced through the gap, churning and thundering northward. Cave-ins near the washout widened the outlet. In less than a year, vast Lake Bonneville, which had covered some 20,000 square miles (51,800 square km), lowered by 375 feet (114 m), until a new threshold of resistant bedrock held the water at Red Rock Pass, just across the state border on US 91 in Idaho.

The lake stayed at this new level for 2,500 years, carving the Provo Shoreline. Often the most obvious of the shorelines, this level includes beaches, spits, and a wide gravel-covered bench fringed with deposits of tufa, calcium carbonate that precipitated from lake water. Incoming rivers built large deltas out into the lake. The bench and the flat-topped deltas have made handy, scenic building sites.

Gradually, the threshold at Red Rock Pass was eroded and the lake dropped again; a later set of beach ridges occurs about 12 feet (3.7 m) below the main Provo level. Then, with an increasingly dry climate that developed as the Pleistocene ice ages waned, the lake dropped below the Provo level and stopped flowing into the Snake River. Since then, periodic stable levels have been reached when inflow balances evaporation. The lowest shoreline above the current level of the Great Salt Lake, the Gilbert Shoreline, developed around 11,500 years ago. It probably took less than 100 years to cut its fairly obscure low-relief benches, ridges, and gravel bars.

The lake dried up completely at least once. Air photos show huge mud cracks on the Great Salt Lake's lake bed—330-foot (100 m) versions of the small ones seen in dry mud puddles.

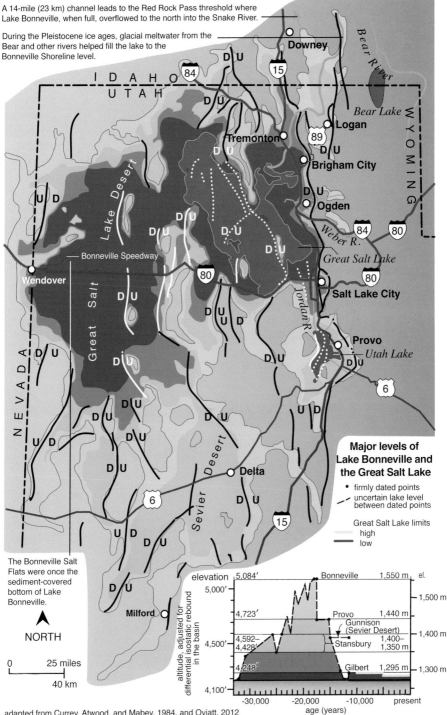

A 14-mile (23 km) channel leads to the Red Rock Pass threshold where Lake Bonneville, when full, overflowed to the north into the Snake River.

During the Pleistocene ice ages, glacial meltwater from the Bear and other rivers helped fill the lake to the Bonneville Shoreline level.

I D A H O
U T A H

W Y O M I N G

Downey

Bear River

Bear Lake

Logan

Tremonton

Brigham City

Lake Desert

Ogden

Weber R.

Great Salt Lake

Wendover

Bonneville Speedway

Great Salt

Salt Lake City

Jordan R.

Provo
Utah Lake

N E V A D A

Delta

Sevier Desert

Major levels of Lake Bonneville and the Great Salt Lake

• firmly dated points
∕ uncertain lake level between dated points

Great Salt Lake limits
high
low

The Bonneville Salt Flats were once the sediment-covered bottom of Lake Bonneville.

Milford

NORTH

0   25 miles
40 km

elevation   5,084′   Bonneville   1,550 m   el.

altitude, adjusted for differential isostatic rebound in the basin

5,000′

4,723′   Provo   1,440 m
    Gunnison (Sevier Desert)
4,592–   1,400–
4,428′   Stansbury   1,350 m
4,500′

4,248′   Gilbert   1,295 m

1,500 m

1,400 m

1,300 m

4,100′
-30,000   -20,000   -10,000   present
age (years)

adapted from Currey, Atwood, and Mabey, 1984, and Oviatt, 2012

The Great Salt Lake's ancestor was Lake Bonneville, a much-larger freshwater lake that developed during Pleistocene time. Abundantly fed by ice age rain, snow, and glacial meltwater, it was about 1,200 feet (370 m) deep. The most noticeable traces of it now are the old shorelines that stair-step up slopes all over northwestern Utah. Many of the present mountain ranges were islands in the lake's broad expanse.

Since 1850 the lake level has been between 4,191 feet and 4,212 feet (1,277 m and 1,284 m). Normally, it fluctuates 2 to 3 feet (0.6 to 0.9 m) with seasonal changes in precipitation. About 2.9 million acre-feet (3.6 cubic km) of water evaporate from it each year.

Between 1983 and 1987, the Great Salt Lake rose 12 feet (3.7 m). To combat the flooding, part of the causeway was opened so more water could drain into the lower north arm. That wasn't enough, so a station was built on the western shore to pump lake water to a dry desert basin to increase the lake's evaporating area. This helped, and so did dryer years that followed. Used for twenty-six months, the system is maintained in case of future need.

Rivers flowing into the lake build broad mudflats and deltas. They also bring in around 2 million tons (1.8 million metric tons) of dissolved minerals each year, adding to those handed down from Lake Bonneville ever since the great lake dropped below its outlet. Salt flats are most common on the western side of the lake where there are no rivers to carry in sediment to dilute the salt.

The lake has many uses. Early people used its salt and ate the roots of cattails that grow where freshwater enters the lake. Today, people use its minerals and ship its plentiful brine shrimp worldwide for use as fish food and as pets called "sea monkeys." The brine shrimp are also an abundant and important food for

Children venture into the warm water of the Great Salt Lake to collect brine shrimp during a science field trip to Antelope Island. The lake is so salty that few species can survive in it. Brine shrimp, which flourish here, are an exception. —Lucy Chronic photo

Oolitic sand borders much of the Great Salt Lake. Each individual oolite grows in agitated water as calcium and magnesium carbonates build layers around a tiny grain. Here, the grain is usually the fecal pellet of a brine shrimp. —Felicie Williams photo

migratory birds on the great Central Flyway. No fish live in the saltwater; only brine shrimp, algae, and brine flies do. The Great Salt Lake is greatly valued for its wildlife preserves, recreational uses, and, of course, its scenery.

Evaporite mineral production is a major industry at the lake. Water from it is first pumped into settling ponds to let sediment settle out. Then the water is moved to solar evaporating pans, where it is concentrated until calcium and magnesium carbonates (calcite and dolomite) crystallize and settle out. At this stage, called the salt point, the remaining brine is pumped into other ponds where, after more evaporation, common salt (sodium chloride) precipitates. The salt goes into water softener pellets and salt lick blocks and is used in industrial processes and for salting highways in winter. Surprisingly, none of it is used as table salt; purer salt deposits are available in other places. After the common salt has been removed, the brine is pumped to yet other evaporation ponds in order to precipitate magnesium and potassium salts, which are used to make magnesium metal, chlorine gas, fertilizer, and magnesium chloride to treat road dust and ice.

ROAD GUIDES OF THE
GREAT BASIN AND WASATCH FRONT

I-15 and I-84
Idaho State Line—Salt Lake City
91 miles (146 km), 115 miles (185 km)

I-84

I-84 enters Utah on a gentle slope that was covered by Lake Bonneville during Pleistocene time. The terraces of Lake Bonneville can be seen as rings around many of the hills, in places covered by younger sediment. The landscape along the interstate is typical of the Basin and Range, with long north-south-trending mountain ranges separated by vast basins of sediment. The ranges expose thick sequences of Paleozoic sedimentary rock, mostly limestone, deposited in the Oquirrh Basin. I-84 joins I-15 at Tremonton.

I-15

The Malad Range, east of I-15 where it crosses into Utah, looks like a bunch of broken steps. Its Paleozoic limestone fragments angle every which way but generally drop down to the west, offset along normal faults. The east side of Malad River valley, fitting the Basin and Range pattern, has dropped down along the Wasatch Fault. Though we refer to it as a singular fault, the Wasatch is, in detail, a zone of many smaller faults.

Thrust faulting during the Sevier Orogeny moved the Paleozoic sedimentary rock of the Malad Range eastward 20 to 30 miles (32 to 48 km) from where it had been deposited, stacking it into mountains; later, Basin and Range extension broke the mountains apart, and some of the rocks were carried partway back (west) along the same faults.

The Bear River, after looping northward from Bear Lake, which is on the east side of the Bear River Range, enters the Malad River valley southeast of Plymouth but does not join the Malad River. It turns south on the extremely flat valley floor—a heritage of flat-lying sediments deposited in Lake Bonneville. The Bear River parallels the Malad for some distance, winding back and forth in intricate meanders.

The Wellsville Mountains, southeast of Collinston, are a single, large wedge of upper Precambrian through Paleozoic sedimentary rocks. The layers dip northeastward dramatically, and many landslides on the far side of the mountains have ridden the slanting bedding planes downward. These mountains were thrust some 37 miles (60 km) eastward between 140 and 100 million years ago, during the Sevier Orogeny, as part of the Willard-Paris Thrust Sheet.

Because of their northeast dip, we pass the youngest rocks—Permian to Pennsylvanian in age—first as we sweep south past the Paleozoic section.

IDAHO
UTAH

NORTH

30 Snowville

Malad River
Malad Range
Cache Valley

84
Plymouth

U D

15 Collinston

Cutler
Reservoir

The Salt Lake
Formation includes
lake and river deposits
and volcanic ash layers.

U D

30

Tremonton
D U

Bear River

Wellsville Mountains

The valleys in this region are
floored with deposits of Lake
Bonneville, which was 1,290 feet
(393 m) deeper than the present
Great Salt Lake. More recent alluvium
covers the old lake bed in places.

Thiokol rocket display

89

83
Honeyville

mudflat

Brigham City

The Bear and Malad Rivers lose
themselves in the marshes of Bear
River Bay, an arm of Great Salt Lake.
When the lake water is high it floods
the marshy areas and mudflats.

Bear River
Migratory
Bird Refuge

Willard-Paris Thrust Fault

mudflat

Willard

Bear
River
Bay

Willard Bay
State Park

Wasatch Fault

Huntsville

39
Pineview
Reservoir

Great

Weber River

Ogden

Quaternary

alluvium

landslides

Lake Bonneville deposits

Tertiary

Salt Lake Fm.

Wasatch Fm. congl. and sandstone

Jurassic

Twin Creek Limestone

Triassic-Jurassic

Nugget Sandstone

Triassic

limestone, shale, sandstone

Permian

Park City Fm.

Paleozoic

Pennsylvanian-Permian Oquirrh Gp.
limestone and sandstone

Mississippian limestone/dolomite

Devonian limestone/dolomite

Silurian limestone/dolomite

Ordovician limestone/dolomite

younger Cambrian dolomite, shale

Cambrian Geertsen Canyon Quartzite

Precambrian

younger Proterozoic metasediments

Proterozoic metamorphic rocks

Salt

Lake

108
Layton

Antelope
Island
State
Park

Kaysville

Farmington

Farmington
Bay

15

84

Wasatch Range

Wasatch Front

Jordan

Bountiful
D U

tailings

80
Salt Lake City
80

Igneous rocks

Oligocene volcanic rocks

Tertiary basalt

0   10 miles

10 km

Geology along I-15 and I-84 between the Idaho state line and Salt Lake City.

Because the entire Pennsylvanian-Permian section was deposited in a marine setting in the gradually subsiding Oquirrh Basin west of the Utah Hingeline, it is thick. The layers are mostly gray and tan limestone and dolomite.

The resistant cliff-forming Geertsen Canyon Quartzite crops out between Honeyville and Brigham City. This hard, tightly cemented, pale-tan rock was deposited mostly in nearshore beaches and sandbars as the sea crept eastward across the continent during latest Precambrian to Cambrian time. The Geertsen Canyon correlates with Cambrian sandstones in states south, north, and east of Utah, giving geologists a sense of the extent of this sea's coverage. Because the sea's encroachment was gradual, the sand deposits become younger eastward.

Lake Bonneville shorelines crease the mountain slopes all along I-15. The uppermost shorelines, the Bonneville and the Provo, were both long enduring, so their shorelines are highly visible. Near Honeyville, the gentle, smooth slope of a delta that once fed from the mountains into Lake Bonneville reaches out into the valley. Similar deltas can be seen farther south.

Both railroad and highway run for some distance barely above lake level near Ogden. During the high water levels of the mid-1980s, the lake lapped against the railroad and highway berms. The marshy region west of the highway, part of it the Bear River Migratory Bird Refuge, includes the deltas of the Bear and Weber Rivers, where the sand and silt their waters carry have partly filled Bear River Bay.

The uppermost shorelines of Lake Bonneville are etched into the mountain front along I-15. —Lucy Chronic photo

plane of the Willard-Paris Thrust Fault

Near Willard, tilted, light-colored Cambrian Tintic Quartzite, which correlates with the Geertsen Canyon Quartzite to the north, overlies dark, metamorphosed Precambrian rocks. More Precambrian rocks were pushed above the quartzite along the Willard-Paris Thrust Fault. All the rocks we have passed since the Idaho border are on the upper, or hanging wall, of this fault; they were pushed here from the west as part of the gigantic Willard-Paris Thrust Sheet. Though originally the fault sloped very gently westward, it now dips east, a result of later, deeper thrust faulting. —Lucy Chronic photo

recent
fault scarps

triangular facets

Lake Bonneville
shorelines

Mountain spurs near Odgen, as well as in most places along the Wasatch Front, end in triangular faces, or facets. These facets are the uneroded surface of the Wasatch Fault, which parallels I-15 for much of its distance across Utah. They show the fault's exact position and, because they have eroded little, indicate quite recent movement on the fault.

This boulder of Precambrian Mineral Fork Formation from Antelope Island shows large cobbles suspended in a fine-grained black matrix. The original sediment was deposited when an ice sheet melted. This dark layer of rock is evidence of a time when ice and snow covered all the continents, as well as most of the oceans. Geologists have coined the name Snowball Earth for this episode of Earth-wide glaciation. —Lucy Chronic photo

Several miles south of Ogden, the highway rises onto Lake Bonneville's Weber River delta. Layton and Kaysville are on this delta, quarries produce gravel from it, and I-84 crosses it between I-15 and the mountains after branching eastward off I-15.

Utah 108 to Antelope Island State Park turns off I-15 between Clearfield and Layton. The park's geology makes it well worth visiting. Superb exposures of Precambrian rocks, including ancient glacial deposits, are juxtaposed with recent geological processes, such as the formation of oolites, which are being created by the Great Salt Lake's salt-concentrated water.

Near Farmington, debris from a large landslide complex extends out into the Great Salt Lake. The slide has moved several times in the past, triggered by earthquakes of magnitude 7 or greater. Such quakes are likely to occur again along the active Wasatch Fault, which separates the Wasatch Range from the Basin and Range country to the west.

The Precambrian rocks in the mountains right above Farmington are very resistant. The streams that cut into the mountains between Farmington and Bountiful are short, and little sediment is carried down to the lake plain or into the lake. Because of this, the easternmost tip of the Great Salt Lake reaches almost to the mountains.

South of Bountiful, in more easily eroded Paleozoic sedimentary rocks, streams have built an apron of merging alluvial fans and deltas that extends well south of Salt Lake City. Like towns farther north, Salt Lake City was initially built on delta deposits; it has now spread west across what was once the floor of Lake Bonneville.

The Wasatch Fault swings west and crosses beneath these delta deposits south of the main part of the city. Hot salty springs occur along this part of the fault. The hot water is another indication that the region is still geologically active, because it means there is a heat source not far beneath the surface.

Salt Lake City, like other cities just west of Utah's major mountains, suffers from a perennial shortage of freshwater. Streams that drain the west face of the Wasatch Range and those that naturally cut through it do not furnish enough for the wants of the urban area. A network of dams, canals, pipelines, and tunnels—the Central Utah Project—augments the flow of several rivers, the Spanish Fork and Provo in particular. The project diverts water that formerly flowed off the Uinta Mountains into the Green River and then, via the Colorado River, across the Southwest to Mexico and the Gulf of California.

I-15
Salt Lake City—Spanish Fork
51 miles (82 km)

Backed by the steep western face of the Wasatch Range, with the waters of the Great Salt Lake stretching far to the west, Salt Lake City's geology makes its scenery, but it also makes problems: Rising and falling lake levels sometimes threaten lakeside industries and recreation. And the active Wasatch Fault, right at the base of the mountains, holds the threat of earthquakes.

The steepness of the mountain front makes it ripe for earthquake-triggered landslides. Thick layers of uncemented sediment beneath the flats could amplify earthquake motion because the sediment can liquefy during a quake. East of Midvale and Sandy, Lake Bonneville shoreline features are offset along the Wasatch Fault by as much as 50 feet (15 m), indicating the fault has continued to be active in the last 15,000 years—the age of the features.

A branch of the Wasatch Fault, the East Bench Fault, runs through residential districts east of I-15. Water in the artesian springs here comes from the mountains, flowing through alluvial fans that slope under the clayey sediment left by Lake Bonneville. Confined by the clay, the water builds hydrostatic pressure. Where the clay layers are broken by the fault, the water flows out, making soft, boggy ground at the artesian springs.

The Oquirrh Mountains rise west of Midvale. They are composed largely of some 25,000 feet (7,620 m) of the Oquirrh Group marine limestone and sandstone that filled the deeply subsiding Oquirrh Basin during Pennsylvanian and Permian time. About 50 miles (80 km) to the east is the stable Colorado Plateau, to the west, the Great Basin. The Utah Hingeline, which runs roughly along the mountain front, marks the transition between these tectonically different areas. For its entire length, I-15 is close to the Utah Hingeline. During most of

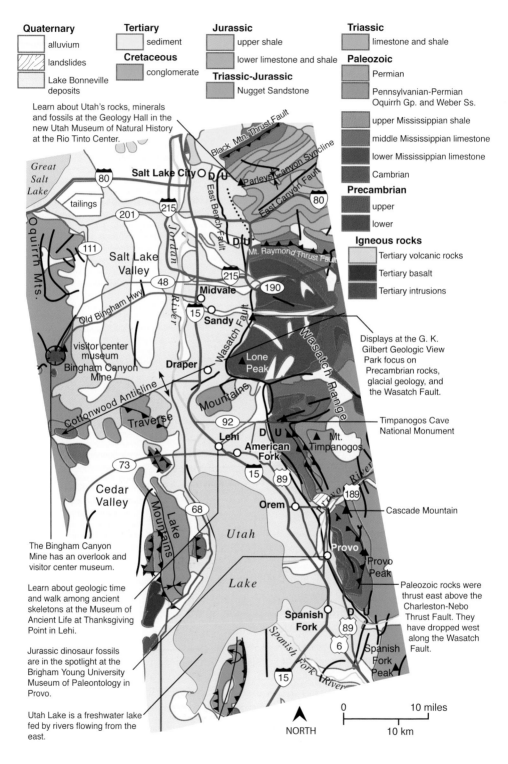

Quaternary
- alluvium
- landslides
- Lake Bonneville deposits

Tertiary
- sediment

Cretaceous
- conglomerate

Jurassic
- upper shale
- lower limestone and shale

Triassic-Jurassic
- Nugget Sandstone

Triassic
- limestone and shale

Paleozoic
- Permian
- Pennsylvanian-Permian Oquirrh Gp. and Weber Ss.
- upper Mississippian shale
- middle Mississippian limestone
- lower Mississippian limestone
- Cambrian

Precambrian
- upper
- lower

Igneous rocks
- Tertiary volcanic rocks
- Tertiary basalt
- Tertiary intrusions

Learn about Utah's rocks, minerals and fossils at the Geology Hall in the new Utah Museum of Natural History at the Rio Tinto Center.

Great Salt Lake

Salt Lake City

Black Mtn. Thrust Fault

Parleys Canyon Syncline

East Canyon Fault

East Bench Fault

Jordan River

tailings

Salt Lake Valley

Mt. Raymond Thrust Fault

Midvale

Sandy

Old Bingham Hwy

Wasatch Fault

Wasatch Range

visitor center museum
Bingham Canyon Mine

Draper

Lone Peak

Cottonwood Anticline

Traverse

Mountains

Displays at the G. K. Gilbert Geologic View Park focus on Precambrian rocks, glacial geology, and the Wasatch Fault.

Timpanogos Cave National Monument

Lehi

American Fork

Mt. Timpanogos

Cedar Valley

Lake Mountains

Utah

Orem

Provo River

Cascade Mountain

Provo

Provo Peak

Paleozoic rocks were thrust east above the Charleston-Nebo Thrust Fault. They have dropped west along the Wasatch Fault.

The Bingham Canyon Mine has an overlook and visitor center museum.

Learn about geologic time and walk among ancient skeletons at the Museum of Ancient Life at Thanksgiving Point in Lehi.

Lake

Spanish Fork

Spanish Fork Peak

Jurassic dinosaur fossils are in the spotlight at the Brigham Young University Museum of Paleontology in Provo.

Spanish Fork River

Utah Lake is a freshwater lake fed by rivers flowing from the east.

0   10 miles

NORTH

10 km

Geology along I-15 between Salt Lake City and Spanish Fork.

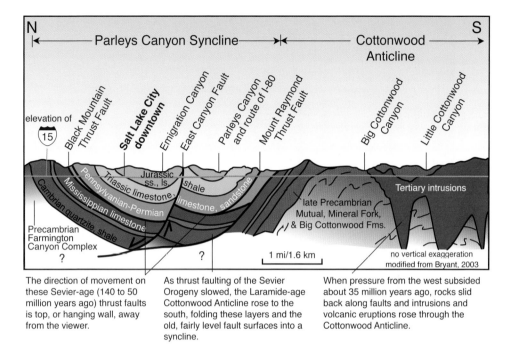

N — Parleys Canyon Syncline — | — Cottonwood Anticline — S

Black Mountain Thrust Fault

Salt Lake City downtown

Emigration Canyon

East Canyon Fault

Parleys Canyon and route of I-80

Mount Raymond Thrust Fault

Big Cottonwood Canyon

Little Cottonwood Canyon

elevation of (15)

Jurassic ss., ls

shale

Triassic limestone

Pennsylvanian-Permian

Mississippian limestone

Cambrian quartzite, shale

limestone, sandstone

Tertiary intrusions

late Precambrian Mutual, Mineral Fork, & Big Cottonwood Fms.

Precambrian Farmington Canyon Complex ?

?

1 mi/1.6 km

no vertical exaggeration modified from Bryant, 2003

The direction of movement on these Sevier-age (140 to 50 million years ago) thrust faults is top, or hanging wall, away from the viewer.

As thrust faulting of the Sevier Orogeny slowed, the Laramide-age Cottonwood Anticline rose to the south, folding these layers and the old, fairly level fault surfaces into a syncline.

When pressure from the west subsided about 35 million years ago, rocks slid back along faults and intrusions and volcanic eruptions rose through the Cottonwood Anticline.

North-south cross section of the Wasatch Range parallel to I-15 near Salt Lake City showing the large-scale folding and faulting that resulted from both the Sevier and Laramide Orogenies. The section runs perpendicular to most motion on the faults.

Paleozoic time, long before faulting reshaped the land to the west, much thicker layers of sediment accumulated west of the hingeline than to the east. Rifting during the Precambrian thinned the crust, so even though the sedimentary layers to the west are thicker, the crust is thinner than that east of the hingeline.

The Bingham Canyon Mine, one of the world's largest open-pit mines, lies in the central Oquirrh Mountains. Its dumps are visible from the highway. The mine is in porphyry that contains copper. Over the years geologists have figured out that these porphyries developed in magma chambers beneath volcanoes.

In 1995 Kennecott Utah Copper Corporation, now a division of Rio Tinto, agreed with the EPA to clean up the Bingham Canyon Mine area and keep their mining environmentally sustainable, so the area has never been listed as a Superfund site.

Large cement and gravel works near Draper offer a good look at the delta gravel along the Wasatch Range. The pebbles and cobbles were rounded and well sorted by ice-fed streams and the lapping waters of Lake Bonneville. I-15 curves around and passes through the Traverse Mountains, which owe their unusual east-west orientation to their location on the Cottonwood Anticline, a gentle east-west-trending anticline that is the westward continuation of the huge Uinta Anticline. The mountains divide the valleys of the Great Salt Lake and Utah Lake.

The Bingham Canyon Mine has been the world's single most productive copper source and has, in addition, produced molybdenum, gold, lead, and zinc. About 500,000 tons (453,600 metric tons) of rock, of which about one-fifth is usable ore, were excavated at the mine every day until a large landslide occurred in the pit on April 10, 2013. Production has slowed since then. —Felicie Williams photo

Like the ranges east and west of Utah Lake, they are composed of Paleozoic—primarily Oquirrh Group—sedimentary rocks that were deposited west of the Utah Hingeline. Much later, during the Sevier Orogeny, they were pushed here above the Charleston-Nebo Thrust Fault. The Traverse Mountains are adjacent to a Tertiary granitic intrusion (to the east) that rose up beneath the Cottonwood Anticline. Lone Peak is part of the intrusion, and the Wasatch Fault curves sharply around it.

Utah Lake occupies a graben that dropped down along faults between the Wasatch Range and the Lake Mountains. Some 6,000 feet (1,830 m) of lake sediment lie under the central part of this valley. The Wasatch Fault dips steeply here. Look for the truncated spurs at the front of the Wasatch Range; they mark the plane of the fault. Because the spurs haven't eroded much, they are evidence of geologically recent movement along the fault.

The cities east of Utah Lake are all built on deposits of Lake Bonneville, which include the deltas of the Provo and Spanish Fork Rivers. Lake Bonneville's ancient shorelines show very well on the mountainsides above Orem. The delta of the Spanish Fork River is particularly large, spreading well out into the valley and filling part of Utah Lake. The river's tributaries flow through soft, slide-prone Tertiary layers, sources of ample sediment for the large delta. In gravel pits near Spanish Fork, the delta gravels are poorly sorted. The poor sorting is a sign of very rapid deposition—evidence that huge amounts of water washed down the Spanish Fork River during the ice ages. Since Lake Bonneville retreated, sloping alluvial fan deposits have covered some of the delta surface, where rivers and creeks exit the mountains.

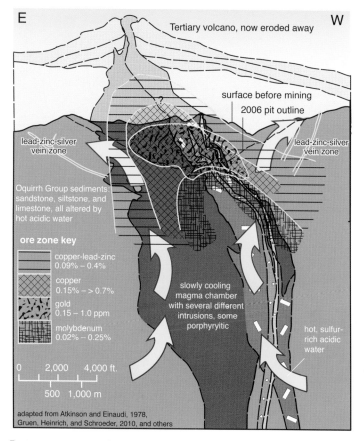

E   Tertiary volcano, now eroded away   W

surface before mining

2006 pit outline

lead-zinc-silver vein zone

lead-zinc-silver vein zone

Oquirrh Group sediments: sandstone, siltstone, and limestone, all altered by hot acidic water

ore zone key

	copper-lead-zinc 0.09% – 0.4%
	copper 0.15% – > 0.7%
	gold 0.15 – 1.0 ppm
	molybdenum 0.02% – 0.25%

slowly cooling magma chamber with several different intrusions, some porphyritic

hot, sulfur-rich acidic water

0   2,000   4,000 ft.

500 1,000 m

adapted from Atkinson and Einaudi, 1978, Gruen, Heinrich, and Schroeder, 2010, and others

East-west cross section of the Bingham Canyon Mine. Porphyry copper deposits are large and very low-grade. They form where magmas intrude with a lot of water and then cool very slowly. The water convects through the cooling intrusion and surrounding rock for quite a while. Very acidic, the water alters the minerals in the rock and deposits ore minerals. Zones of alteration and ore minerals are determined by depth (pressure) and distance from the cooling magma (temperature). The zone of altered rock can extend outward as much as 3 miles (5 km) from the intrusion.

The east-west-trending Traverse Mountains are clearly marked with Lake Bonneville shorelines (foreground, with buildings atop them). —Lucy Chronic photo

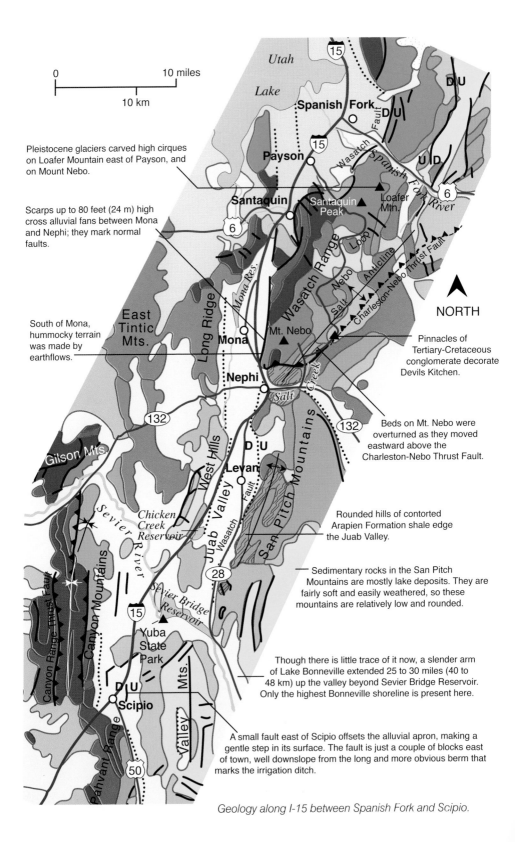

0 ——— 10 miles
——— 10 km

Utah
Lake

🛣️ 15

Spanish Fork ○
D/U

🛣️ 15

Payson ○

D/U

U/D

🛣️ 6

Pleistocene glaciers carved high cirques
on Loafer Mountain east of Payson, and
on Mount Nebo.

Santaquin ○
Santaquin
Peak ▲

Loafer
Mtn. ▲

Spanish Fork River

Scarps up to 80 feet (24 m) high
cross alluvial fans between Mona
and Nephi; they mark normal
faults.

🛣️ 6

Wasatch Fault

Loop

Nebo
Anticline

Salt

Charleston-Nebo Thrust Fault

**East
Tintic
Mts.**

Long Ridge

Mona Res.

Mona ○

Mt. Nebo ▲

South of Mona,
hummocky terrain
was made by
earthflows.

Pinnacles of
Tertiary-Cretaceous
conglomerate decorate
Devils Kitchen.

NORTH ▲

Nephi

Salt

Creek

🛣️ 132

Beds on Mt. Nebo were
overturned as they moved
eastward above the
Charleston-Nebo Thrust Fault.

🛣️ 132

D/U

Levan

San Pitch Mountains

Wasatch Fault

Rounded hills of contorted
Arapien Formation shale edge
the Juab Valley.

Gilson Mts.

West Hills

Chicken
Creek
Reservoir

Juab Valley

🛣️ 28

Sedimentary rocks in the San Pitch
Mountains are mostly lake deposits. They are
fairly soft and easily weathered, so these
mountains are relatively low and rounded.

Sevier River

Canyon Range Thrust Fault

Canyon Mountains

Sevier Bridge
Reservoir

🛣️ 15

Yuba
State
Park ▲

Mts.

Though there is little trace of it now, a slender arm
of Lake Bonneville extended 25 to 30 miles (40 to
48 km) up the valley beyond Sevier Bridge Reservoir.
Only the highest Bonneville shoreline is present here.

D/U

Scipio ○

Pahvant Range

Valley

🛣️ 50

A small fault east of Scipio offsets the alluvial apron, making a
gentle step in its surface. The fault is just a couple of blocks east
of town, well downslope from the long and more obvious berm that
marks the irrigation ditch.

Geology along I-15 between Spanish Fork and Scipio.

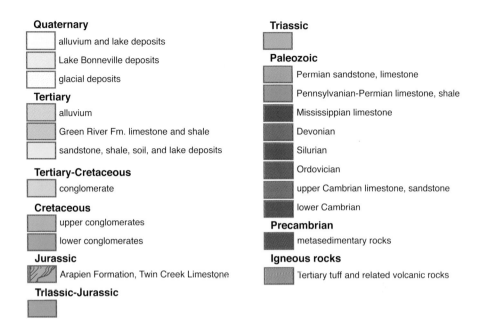

Quaternary
- alluvium and lake deposits
- Lake Bonneville deposits
- glacial deposits

Tertiary
- alluvium
- Green River Fm. limestone and shale
- sandstone, shale, soil, and lake deposits

Tertiary-Cretaceous
- conglomerate

Cretaceous
- upper conglomerates
- lower conglomerates

Jurassic
- Arapien Formation, Twin Creek Limestone

Triassic-Jurassic

Triassic

Paleozoic
- Permian sandstone, limestone
- Pennsylvanian-Permian limestone, shale
- Mississippian limestone
- Devonian
- Silurian
- Ordovician
- upper Cambrian limestone, sandstone
- lower Cambrian

Precambrian
- metasedimentary rocks

Igneous rocks
- Tertiary tuff and related volcanic rocks

I-15
Spanish Fork—Scipio
69 miles (111 km)

The abrupt mountain front east of Spanish Fork, part of the Wasatch Front, marks the location of the Wasatch Fault. Sharp-edged ridges along the front are cut off by the fault, leaving large triangular facets along their toes. These facets are evidence of fairly recent movement along the fault; a long period of erosion would have rounded them. There has been close to 3 miles (5 km) of displacement along the fault here.

Wave-cut shorelines of Pleistocene Lake Bonneville show up well near Spanish Fork. Each shoreline developed during a time when the lake level remained the same. When the lake was at the Provo Shoreline level, the Spanish Fork River built a sizeable delta out into the ancient lake. This delta is readily identifiable as an elevated terrace between I-15 and the mountains. Finer lake sediments form the broad slope that descends to Utah Lake.

The Wasatch Range between Spanish Fork and Nephi is almost entirely Pennsylvanian and Permian limestone and shale of the Oquirrh Group. Deposited in the marine Oquirrh Basin farther west, in Cretaceous time the rock moved 60 miles (100 km) eastward above the Charleston-Nebo Thrust Fault, which is part of the Sevier Thrust Belt. The fault is immense: its hanging wall includes all of the Wasatch Range from well north of Provo clear to Nephi—even the highest mountains you can see between Spanish Fork and Nephi.

South of Payson older strata appear on the lower mountain slopes, from Mississippian down to Proterozoic in age. The rocks were faulted extensively by thrust faults and later by extensional movement along the Wasatch Fault.

Ores in the East Tintic Mountains west of Santaquin produced lead, silver, gold, copper, and zinc. The ore-bearing minerals were deposited where fluids from a Tertiary intrusion interacted with the surrounding limestone and dolomite. The district was mined from 1868 until 2002, and exploration continues.

The highway crosses another Pleistocene delta and a more recently deposited alluvial fan where Santaquin Canyon exits the mountains. Large quarries near exit 244 produced Paleozoic limestone for use as flux in steel mills near Provo. Ground limestone added to smelter furnaces promotes the melting of ore and the separation of iron from it. The mills closed in 2001 and 2009.

At the Utah-Juab County line, near milepost 241, the highway crosses a sand and gravel bar that formed when Lake Bonneville was at its highest. Such bars form when the turbulence of breaking waves piles up sand offshore. They exist, along with other lakeshore features, around the entire perimeter of the former lake.

The huge mass of Mt. Nebo stands east of Mona. Composed of limestone and shale of Pennsylvanian and Permian age, the mass of rock was pushed here from the west as part of the Charleston-Nebo Thrust Sheet. The older rock overlies Triassic and Jurassic sedimentary rocks that were deposited here.

9,000 feet (2,740 m)

Pleistocene glaciers carved cirques on the high summits of Mt. Nebo, east of Mona. Glaciers extended down to about 9,000 feet (2,740 m) here, straightening canyons and rounding them into U-shaped valleys. Lower, unglaciated stream canyons are narrower and V-shaped. —Lucy Chronic photo

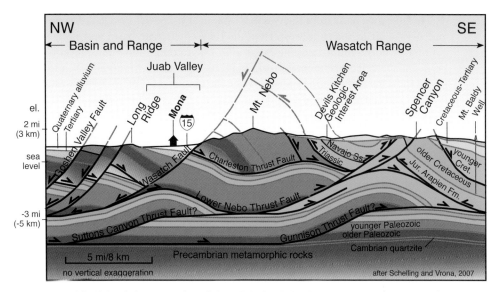

Mt. Nebo is just a portion of the original culmination, or highest part, of the Sevier Thrust Belt. The western side of this high range dropped down along the Wasatch Fault. The dashed lines show how much higher the mountains once were.

Near Salt Creek, gypsum-rich masses of Jurassic Arapien Formation weather into rounded gray-white hills. These beds were contorted by the Sevier Orogeny, and possibly by movement of the gypsum itself. Along the Wasatch Fault, where there was less pressure to confine the material, the gypsum may have flowed upward. Nearer the interstate, alluvial fans contain sand and gravel washed from the San Pitch Mountains. The highway climbs onto one of these fans at milepost 222.

The thinly bedded gray rock on the flanks of the San Pitch Mountains above Levan is Jurassic Twin Creek Limestone, a fossiliferous marine formation deposited in a shallow sea. The San Pitch Mountains contain Cretaceous conglomerate and sandstone that was eroded from the mountains of the Sevier Orogeny. The rocks were part of an alluvial apron much larger than the one that edges the mountain front now. It was 15,000 feet (4,570 m) thick, attesting to the great heights that have since been faulted and eroded into oblivion.

These Cretaceous and younger sedimentary rocks, which became finer grained as the mountains wore away, are exposed in the West Hills west of Levan: the fossiliferous lake limestone of the Flagstaff Formation, the river mudstones of the Colton Formation, and the lake shale and limestone of the Green River Formation. All were deposited in a large basin adjacent to the old and eroding Sevier highlands.

The Sevier River flows northward from the Sevier Bridge Reservoir, meandering gently. Rounding the north edge of the Canyon Mountains, west of the interstate, it curves west and then southwest toward the town of Delta and the dry Sevier Lake. Most years the river sinks into the desert before reaching the lake.

The interstate parts from the Wasatch Fault near the Sevier Bridge Reservoir and swings off to the southwest. Here the road enters true Basin and Range country as north-south-trending range after range comes into view. Precambrian and Devonian sedimentary rocks make up the Canyon Mountains. Scipio lies near the head of the long valley between the Canyon Mountains to the west, the Pahvant Range to the south, and the Valley Mountains to the east-southeast.

I-15
Scipio—Cove Fort
55 miles (89 km)

Southwest of Scipio, I-15 continues to roughly parallel the Utah Hingeline. The north-south ranges of the Great Basin meet the hingeline at an angle, so the ranges along this stretch of highway are staggered.

West of Scipio, in the Canyon Mountains, the Canyon Range Thrust Fault brought red upper Precambrian quartzite and Cambrian layers over gray Paleozoic limestone. Both rocks were then folded into a syncline by younger thrust faults beneath the mountain range: the Pahvant Thrust Fault, which surfaces in the Pahvant Range east of the highway near Scipio, and the Paxton Thrust Fault, which has only been identified in seismic profiles. These thrusts are part of the Sevier Thrust Belt, a series of Cretaceous-age thrust faults along which rocks were shoved eastward a total of some 130 miles (210 km). Paleozoic formations in the Canyon Mountains can be traced across an anticline at Scipio Pass and south into the Pahvant Range. The Cretaceous-age

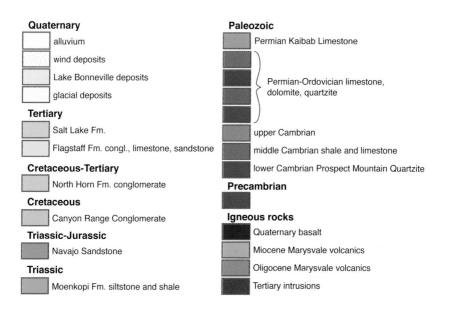

Quaternary
- alluvium
- wind deposits
- Lake Bonneville deposits
- glacial deposits

Tertiary
- Salt Lake Fm.
- Flagstaff Fm. congl., limestone, sandstone

Cretaceous-Tertiary
- North Horn Fm. conglomerate

Cretaceous
- Canyon Range Conglomerate

Triassic-Jurassic
- Navajo Sandstone

Triassic
- Moenkopi Fm. siltstone and shale

Paleozoic
- Permian Kaibab Limestone
- Permian-Ordovician limestone, dolomite, quartzite
- upper Cambrian
- middle Cambrian shale and limestone
- lower Cambrian Prospect Mountain Quartzite

Precambrian

Igneous rocks
- Quaternary basalt
- Miocene Marysvale volcanics
- Oligocene Marysvale volcanics
- Tertiary intrusions

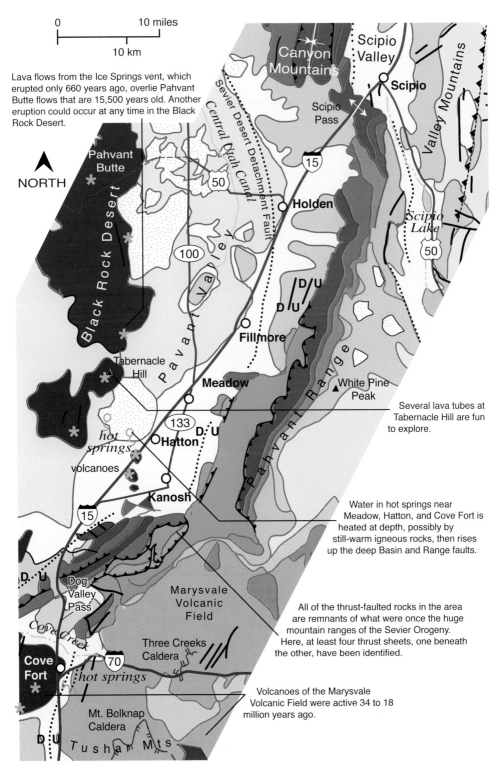

0   10 miles

10 km

Lava flows from the Ice Springs vent, which erupted only 660 years ago, overlie Pahvant Butte flows that are 15,500 years old. Another eruption could occur at any time in the Black Rock Desert.

NORTH

Pahvant Butte

Black Rock Desert

Canyon Mountains

Scipio Valley

Scipio

Scipio Pass

Scipio Lake

Valley Mountains

Sevier Desert Detachment Fault

Central Utah Canal

50

100

Holden

D/U
D/U

Pahvant Valley

Fillmore

Meadow

White Pine Peak

Pahvant Range

Several lava tubes at Tabernacle Hill are fun to explore.

Tabernacle Hill

hot springs

volcanoes

133   D/U

Hatton

Kanosh

15

Water in hot springs near Meadow, Hatton, and Cove Fort is heated at depth, possibly by still-warm igneous rocks, then rises up the deep Basin and Range faults.

D. U

Dog Valley Pass

Marysvale Volcanic Field

Cove Creek

Three Creeks Caldera

Cove Fort

70

hot springs

Mt. Belknap Caldera

D U   Tushar Mts.

All of the thrust-faulted rocks in the area are remnants of what were once the huge mountain ranges of the Sevier Orogeny. Here, at least four thrust sheets, one beneath the other, have been identified.

Volcanoes of the Marysvale Volcanic Field were active 34 to 18 million years ago.

Geology along I-15 between Scipio and Cove Fort.

Thrust faults like the Canyon Range Thrust Fault, shown here in the Canyon Mountains, were almost horizontal when they were active. The thrust fault is between the reddish Precambrian rocks at the top of the slope and the gray Paleozoic rocks of the lower slopes. —Lucy Chronic photo

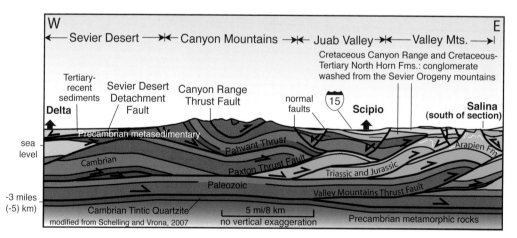

Thrust sheets that were moved during the Sevier Orogeny (145 to 65 million years ago) form a bizarre-looking pileup in the Canyon Mountains. The thrust sheets, in turn, were cut by much younger Basin and Range normal faults and the Sevier Desert Detachment Fault (all younger than 17 million years).

Canyon Creek Conglomerate tops the Pahvant Range; it is part of a thick wedge of coarse sedimentary rock that made up an alluvial apron on the east slope of the mountains of the Sevier Orogeny. Waves of mountain building are recorded in the conglomerate: coarser sediment was deposited when faulting occurred and invigorated rivers, and the newly deposited layers were folded by movement of later thrust faults beneath them.

South of Scipio Pass the highway drops down into a broad, flat-floored valley, part of the bed of Lake Bonneville, though younger sediment covers the old lake deposits. To the southwest, the Cricket Mountains are composed of upper Precambrian and Cambrian rocks similar to those in the Canyon Mountains; they may also have been pushed east above the Canyon Range Thrust Fault.

To the west are lava flows centered around Pahvant Butte, a small volcano that erupted explosively 15,500 years ago and has scarcely eroded since. Both Pahvant Butte and Tabernacle Hill, to the south, erupted into Lake Bonneville. The evidence for this is clear: lava flows and lake sediments are interfingered, and there are pillow-shaped lumps of basalt that formed when hot lava encountered lake water and cooled quickly.

The Black Rock Desert is named for its basalt cinder cones and flows. In patches, these young volcanic rocks extend south beyond Cove Fort across an area about 10 miles (16 km) wide and 60 miles (100 km) long. Their source was deep: their melted rock rose up Basin and Range faults from the top of Earth's mantle. Another eruption could occur in this desert at any time.

Cinder cones form where basalt magma contains quite a bit of steam. Escaping steam flings frothy blobs of lava from a volcanic vent, and the blobs solidify as cinders and accumulate around the vent to build a cone. Cinder cones generally form early in an eruption, and lava flows out through the porous base of the cone once the froth is expended.

Near Fillmore, several cinder cones have been quarried for use as lightweight aggregate for construction purposes; the bubble-filled volcanic cinders are seemingly ready-made for such use.

The entire volcanic area around and in the Black Rock Desert is a geothermal region. There are hot springs west of Holden, near Hatton, and elsewhere. Heated water likely flows up the same faults that were used by the basalt.

Several buttes near Fillmore are of Tertiary sedimentary rock; the buttes are composed of layers of coarse conglomerate laid down by rivers flowing east from the mountains of the Sevier Orogeny.

From Fillmore south, this side of the Pahvant Range contains Cambrian strata thrust over much younger Triassic and Jurassic layers. The rocks of Cambrian age were thrust eastward about 43 miles (70 km) on the Pahvant Thrust Fault 110 to 86 million years ago. The light-colored lower slopes are Navajo Sandstone, a massive rock that formed in dunes and is exposed in spectacular outcrops in the Colorado Plateau.

Above the Cambrian rocks is an unconformity. The North Horn Formation, which spans the boundary between Cretaceous and Tertiary time, was laid down on the eroded Cambrian surface. On the top of the range, White Pine Peak is composed of the North Horn Formation.

Lava tubes, such as this one near Tabernacle Hill, form when a flow cools on the outside and the lava inside continues flowing and eventually drains out, leaving a long, sinuous cave. —Lucy Chronic photo

Low, light-brown hills directly west of Hatton are tufa cones, spongy deposits of calcium carbonate near Hatton hot springs. When the hot-spring water evaporates, it precipitates calcium carbonate that it carried in solution.

A short distance south of Hatton, I-15 passes between two small volcanoes. Reddish soils are derived from the basaltic cinders and lava flows. The Twin Peaks, small light-colored peaks in the distance near the southern end of the desert, are rhyolite domes that are about 3 million years old.

The highway approaches Dog Valley Pass south of Hatton, climbing a stream-formed delta that dates back to the Lake Bonneville days. Slices of crushed, folded, broken, and sometimes completely overturned Paleozoic sedimentary rocks border the highway on the way to the pass. They are below the Pahvant Thrust Fault and were crumpled and broken as rocks moved eastward above them.

South of the pass, another young, flat-topped cinder cone rises above the lava flows of a small shield volcano. I-15 drops into older and quite different Tertiary volcanic rocks: light-colored pinkish tuff, mudflow deposits, and breccia; these deposits are part of a huge swath of volcanic rocks that spread across the West after the Sevier Orogeny but before Basin and Range faulting.

Cove Fort's geothermal area includes hot springs and a small power plant. Sulfur deposits, mined for a time, formed around some hot spring vents. Cove

Creek, which flows through the community, divides the Pahvant Range from the Tushar Mountains to the southeast.

Cove Fort, built of volcanic rock, was a late nineteenth-century stagecoach stop between the towns of Fillmore and Beaver. It provided travelers protection from Native American attacks during the Black Hawk War.

I-15

Cove Fort—Cedar City

76 miles (122 km)

As I-15 heads south from Cove Fort, two Quaternary cinder cones rise atop basalt flows to the west; one is quite near milepost 132.

To the east are older volcanic rocks: the high summits of the Tushar Mountains are eroded Tertiary calderas of the Marysvale Volcanic Field. Exposed by Basin and Range faulting about 10 million years ago, they now form the highest range between the Rocky Mountains of Colorado and the distant Sierra Nevada.

Volcanism came in two separate episodes in Utah: the first, which includes the Marysvale Volcanic Field, was between 34 and 18 million years ago, when the subducted Farallon Plate was passing shallowly beneath this region. Rocks of the first episode are usually andesite or dacite that developed from melted crust. The second episode, which continues today, started around 17 million years ago with the beginning of Basin and Range normal faulting. Its eruptions have been bimodal, meaning they produce both dark basalt and light-colored rhyolite. It is thought that basalt magma, rising up from the mantle

A cinder cone (left-center) of Quaternary age near milepost 132. —Lucy Chronic photo

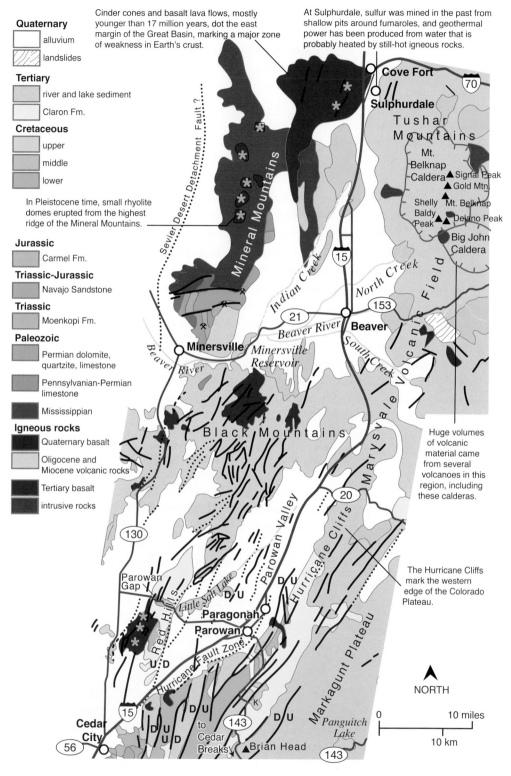

Quaternary
- alluvium
- landslides

Tertiary
- river and lake sediment
- Claron Fm.

Cretaceous
- upper
- middle
- lower

In Pleistocene time, small rhyolite domes erupted from the highest ridge of the Mineral Mountains.

Jurassic
- Carmel Fm.

Triassic-Jurassic
- Navajo Sandstone

Triassic
- Moenkopi Fm.

Paleozoic
- Permian dolomite, quartzite, limestone
- Pennsylvanian-Permian limestone
- Mississippian

Igneous rocks
- Quaternary basalt
- Oligocene and Miocene volcanic rocks
- Tertiary basalt
- intrusive rocks

Cinder cones and basalt lava flows, mostly younger than 17 million years, dot the east margin of the Great Basin, marking a major zone of weakness in Earth's crust.

At Sulphurdale, sulfur was mined in the past from shallow pits around fumaroles, and geothermal power has been produced from water that is probably heated by still-hot igneous rocks.

Huge volumes of volcanic material came from several volcanoes in this region, including these calderas.

The Hurricane Cliffs mark the western edge of the Colorado Plateau.

Cove Fort

Sulphurdale

Tushar Mountains

Mt. Belknap Caldera

Signal Peak
Gold Mtn.
Shelly Baldy Peak
Mt. Belknap
Delano Peak
Big John Caldera

Mineral Mountains

Sevier Desert Detachment Fault ?

Indian Creek

North Creek

Beaver River

Beaver

Minersville

Minersville Reservoir

South Creek

Marysvale Volcanic Field

Black Mountains

Parowan Valley

Hurricane Cliffs

Markagunt Plateau

Parowan Gap

Red Hills

Little Salt Lake

Paragonah

Parowan

Hurricane Fault Zone

Cedar City

to Cedar Breaks

Brian Head

Panguitch Lake

NORTH

0   10 miles

10 km

Geology along I-15 between Cove Fort and Cedar City.

along normal faults, melts some of the crust along the way, generating rhyolitic magma.

Southwest of milepost 130, to the west, is the large granite complex of the Mineral Mountains. The mountains are part of the Marysvale Volcanic Field. As the granite cooled and cracked, the cracks filled with crystals of smoky quartz and feldspar, for which the mountains are named. They have dropped down along normal faults and eroded.

Despite their name, the Mineral Mountains have not been very important in Utah's mining history. A few mines are clustered near Minersville, at the south end of the range. There, hot mineral-rich solutions from the granitic intrusion deposited ore that contains gold, silver, copper, lead, zinc, and tungsten where they met Paleozoic limestone.

As I-15 approaches milepost 126, it climbs onto a terrace of gravel deposited by sediment-burdened rivers of Pleistocene time. The terrace is offset along many small normal faults, proof that faulting occurred here relatively recently and is still going on.

A hilly region of much-faulted Tertiary volcanic rocks—the Black Mountains—can be seen between mileposts 119 and 118, south of the divide. The Black Mountains extend well to the west beyond the Mineral Mountains. South of them lies the Markagunt Plateau, fronted by the Hurricane Cliffs. The plateau's southern edge marks the southern extent of the Marysvale Volcanic Field.

I-15 crosses two more Pleistocene river terraces before reaching the town of Beaver, located in a fertile valley nourished by the Beaver River.

South of Beaver, I-15 crosses the east end of the Black Mountains. These 22-to-19-million-year-old dacite and andesite lava flows, volcanic breccias, and ashflow tuffs were derived from stratovolcanoes. A dense concentration of northeasterly faults continues east from the mountains into the low area between the Tushar Mountains and the Markagunt Plateau. Some of the faults displace upper Tertiary gravel, so there must have been movement along the faults during the last 2.6 million years.

The Parowan Valley, south of milepost 95, is a graben between the Red Hills and the Markagunt Plateau. A Pleistocene lake occupied the valley, but ridges to the west and south isolated the lake, so it was not part of Lake Bonneville. Little Salt Lake, a few miles long and ephemeral, is all that remains. The valley narrows at its southern end between Quaternary lava flows and fault blocks that have dropped down along the Hurricane Cliffs.

East of Parowan the mountain front rises abruptly along the Hurricane Fault Zone. More than 160 miles (260 km) long, this fault zone extends south into Arizona, dividing the Colorado Plateau from the Great Basin. Rocks west of it have dropped several thousand feet (more than 1 km). (See the cross section in the Utah 14: Cedar City—Long Valley Junction road guide in Utah's Backbone: The High Country.)

From the milepost 83 rest area, sculptured pinkish-red rocks of the Tertiary Claron Formation can be seen to the east and in the Red Hills to the west. This soft, silty, limy rock has eroded into several breaks along the margins of Utah's high plateaus, including at Cedar Breaks National Monument and Bryce

The Fremont people etched a spectacular number of petroglyphs through desert varnish on blocks of Navajo Sandstone in Parowan Gap, which is west of Parowan in the Red Hills. —Lucy Chronic photo

Canyon National Park, both farther east. Basalt flows cap the southern part of the Red Hills.

Continuing south to Cedar City, I-15 remains close to the Hurricane Cliffs. Part of Cedar City's economy has long depended on mining. Iron is mined from the mountains to the west. Coal and uranium have been produced from Cretaceous rocks on the Markagunt Plateau to the east.

I-15
Cedar City—Arizona State Line
59 miles (95 km)

West of Cedar City, Tertiary intrusions and volcanic rocks form the low hills of the Iron Springs Mining District. Here, intruding magmas domed Jurassic and Cretaceous strata, and hot water that followed the intrusions leached the iron from their minerals and redeposited it along bedding in adjoining limestone. The resulting pockets and veins of magnetite and hematite are mined for iron.

East of Cedar City along the Hurricane Cliffs, the succession of layers is a signature of the Colorado Plateau. The Vermilion Cliffs are composed of colorful Triassic and Jurassic rocks, including the deep red-brown Chinle and Moenkopi

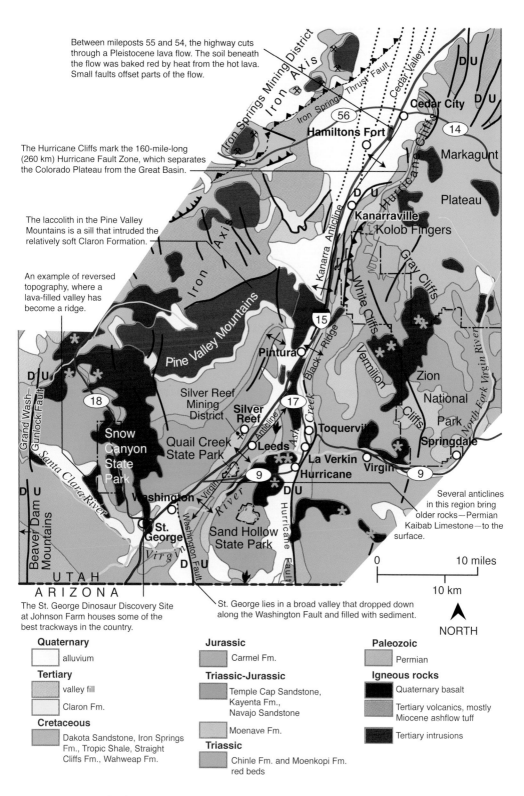

Between mileposts 55 and 54, the highway cuts through a Pleistocene lava flow. The soil beneath the flow was baked red by heat from the hot lava. Small faults offset parts of the flow.

The Hurricane Cliffs mark the 160-mile-long (260 km) Hurricane Fault Zone, which separates the Colorado Plateau from the Great Basin.

The laccolith in the Pine Valley Mountains is a sill that intruded the relatively soft Claron Formation.

An example of reversed topography, where a lava-filled valley has become a ridge.

Iron Springs Mining District

Iron Axis

Iron Springs Thrust Fault

Cedar Valley

D|U

56

Cedar City

14

Hamiltons Fort

Markagunt

Hurricane Cliffs

Plateau

D|U

Kanarra Anticline

Kanarraville

Kolob Fingers

Iron Axis

Gray Cliffs

White Cliffs

15

Vermilion

Pintura

Pine Valley Mountains

Zion

Black Ridge

National

Silver Reef Mining District

18

Silver Reef

17

Ash Creek

Anticline

Toquerville

Cliffs

Park

North Fork Virgin River

Springdale

Snow Canyon State Park

Quail Creek State Park

Leeds

La Verkin

Virgin

9

Grand Wash–Gunlock Fault

Santa Clara River

Washington

9

Hurricane

D|U

Virgin River

Several anticlines in this region bring older rocks—Permian Kaibab Limestone—to the surface.

D U

St. George

Sand Hollow State Park

Washington Fault

Hurricane Fault

Beaver Dam Mountains

Virgin River

D U

0   10 miles

UTAH

A R I Z O N A

10 km

The St. George Dinosaur Discovery Site at Johnson Farm houses some of the best trackways in the country.

St. George lies in a broad valley that dropped down along the Washington Fault and filled with sediment.

NORTH

Quaternary

alluvium

Tertiary

valley fill

Claron Fm.

Cretaceous

Dakota Sandstone, Iron Springs Fm., Tropic Shale, Straight Cliffs Fm., Wahweap Fm.

Jurassic

Carmel Fm.

Triassic-Jurassic

Temple Cap Sandstone, Kayenta Fm., Navajo Sandstone

Moenave Fm.

Triassic

Chinle Fm. and Moenkopi Fm. red beds

Paleozoic

Permian

Igneous rocks

Quaternary basalt

Tertiary volcanics, mostly Miocene ashflow tuff

Tertiary intrusions

Geology along I-15 between Cedar City and the Arizona state line.

Formations. The Navajo Sandstone is the prominent White Cliffs right above town on the mountain front. These layers are faulted and folded. Cretaceous rocks, the Gray Cliffs, lie more or less horizontally above them: forested slopes of shale capped by Dakota and Straight Cliffs sandstones. From high spots you can glimpse even younger Tertiary rock: the Pink Cliffs of the Claron Formation, eroded in one of several breaks that edge the Markagunt Plateau.

Cedar City is at the north end of the Kanarra Anticline, a fold that is the area's easternmost evidence of mountain building related to the Sevier Orogeny. In the last 13 million years, the western side of the anticline has dropped downward along the Hurricane Fault. It is now far below Cedar Valley.

The Pine Valley Mountains, to the west, are a single, huge laccolith that intruded 20 million years ago. It is part of the Iron Axis, a line of laccoliths that roughly parallels thrust faults of the Sevier Orogeny and the Hurricane Fault. Laccoliths form when magma spreads between rock layers and causes them to dome up. Here, the layers of the Claron Formation were bent so steeply they finally slid off the intrusion. This lowered the pressure on the still-hot intrusion, which then erupted violently, blanketing the area with ash and breccia.

East of Pintura, much younger basalt (850,000 years old) appears on Black Ridge. These flows are examples of reversed topography. This type of landform commonly occurs where lava has flowed down a valley and solidified; and then, over time, softer rocks surrounding the flows erode away. Near the center of Black Ridge one flow is quite thick, as seen in the deep gorge at milepost 37.

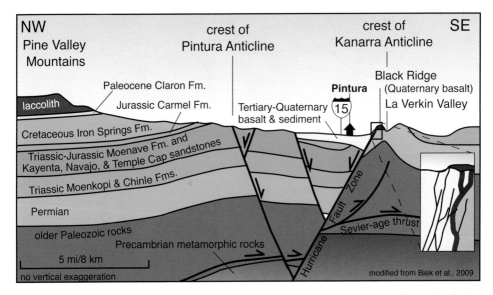

Near Pintura, I-15 rides down a valley that is a graben between two normal faults. Caused by Basin and Range extension, the faults postdate and offset the thrust faults and folds that formed during the earlier Sevier Orogeny. Inset: Each fault line on maps and cross sections actually represents a fault zone composed of many subparallel fractures and faults. Magma travels up these fault zones, which are weaker than surrounding rock.

The basalt down low behind the trees and on the ridgetop are part of the same flow; the flow has been offset across the Hurricane Fault. —Lucy Chronic photo

Pale Permian rocks appear low on the Hurricane Cliffs near Pintura. The Hurricane Fault occurs at the base of the cliffs. The lowest rock exposed in the cliffs, the Queantoweap Sandstone, was deposited in dunes along a coastal plain and in deltas. Next, the nearshore Toroweap Formation, dark shale, limestone, and some gypsum, records a single incursion and withdrawal of the sea. Above it, the lighter-colored cliff of the Kaibab Limestone marks a more extensive marine incursion.

Around milepost 27, I-15 passes a hill of granitelike quartz monzonite that intruded as part of the Pine Valley Laccolith. The Hurricane Fault swings away southeastward. Near Leeds, the road crosses onto the Virgin Anticline. This anticline is said to be doubly plunging because its north and south ends also plunge downward. It is, in a way, a very long dome. Near the exit to Toquerville, you can see the east limb of the Kanarra Anticline to the northeast.

Leeds is in the heart of the Silver Reef Mining District. Here, silver, copper, and uranium minerals, carried by groundwater percolating through river-channel deposits, coated and replaced poorly preserved petrified and carbonaceous plant remains. The deposits are in the Springdale Sandstone Member of the Jurassic-age Kayenta Formation. Most mining here took place between 1876 and 1909.

Mercury, used to process silver ore, is present in the district's tailings; the EPA has cleaned up some of the contaminated sites and warns people away from others. The hundreds of mine openings scattered around the district now

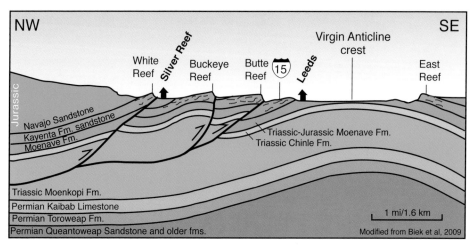

NW   SE

Silver Reef

White Reef   Buckeye Reef   Butte Reef (15) Leeds

Virgin Anticline crest

East Reef

Jurassic

Navajo Sandstone

Kayenta Fm. sandstone

Moenave Fm.

Triassic-Jurassic Moenave Fm.
Triassic Chinle Fm.

Triassic Moenkopi Fm.
Permian Kaibab Limestone
Permian Toroweap Fm.
Permian Queantoweap Sandstone and older fms.

1 mi/1.6 km

Modified from Biek et al, 2009

Simplified cross section of the Virgin Anticline through the Silver Reef Mining District. The anticline and associated back thrusts brought the ore-bearing sandstone to the surface in a repeating pattern. Reef is both a mining term for an ore-bearing layer and a name applied to a long, sharp outcrop of an erosion-resistant bed. Both definitions apply to the rocks in this district.

A swimming dinosaur left long claw marks on the bottom of a lake 200 million years ago, during Jurassic time. These molds of the marks formed from sediment that filled in the claw marks and lithified as rock. They can now be seen, along with many other prints, in the Moenave Formation at the St. George Dinosaur Discovery Site at Johnson Farm. —Felicie Williams photo

house a good population of bats, so the unsafe workings have been sealed with metal grids to both allow the bats to move freely and keep people out.

Dinosaur tracks have been found in the Moenave Formation northeast of Leeds. Thousands more mark the same layers of rock in St. George.

Between Leeds and Washington, I-15 parallels the Virgin Anticline's edge, following a racetrack valley in the Chinle Formation. The valley eroded in tilted layers of soft mudstone and siltstone between cuestas of harder sandstone.

Quail Creek State Park and its reservoir lie in the anticline's center, where soft layers of red Moenkopi Formation have worn down into a valley with walls topped by the cliff-forming Shinarump Conglomerate, the lowest member of the Chinle Formation. The Moenkopi was deposited on tidal flats. It records transgressions and regressions of the nearby ocean: white gypsum beds, from evaporated seawater, are interlayered with mudstone and siltstone. The Chinle, on the other hand, was deposited on a continental plain. The soft and very colorful Petrified Forest Member, which often contains pieces of petrified wood, rests on top of the Shinarump Conglomerate. The Petrified Forest Member's abundant clay, weathered from volcanic ash, swells when wet and shrinks when dry. Beds of it often form spectacular badlands.

On the southern slope of the Pine Valley Mountains to the west, we see again the massive, light-colored Navajo Sandstone that we saw in the Hurricane Cliffs. The Navajo's long crossbeds formed in dunes in a vast Jurassic sea of sand. Highway cuts over the next few miles contain either coarse, cobbly gravel washed down from the Pine Valley Mountains, or lava flows.

The Virgin River, born in the highlands of the Markagunt Plateau, flows south of St. George toward Lake Mead and the Colorado River, meandering where its gradient is low. Crossing the river's floodplain, I-15 rises onto red Triassic layers as it approaches the Arizona state line. The Beaver Dam Mountains, to the west, expose Precambrian metamorphic rock along with Paleozoic layers.

I-80
Wendover—Salt Lake City
122 miles (196 km)

Imagine a vast, waveless ocean stricken dead and turned to ashes; imagine this solemn waste tufted with ash-dusted sage-bushes; imagine the lifeless silence and solitude that belong to such a place; imagine a coach, creeping like a bug through the midst of this shoreless level, and sending up tumbled volumes of dust as if it were a bug that went by steam; . . . imagine team, driver, coach and passengers so deeply coated with ashes that they are all one colorless color.

—Mark Twain's description of the "alkali desert" west of the Great Salt Lake in *Roughing It.*

The Silver Island Mountains, just east of the state line, are the first of many ranges I-80 passes on the way to Salt Lake City. Their steeply tilted Paleozoic limestones and dolomites, intruded by granite and related rocks, are circled by a scenic backcountry drive with hiking trails up some of the peaks. The

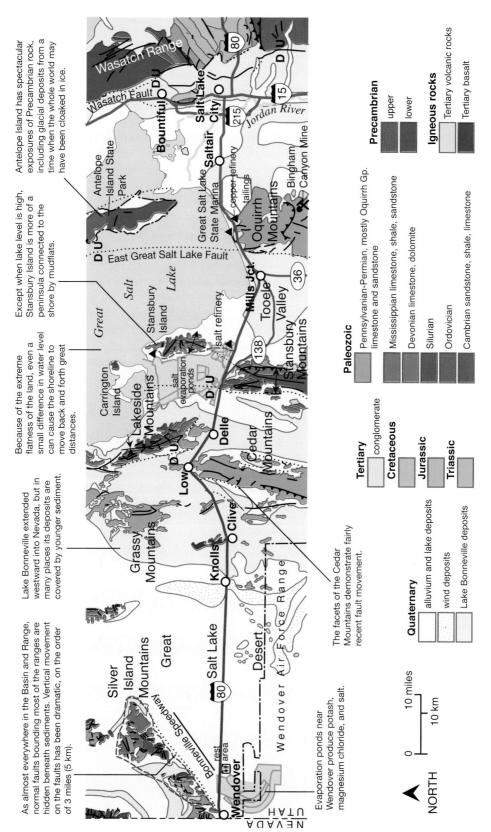

NORTH

0   10 miles

10 km

Antelope Island has spectacular exposures of Precambrian rock, including glacial deposits from a time when the whole world may have been cloaked in ice.

Except when lake level is high, Stansbury Island is more of a peninsula connected to the shore by mudflats.

Because of the extreme flatness of the land, even a small difference in water level can cause the shoreline to move back and forth great distances.

Lake Bonneville extended westward into Nevada, but in many places its deposits are covered by younger sediment.

As almost everywhere in the Basin and Range, normal faults bounding most of the ranges are hidden beneath sediments. Vertical movement on the faults has been dramatic, on the order of 3 miles (5 km).

The facets of the Cedar Mountains demonstrate fairly recent fault movement.

Evaporation ponds near Wendover produce potash, magnesium chloride, and salt.

Quaternary
- alluvium and lake deposits
- wind deposits
- Lake Bonneville deposits

Tertiary
- conglomerate

Cretaceous

Jurassic

Triassic

Paleozoic
- Pennsylvanian–Permian, mostly Oquirrh Gp. limestone, shale, sandstone
- Mississippian limestone, shale, sandstone
- Devonian limestone, dolomite
- Silurian
- Ordovician
- Cambrian sandstone, shale, limestone

Precambrian
- upper
- lower

Igneous rocks
- Tertiary volcanic rocks
- Tertiary basalt

Geology along I-80 between Wendover and Salt Lake City.

intrusions provide evidence of a spell of extension that preceded the eastward sweep of Sevier Orogeny mountain building across Utah. The extension, during Jurassic time, allowed magma from the lower crust or mantle to rise and intrude Paleozoic rocks.

In Cretaceous time, during the Sevier Orogeny, all the rocks along this section of highway were thrust eastward. Then, starting 17 million years ago, they were pulled westward, breaking into isolated ranges—tilted fault blocks separated by westward-dipping normal faults. The down-dropping on these faults has been very dramatic—on the order of 3 miles (5 km). Most of these normal faults are hidden beneath sediment. The faults flatten with depth and probably merge into gently sloping detachment faults. (For a cross section reflecting much of the geology along this stretch of I-80, see The Great Salt Lake and Lake Bonneville section at the beginning of this chapter.)

Part of the smooth, hard surface of the salt flats southeast of the Silver Island Mountains has become one of the world's most famous racetracks: the Bonneville Speedway. About 12 miles (19 km) long and 80 feet (24 m) wide, it is so flat that you can't see both ends of the speedway at once due to the curvature of the Earth. This curvature can be seen quite well from the rest area east of Wendover, where there is an exhibit about the racetrack's history.

I-80 crosses about 40 miles (64 km) of salt flats here. You could ask for no more barren a desert than this. The wind has scoured away the sediment around the few salt-tolerant plants, leaving them on low pedestals held together by their roots. The salt crust of the flats is as much as 5 feet (1.5 m) thick and is 90 percent pure salt, concentrated here by the evaporation of Lake Bonneville. The rest of the crust is gypsum, potash, calcium carbonate, and other minerals. All the minerals were carried into the lake by streams.

Scattered sand dunes edge the highway near milepost 39. Near Knolls, the westerly winds that bounce near-surface grains of gypsum sand along are

Wave-cut shorelines (arrows) stair-step up the lower slopes of the Silver Island Mountains. In 1890 geologist G. K. Gilbert recognized the terraces as the work of an immense ice age lake he named Lake Bonneville, which extended west well into Nevada. —Lucy Chronic photo

Ore from the Bingham Canyon Mine is processed at the smelter near I-80. Even the sulfur is recovered to make sulfuric acid. Tailings east of the Great Salt Lake State Marina build up at a rate of 60 million tons (54 million metric tons) each year. Older tailings have been reclaimed as grassland. —Lucy Chronic photos

slowed as they begin to rise over the mountains, causing the wind to drop the sand. The largest dunes develop on the western slopes of ridges.

East of Knolls the highway rises almost imperceptibly above the salt flats onto older Lake Bonneville deposits. These are mostly composed of fine silt without salt; Lake Bonneville's water was fresh before the extensive evaporation that left the much smaller Great Salt Lake and playas.

The Cedar Mountains, south of Low, consist of east-dipping Oquirrh Group layers, mostly limestone and sandstone. The Oquirrh Basin, which encompassed much of northwestern Utah, filled with as much as 25,000 feet (7,620 m) of sediment during Pennsylvanian and Permian time. The basin subsided slowly, maintaining shallow marine conditions ideal for a great variety of life. Brachiopods, bryozoans, corals, sponges, and fusulinids are among those fossilized in these rocks.

Rocks in the Lakeside Mountains, to the north, are older; these Cambrian to Mississippian rocks were bent into a syncline. Shelflike Lake Bonneville shorelines mark all ranges. The shorelines show up best on smaller buttes, those with minimal watersheds to cause stream erosion.

Near exit 77 the Stansbury Mountains, to the south, and Stansbury Island, to the northeast, are parts of a long, narrow fault block in which Paleozoic rocks arch into an anticline. Light-colored Cambrian quartzite at the core of the anticline forms the central ridge of the Stansbury Mountains.

The north end of the Oquirrh Mountains ahead is, like the Cedar Mountains, composed of Pennsylvanian and Permian Oquirrh Group rocks. The range includes several Tertiary granitic intrusions, one of which is responsible for the rich ore at the Bingham Canyon Mine. The nation's largest open-pit copper mine, Bingham Canyon Mine also contains molybdenum, silver, gold, and lead. Prominent Lake Bonneville shorelines, some with porous white lake deposits called tufa, crease the slopes of the Oquirrh Mountains.

Antelope Island, north of the Oquirrh Mountains and in the lake, is another faulted mountain ridge. Its rocks are Precambrian, so they are much older than those we've seen so far. Precambrian rocks underlie all of the state but are rarely exposed.

Much of Salt Lake City lies on deltas that were built out into Lake Bonneville by rivers of glacial meltwater during the Pleistocene ice ages. The western part of the city spreads onto the lower, younger alluvial floodplain of the Jordan River. The Jordan drains Utah Lake, a freshwater lake that, like the Great Salt Lake, occupies a part of former Lake Bonneville's immense basin.

US 6
Santaquin—Delta
71 miles (114 km)

US 6 leaves Santaquin on the wide Provo Shoreline of ancient Lake Bonneville. Between the highway and Utah Lake is West Mountain. Its quarries long provided limestone and dolomite to a steel plant near Orem. Water from springs to the south, warmed at depth, flows toward Utah Lake but evaporates on the way,

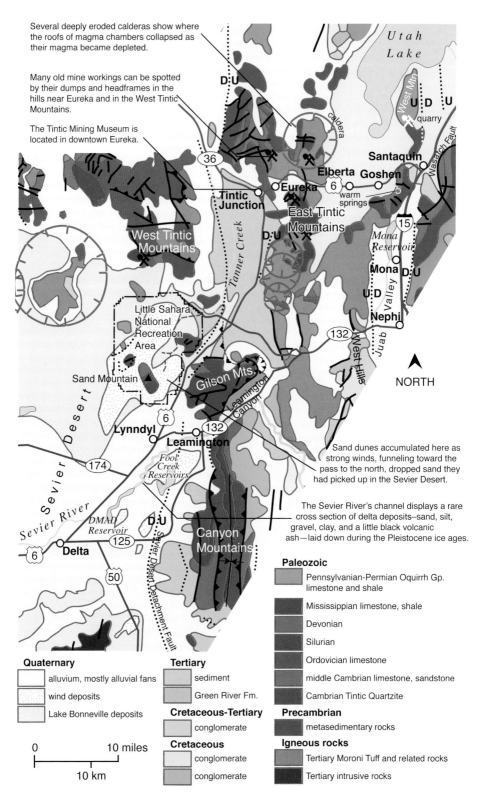

Several deeply eroded calderas show where the roofs of magma chambers collapsed as their magma became depleted.

Many old mine workings can be spotted by their dumps and headframes in the hills near Eureka and in the West Tintic Mountains.

The Tintic Mining Museum is located in downtown Eureka.

Utah Lake

caldera

quarry

Wasatch Fault

West Mtn

D:U

U:D U

Santaquin

36

Elberta Goshen

Eureka

6

warm springs

Tintic Junction

15

East Tintic Mountains

D:U

Mona Reservoir

West Tintic Mountains

Mona D:U

U:D

Nephi

Little Sahara National Recreation Area

132

Juab Valley

West Hills

NORTH

Sand Mountain

Gilson Mts.

Leamington Canyon

Tanner Creek

Desert

6

Lynndyl

132

Leamington

Fool Creek Reservoirs

Sand dunes accumulated here as strong winds, funneling toward the pass to the north, dropped sand they had picked up in the Sevier Desert.

174

Sevier River

The Sevier River's channel displays a rare cross section of delta deposits–sand, silt, gravel, clay, and a little black volcanic ash—laid down during the Pleistocene ice ages.

DMAD Reservoir

D:U

125

D

Delta

Canyon Mountains

Sevier Desert Detachment Fault

6

50

Paleozoic

Pennsylvanian-Permian Oquirrh Gp. limestone and shale

Mississippian limestone, shale

Devonian

Silurian

Ordovician limestone

middle Cambrian limestone, sandstone

Cambrian Tintic Quartzite

Quaternary

alluvium, mostly alluvial fans

wind deposits

Lake Bonneville deposits

0   10 miles

10 km

Tertiary

sediment

Green River Fm.

Cretaceous-Tertiary

conglomerate

Cretaceous

conglomerate

conglomerate

Precambrian

metasedimentary rocks

Igneous rocks

Tertiary Moroni Tuff and related rocks

Tertiary intrusive rocks

Geology along US 6 between Santaquin and Delta.

leaving white salt in flats near Goshen. Since its separation from its big sister to the north, Utah Lake has been fresh; an outlet via the Jordan River, flowing into the Great Salt Lake, prevents salt from building up. Most of the lake's water comes from the Provo River, with its headwaters in the Uinta Mountains.

West of Elberta US 6 climbs steadily across a broad alluvial fan dotted with boulders eroded from the East Tintic Mountains. After milepost 146, look for basalt capping a ridge ahead; below it are several layers of welded rhyolite tuff, part of the Tertiary Tintic Mountain Volcanic Complex. Immense volcanoes stood in this region 31 to 30 million years ago. Most of their eruptions were violent, piling up layer on layer of breccia and ashflow tuff with only occasional lava flows.

Rhyolite tuff near milepost 144 is colored by iron oxide from weathered sulfide minerals. The Tintic Mining District has had some of the richest and most productive mines in the country: by 1976, 18.4 million tons (16.7 million metric tons) of ore had been mined to produce 2.8 million ounces (79,400 kg) of gold, 275 million ounces (7.8 million kg) of silver, 125,000 tons (113,000 metric tons) of copper, 1,160,000 tons (1,052,000 metric tons) of lead, and 196 tons (178 metric tons) of zinc. Like many mineral deposits, those in the Tintic area are zoned: copper and gold are found with coarse quartz crystals that formed

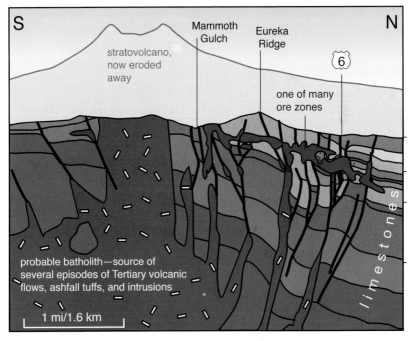

Cross section through the Tintic Mining District in the East Tintic Mountains near Eureka. Hot, acidic metal-rich solutions flowed along the faults and along bedding planes, interacting with and replacing surrounding limestones to form deposits of gold, silver, copper, lead, and zinc. Deposits are called veins when they are in fault zones and replacements when they replace bedrock.

Eurekadumpite, containing copper, zinc, tellurium, arsenic, and chlorine, was approved as an official new mineral in 2009. Individual spheres are around 0.02 inch (0.5 mm) across. —Tom Loomis photo, Dakota Matrix

from the hottest fluids—those circulating closest to the magma bodies; zones with silver and lead associated with finer-grained quartz are from less-hot fluids; and zones with zinc and lead associated with very fine-grained quartz are from the coolest fluids. Mining had tapered down here by the mid 1960s, but interest in the deposits continues today.

This district is exceptional for its unusual minerals, quite a few of them first discovered here. Eighty-five different minerals have been identified from Eureka's Centennial Eureka Mine alone.

Limestones show up at the top of the canyon after milepost 143. On the right a large quarry provides gravel and rock for the district's EPA Superfund projects. Gray limestone gravel is used to fill in and cover areas contaminated by mine tailings in Eureka. During the early mining days, nobody thought of the poisonous lead and arsenic in tailings, so tailings were used as fill in towns. In order to clean up these toxic metals, the tailings were dug out from around the houses to a depth of 18 inches (46 cm). Fabric was put down to prevent dust from moving upward, and then the excavation was filled with limestone gravel. If the ground level was unimportant, fabric and fill were put on top of contaminated fill without excavating. The EPA funded most of the work, and mining companies helped finance the cleanup of specific mines.

At Tintic Junction, US 6 turns south down the broad valley of Tanner Creek. The valley is floored with upper Tertiary sedimentary rocks: sandstone, mudstone, siltstone, volcanic ash, and salt, in places thousands of feet (more than 1 km) thick.

US 6 continues southwest into the Sevier Desert, a broad, open basin that was part of Lake Bonneville in Pleistocene time. Much of it is surfaced with fine lake-bottom sediment. At the desert's northeast end is a field of dunes known as Little Sahara or Lynndyl Sand Dunes. Dunes develop where there is plenty of both sand and wind. Here, the sand comes from the Pleistocene delta of the Sevier River, which we will see farther south. Vegetation has stabilized many dunes, especially around the field's margins.

The Sand Hills, dunes anchored by bedrock, are the highest dunes in Little Sahara National Recreation Area. —Felicie Williams photo

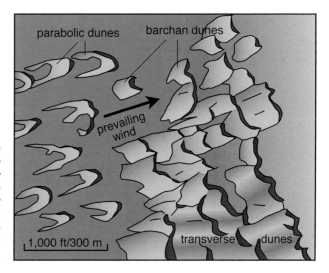

Low parabolic dunes trail long arms stabilized by vegetation, barchan dunes are crescent shaped, and transverse dunes are formed of coalescing barchans.

Much-faulted Devonian and Mississippian rocks make up most of the Gilson Mountains northeast of Lynndyl. The thrust faults date to the Sevier Orogeny of 145 to 65 million years ago.

At Lynndyl, US 6 meets up with the Sevier River, which flows into the desert through Leamington Canyon. During the Pleistocene ice ages, this river, laden with sediment from the glacier-capped mountains of central Utah, built a delta more than 20 miles (32 km) across out into Lake Bonneville. The delta extends to beyond the town of Delta.

A plant a few miles south of Lynndyl processes beryllium mined from an altered layer of tuff on Spor Mountain, 48 miles (77 km) to the northwest.

Beryllium is stiff, lightweight, nonmagnetic, and transparent to X-rays and has a very high melting point of 2,349°F (1,287°C).

The Canyon Mountains, east of Delta, are built of Precambrian and Cambrian sedimentary rock that was faulted and folded into a long syncline during the Sevier Orogeny. (See the cross section of the range in the I-15: Scipio to Cove Fort road guide.)

Along the west flank of the Canyon Mountains is the breakaway zone for the very extensive Sevier Desert Detachment Fault. It plunges westward and then gradually flattens until it slopes at about 10 degrees. It extends at least into Nevada. The different crustal blocks of the Basin and Range country west of here have slid westward along this fault as the continent stretches. And this surface may have been recycled: it may be a ramp up which thrust sheets traveled eastward during the Sevier Orogeny.

The town of Delta is a true oasis in the desert: water from the river and water pumped from its Pleistocene delta are used to farm the delta's level, fertile surface.

US 6/US 50
Delta—Nevada State Line
89 miles (143 km)

From Delta, US 6/US 50 runs southwest across the Sevier Desert on Lake Bonneville deposits: sandy delta sediments near town and then finer lake-bottom silt. The Sevier River passes beneath the highway just after town then twists across the nearly level desert floor. The river flows here, but farther out in the desert its channel is usually dry. Some of its water is diverted for irrigation. Remaining water sinks into the valley sediment or evaporates in the dry climate.

Between Delta and Nevada, US 6/US 50 crosses multiple ranges of the Basin and Range. Starting about 17 million years ago, tension across the West caused near-surface crust, which had previously been pushed into ranges of mountains, to slide westward down the long, gently dipping Sevier Desert Detachment Fault, named for its presence below the Sevier Desert. As the crust slid, it broke along normal faults, and the resulting pieces (individual fault blocks) of crust tipped and folded with the motion to form a succession of valleys and mountain ranges. Movement on the fault has ranged between 18 and 36 miles (30 and 60 km).

The faulting has caused there to be repeated outcrops of the thick Paleozoic layers. The fossil sequence here is famous worldwide for several reasons: the exposures are superb and easily accessed; this edge of the continent subsided steadily during Paleozoic time, allowing complete time sequences of sediment and fossils to accumulate; and rapid burial of the fossils and, at times, oxygen-depleted conditions, aided in their preservation.

During Cambrian time, the continent Laurentia (the precursor to North America) was close to the equator and rotated about 90 degrees clockwise of North America's present position. A sea covered much of the West. An embayment in southwest Utah provided habitat for numerous shallow-water species.

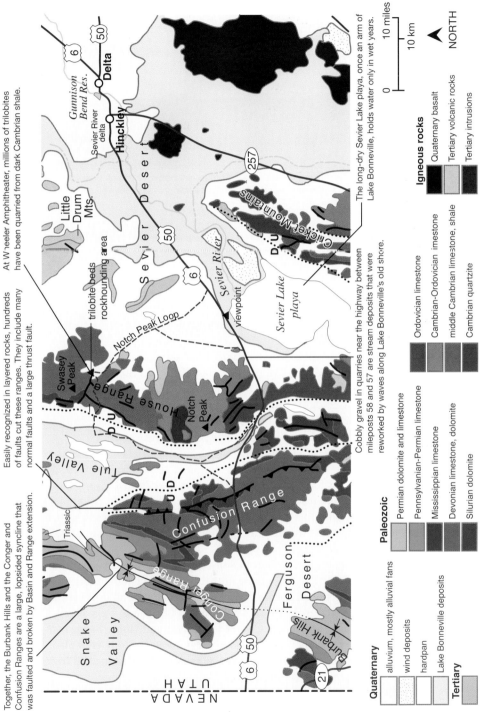

At Wheeler Amphitheater, millions of trilobites have been quarried from dark Cambrian shale.

Easily recognized in layered rocks, hundreds of faults cut these ranges. They include many normal faults and a large thrust fault.

Together, the Burbank Hills and the Conger and Confusion Ranges are a large, lopsided syncline that was faulted and broken by Basin and Range extension.

Cobbly gravel in quarries near the highway between mileposts 58 and 57 are stream deposits that were reworked by waves along Lake Bonneville's old shore.

The long-dry Sevier Lake playa, once an arm of Lake Bonneville, holds water only in wet years.

trilobite beds rockhounding area

Notch Peak Loop

Swasey Peak

Notch Peak

House Range

Tule Valley

Confusion Range

Conger Range

Snake Valley

Ferguson Desert

Burbank Hills

Triassic

Sevier Desert

Sevier River delta

Little Drum Mts.

Gunnison Bend Res.

Delta

Hinckley

Sevier River

viewpoint

Sevier Lake playa

Cricket Mountains

NEVADA
UTAH

50
6
257
50
6
50
6
21

Quaternary
- alluvium, mostly alluvial fans
- wind deposits
- hardpan
- Lake Bonneville deposits

Tertiary

Paleozoic
- Permian dolomite and limestone
- Pennsylvanian-Permian limestone
- Mississippian limestone
- Devonian limestone, dolomite
- Silurian dolomite
- Ordovician limestone
- Cambrian-Ordovician limestone
- middle Cambrian limestone, shale
- Cambrian quartzite

Igneous rocks
- Quaternary basalt
- Tertiary volcanic rocks
- Tertiary intrusions

0   10 miles

10 km

NORTH

Geology along US 6/US 50 between Delta and the Nevada state line.

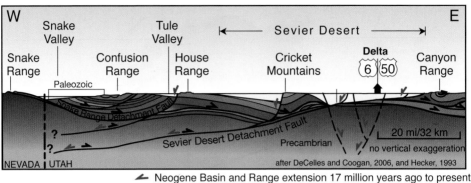

Neogene Basin and Range extension 17 million years ago to present
Cretaceous Sevier Orogeny compression 145–65 million years ago

This simplified cross section shows the Sevier Desert Detachment Fault as it crosses beneath the Basin and Range. Only a few of the steep normal faults that curve to meet the detachment are shown. Some faults extend below the detachment and have become conduits for relatively recent volcanism. To the west, we do not know what happens: are the Snake Range and Sevier Desert Detachment Faults somehow connected?

The Drum Mountains, northwest of Delta and north of the Little Drum Mountains, have an incredibly thick, complete, and unfaulted section of middle Cambrian limestone, for which the Drumian stage of Cambrian time is named. Its agnostid trilobites are cosmopolitan, meaning they have been found in many parts of the world. Geologists use them to correlate rocks of this age on most continents. Volcanic rocks cover much of the Little Drum and Drum Mountains. They erupted in Tertiary time from at least four large volcanoes.

The northern part of the House Range ahead, and the Cricket Mountains to the south, also have very thick, complete Cambrian sections. The House Range, in particular, is famous for its Cambrian fauna. You have probably seen specimens of *Elrathia kingii,* a trilobite so abundant in the Wheeler Formation there that millions of them have been quarried and sold commercially. Regrettably, it is likely that the quarrying inadvertently demolished less-obvious trace fossils of soft-bodied life.

These ranges contain extraordinary fossil-rich deposits that are significant to our knowledge of Earth's past because of their unusually good preservation, because they represent the great diversity of fauna of their time, and/or because they preserve the very unusual traces of soft-bodied organisms. Deposits such as these are called lagerstätten. The Cricket Mountains and House Range contain Cambrian-age soft-bodied animals. They are preserved in only a few places in the world, and the fauna includes groups of animals that disappeared from the fossil record after Cambrian time.

Hills and mountains here bear the shoreline benches carved by Lake Bonneville; this valley, too, was part of the great Pleistocene lake. Interestingly, the benches are no longer level because of isostatic rebound. When the lake was full, the weight of its 1,200 feet (370 m) of water depressed Earth's crust, making the crust "float" lower in the mantle. Since the lake's demise, the land has rebounded.

From milepost 60, Sevier Lake can be seen to the south. During the wet years of the early 1980s the Sevier River carried enough water into the glaring white playa to turn it into a shimmering lake for the first time in living memory.

Silurian and Devonian limestone and dolomite are exposed in the Confusion Range and small hills in the valley east of it. Mississippian through Permian limestone and dolomite at the range's northwest end and in the smaller Conger Range still farther west are also well exposed and contain many fossils.

We know very little about the soft-bodied animals that made up the majority of life-forms during Cambrian time. Fossils like the one on the right were at first thought to be a jellyfish-like animal. Later finds proved them to be the grabbing front limbs of the largest known Cambrian predator, Anomalocaris *(top).*
—Photo courtesy of Virtual Fossil Museum, www.fossilmuseum.net

1 foot/30 cm

Slightly dipping layers of upper Cambrian and Ordovician dolomite of the Notch Peak Formation form Notch Peak at the crest of the House Range. The pink rock at left is part of a Jurassic intrusion. —Lucy Chronic photo

Approaching the Confusion Range, US 6/US 50 first crosses the southern end of landlocked Tule Valley. Water entering Tule Valley sinks into the broad bajadas around its margin or flows down its washes into a small playa. The valley is a graben, having dropped down along faults on both of its sides. At about milepost 26, US 6/US 50 reaches the edge of the Confusion Range, crossing a thrust fault that brought Silurian rocks east over younger Devonian rocks during the Sevier Orogeny.

Once across the summit, broad alluvial fans, so typical of the Basin and Range country, surround the little Conger Range to the northwest. The highway descends into the Ferguson Desert, which drains northwest into the Snake Valley, which, during Pleistocene time, was yet another arm of Lake Bonneville. The lake reached about 10 miles (16 km) farther north into Nevada.

US 50
Delta—Salina
70 miles (113 km)

The first part of US 50 east of Delta crosses the nearly level surface of the Sevier River delta, which built into Lake Bonneville during the Pleistocene ice ages. Composed of Pleistocene-age through recent alluvium, the delta makes good soil. Water carried in ditches from reservoirs along the Sevier River is used for irrigation.

By the time the road turns southeast, it is on Lake Bonneville sediment, much of which is covered with sand that the wind has carried from the delta.

The Canyon Mountains, to the east, are part of a syncline; its rocks were pushed eastward and folded during the Sevier Orogeny. Precambrian and Cambrian rock form the center of the syncline. This older rock was thrust over Silurian, Ordovician, and Cambrian layers before all were folded into the syncline. Range after range of mountains once marched west of here, resembling the Canadian Rockies. They are now fragmented, having been stretched apart by Basin and Range faulting, which began in late Tertiary time and continues today.

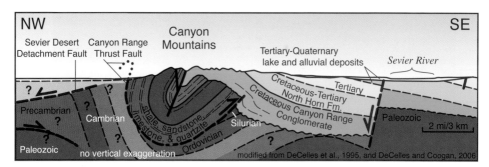

Cross section of the northern part of the Canyon Mountains showing how Cretaceous alluvial fans (Canyon Range Conglomerate and North Horn Formation) that built out along the east side of the Sevier Orogeny mountains were folded as the mountains continued to rise. Question marks are in areas where the geology is theoretical.

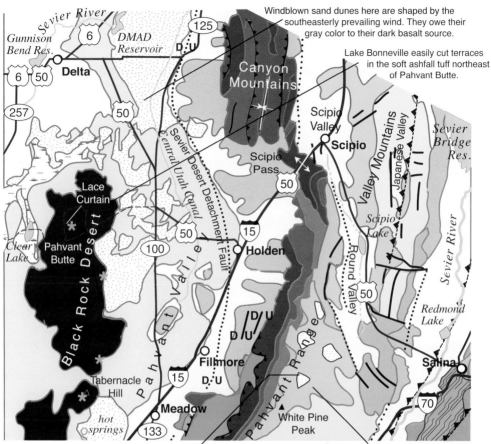

Windblown sand dunes here are shaped by the southeasterly prevailing wind. They owe their gray color to their dark basalt source.

Lake Bonneville easily cut terraces in the soft ashfall tuff northeast of Pahvant Butte.

The summits of the Pahvant Range reach a little over 10,000 feet (3,050 m) and were glaciated during the Pleistocene ice ages; the glaciers left moraines above 8,500 feet (2,600 m).

The eastern Pahvant Range is composed of Cretaceous and Tertiary sediment washed from the range and from the mountains of the Sevier Orogeny that were west of it.

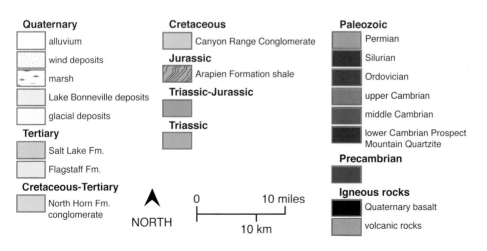

Quaternary
- alluvium
- wind deposits
- marsh
- Lake Bonneville deposits
- glacial deposits

Tertiary
- Salt Lake Fm.
- Flagstaff Fm.

Cretaceous-Tertiary
- North Horn Fm. conglomerate

Cretaceous
- Canyon Range Conglomerate

Jurassic
- Arapien Formation shale

Triassic-Jurassic

Triassic

Paleozoic
- Permian
- Silurian
- Ordovician
- upper Cambrian
- middle Cambrian
- lower Cambrian Prospect Mountain Quartzite

Precambrian

Igneous rocks
- Quaternary basalt
- volcanic rocks

NORTH

0 — 10 miles

10 km

Geology along US 50 between Delta and Salina.

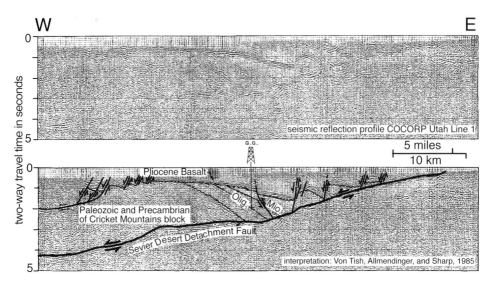

W   E

two-way travel time in seconds

0

5

seismic reflection profile COCORP Utah Line 1

5 miles

10 km

G.G.

0

Pliocene Basalt

Olig.

Mio.

Paleozoic and Precambrian of Cricket Mountains block

Sevier Desert Detachment Fault

interpretation: Von Tish, Allmendinger, and Sharp, 1985

5

Revealed in drill holes and in seismic reflection profiles like this one, the Sevier Desert Detachment Fault is probably active. It is hard to tell, since there have been no earthquakes that can be tied to it. Instead, fault blocks may move along the fault in what geologists call aseismic creep—movement so gradual it doesn't register on seismometers.

The Lace Curtain on the north side of Pahvant Butte was revealed when Lake Bonneville's waves cut into volcanic ash layers. Mineralized groundwater had previously circulated in erratic paths through the still-warm cinder cone, cementing part of the ash in a lacelike pattern.
—Felicie Williams photo

As the highway follows the Pahvant Valley, it parallels a shallow west-dipping fault that is thought to be the breakaway zone of the Sevier Desert Detachment Fault, which underlies the Sevier and Black Rock Deserts and may extend westward into Nevada. The detachment is believed to be the surface along which at least some of the ranges here have gradually moved westward due to tectonic extension. Probably active since late Oligocene time, its hanging wall has slipped as much as 29 miles (47 km) down to the west, greatly extending and thinning the crust in the Basin and Range.

Pahvant Butte appears west of the Pahvant Valley, near where Utah 100 turns off US 50. It formed above a steep Basin and Range fault that tapped Earth's upper mantle. The butte is a crater: its basalt lava erupted into the water of Lake Bonneville 15,500 years ago, causing huge steam explosions. Between Holden and Scipio US 50 follows I-15, climbing over Scipio Pass, a gentle downwarp between the Canyon Mountains and the Pahvant Range. There are Ordovician rocks at the pass: sandstone, shale, and conglomerate with thinly bedded fossiliferous limestone in the upper layers.

Turning southeast at Scipio, US 50 follows Round Valley to Salina. The valley has dropped down along Basin and Range faults. Cretaceous and younger sedimentary rocks form the low Valley Mountains to the northeast; their sediment washed down from the Sevier Orogeny mountains. Japanese Valley is a graben that developed in the range.

Salina lies in another basin that dropped down between ranges. Arapien Formation shale makes badlands of white and tan hills. The salt- and gypsum-rich Arapien was deposited in a long arm of the sea that stretched from the open ocean to the north during Jurassic time. Soft and pliable, the Arapien layers were easily folded during the Sevier Orogeny. Now, in places, the evaporite minerals have flowed (slowly, of course) up to the surface.

Utah 18
St. George—Beryl Junction
50 miles (80 km)

Leaving St. George, Utah 18 leads first to starkly beautiful examples of reversed topography: The highway ascends onto and follows a basalt flow that, when it erupted between 2 and 1 million years ago, flowed down a stream channel from north to south. The channel walls kept the lava confined. The basalt was stronger than the surrounding sedimentary beds, so they eroded away and the basalt remains—a valley floor that became a ridge.

At milepost 4, at the top of the rise, the Beaver Dam Mountains are visible to the west. The oldest rocks in the range were part of an island chain that collided with the early North American continent 1.78 billion years ago. The collision buried them 10 to 15 miles (16 to 24 km) deep, causing intense metamorphism of their sedimentary and igneous rocks.

Soon after passing the last of St. George's houses, Utah 18 enters Snow Canyon State Park, named for the white Navajo Sandstone at its north end. Here, the Navajo is close to its maximum thickness of 2,500 feet (760 m). Deposited

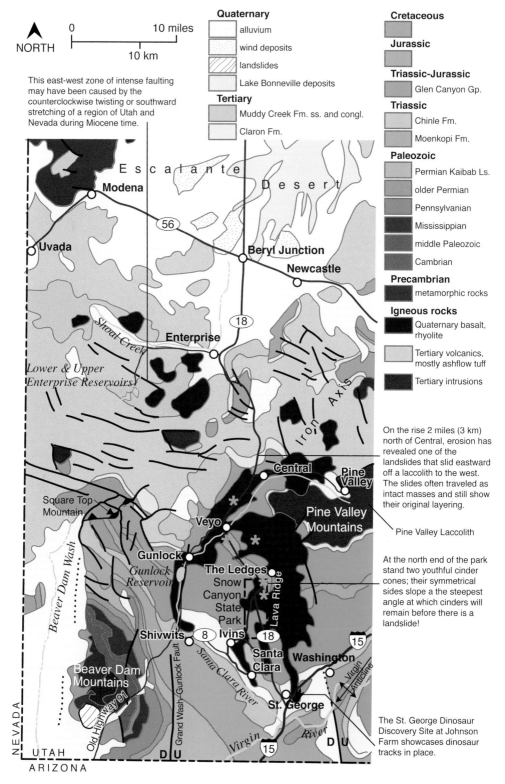

Quaternary
- alluvium
- wind deposits
- landslides
- Lake Bonneville deposits

Tertiary
- Muddy Creek Fm. ss. and congl.
- Claron Fm.

Cretaceous

Jurassic

Triassic-Jurassic
- Glen Canyon Gp.

Triassic
- Chinle Fm.
- Moenkopi Fm.

Paleozoic
- Permian Kaibab Ls.
- older Permian
- Pennsylvanian
- Mississippian
- middle Paleozoic
- Cambrian

Precambrian
- metamorphic rocks

Igneous rocks
- Quaternary basalt, rhyolite
- Tertiary volcanics, mostly ashflow tuff
- Tertiary intrusions

NORTH

0 — 10 miles

10 km

This east-west zone of intense faulting may have been caused by the counterclockwise twisting or southward stretching of a region of Utah and Nevada during Miocene time.

Escalante Desert

Modena

Uvada

56

Beryl Junction

Newcastle

Shoal Creek

18

Enterprise

Lower & Upper Enterprise Reservoirs

Iron Axis

On the rise 2 miles (3 km) north of Central, erosion has revealed one of the landslides that slid eastward off a laccolith to the west. The slides often traveled as intact masses and still show their original layering.

Central

Pine Valley

Square Top Mountain

Veyo

Pine Valley Mountains

Pine Valley Laccolith

Gunlock

Gunlock Reservoir

The Ledges

Snow Canyon State Park

Lava Ridge

At the north end of the park stand two youthful cinder cones; their symmetrical sides slope a the steepest angle at which cinders will remain before there is a landslide!

Beaver Dam Wash

Shivwits

8

Ivins

18

Santa Clara

Washington

15

Beaver Dam Mountains

Grand Wash–Gunlock Fault

Old Highway 91

Santa Clara River

St. George

Virgin Anticline

D U

NEVADA

UTAH

ARIZONA

D U

Virgin River

15

D U

The St. George Dinosaur Discovery Site at Johnson Farm showcases dinosaur tracks in place.

Geology along Utah 18 between St. George and Beryl Junction.

Hemmed in by lava flows, St. George is surrounded by a rugged landscape. Not so long ago a basalt flow spilled down Navajo Sandstone cliffs (middle foreground) and spread across the valley floor. The Pine Valley Mountains, light-colored summits composed of a very large igneous intrusion, rise in the distance. —Lucy Chronic photo

A side trip down into Snow Canyon shows magnificent views of the basalt that flowed down Snow Canyon just 27,500 to 27,000 years ago. The Santa Clara River had cut its canyon here, where strong jointing had weakened the sandstone. The lava filled the broad valley of the river, which now flows along the side of the flow. —Lucy Chronic photo

The cinder cones erupted at Veyo 690,000 years ago; they are so young that their black rock shows little evidence of weathering or erosion. —Lucy Chronic photo

in a great field of Jurassic dunes, the rock was originally colored by red hematite. During either the Laramide Orogeny or the Tertiary volcanic episode, groundwater moved through the sandstone, dissolving and carrying off the hematite. The hematite was redeposited as fillings in fractures, as irregular bodies, and as spherical concretions called Moqui marbles.

Utah 18 continues to ride on basalt until past Veyo, where cinder cones tower above local farms. The light-colored Pine Valley intrusion, to the east on the flanks of the Pine Valley Mountains, and other intrusions nearby are much older than the basalt; they were emplaced between 24 and 20 million years ago. These intrusions make up a grouping of laccoliths—called the Iron Axis—that bowed up mountains along a northeasterly trend. They are so named because hydrothermal water from the laccoliths deposited iron within the laccoliths and surrounding rocks.

The magma of the Pine Valley intrusion spread out into the relatively weak Claron Formation and caused the overlying layers to dome upward. When the sides of the dome became too steep, huge chunks of the Claron slid off. Uncovered, the magma erupted violently. Erosion has since revealed both the laccolith and landslides. From close to Central, you can see the laccolith's great light-colored mass, with its almost horizontal base.

On Square Top Mountain, to the west, Paleozoic strata were thrust over Mesozoic layers during the Sevier Orogeny. The Paleozoic strata are part of the easternmost large thrust sheet of the Sevier Thrust Belt in southern Utah.

The basalt flows and the intrusions belong to two different waves of igneous activity that swept across Utah. The first wave had magma of a composition intermediate between basalt and rhyolite and typically was fairly thick. When it reached the surface, gases in it often made it very explosive. The resulting volcanic rock often has phenocrysts, large crystals that grew while the magma was still in a chamber underground.

The second wave of volcanism accompanied Basin and Range faulting. Very hot, fluid, dark-colored basaltic magma rose up from the mantle along deep normal faults, melting some of the crust on the way up. The basalt and molten crust did not tend to mix; the eruptions of this phase are called bimodal because they brought both basalt and rhyolite to the surface, two very different types of magma.

With few exceptions, the hills north of Central are volcanic or intrusive rocks of the older wave—tuffs (welded ash) and breccias (ash with fragments of volcanic rock). The few younger volcanic rocks are easy to pick out: conical cinder cones and dark basalt flows.

About 6 miles (10 km) north of Central, Utah 18 briefly crosses red and tan beds of the Claron Formation. These soil and limy lake sediments were laid down on a wide plain between mountain ranges. After Enterprise the highway rides down a broad alluvial fan to the Escalante Desert. Irrigation water here comes from the nearby mountains or is pumped from below the valley floor.

Utah 21
Nevada State Line—Beaver
107 miles (172 km)

At the start of Utah 21, alluvial fans flank the ranges. A wide slope of merged fans is called an alluvial apron, or bajada. In places, these slopes are truncated at their lower ends by stream channels.

From Garrison, Utah 21 diagonals southeastward along a valley that crosses a northeast-southwest-trending syncline. The syncline is one of many folds that formed during the Sevier Orogeny, between 145 and 65 million years ago. The

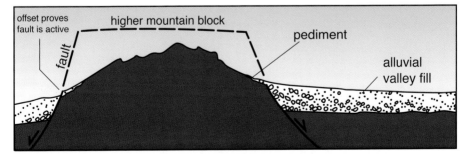

Pediments are erosional surfaces cut into the bedrock of mountains but usually covered with gravel. They are often continuous with valley fill and veneered with alluvial fans—it may be difficult to know where pediment ends and valley fill begins.

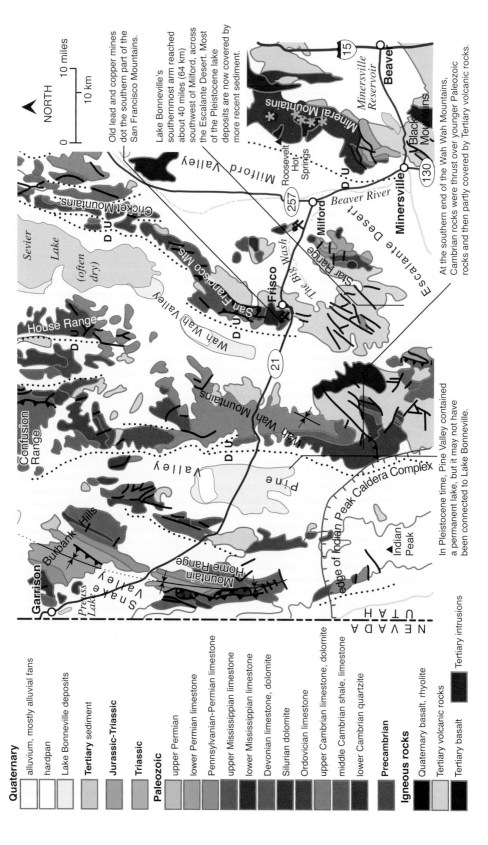

Geology along Utah 21 between the Nevada state line and Beaver.

Quaternary
- alluvium, mostly alluvial fans
- hardpan
- Lake Bonneville deposits

Tertiary sediment

Jurassic-Triassic

Triassic

Paleozoic
- upper Permian
- lower Permian limestone
- Pennsylvanian-Permian limestone
- upper Mississippian limestone
- lower Mississippian limestone
- Devonian limestone, dolomite
- Silurian dolomite
- Ordovician limestone
- upper Cambrian limestone, dolomite
- middle Cambrian shale, limestone
- lower Cambrian quartzite

Precambrian

Igneous rocks
- Quaternary basalt, rhyolite
- Tertiary volcanic rocks
- Tertiary basalt
- Tertiary intrusions

NORTH

0   10 miles

10 km

Old lead and copper mines dot the southern part of the San Francisco Mountains.

Lake Bonneville's southernmost arm reached about 40 miles (64 km) southwest of Milford, across the Escalante Desert. Most of the Pleistocene lake deposits are now covered by more recent sediment.

At the southern end of the Wah Wah Mountains, Cambrian rocks were thrust over younger Paleozoic rocks and then partly covered by Tertiary volcanic rocks.

In Pleistocene time, Pine Valley contained a permanent lake, but it may not have been connected to Lake Bonneville.

Milford Valley

Cricket Mountains

Sevier Lake (often dry)

House Range

Confusion Range

Wah Wah Valley

San Francisco Mtns

Frisco

The Big Wash

Star Range

Milford

Beaver River

257

21

Minersville Desert

Minersville

130

Escalante Desert

Roosevelt Hot Springs

Minersville Reservoir

Mineral Mountains

Black Mountains

Beaver

15

Pine Valley

Wah Wah Mountains

Indian Peak

edge of Indian Peak Caldera Complex

Mountain Home Range

Burbank Hills

Snake Valley

Pre-Lake Lake

Garrison

NEVADA   UTAH

folds can often be traced between mountain ranges despite later offset by Basin and Range normal faulting. This particular fold includes the Burbank Hills north of Utah 21 and the Mountain Home Range south of the highway. Parts of the folds now lie deep beneath the sediment-filled basins that developed as a result of Basin and Range extension.

Utah 21 passes oldest layers to youngest as it heads into the syncline: Devonian strata appear near Garrison, Mississippian near Pruess Lake, and then a wide band of light-gray Pennsylvanian and Permian limestone and shale in the Burbank Hills—quite close to the highway near mileposts 16 and 17, where the road bends slightly east. Southeast of the pass, between the two ranges, rocks become older again as the road crosses the other limb of the syncline, though near the highway they lie beneath bajada gravels.

Practically all the Paleozoic rocks here are marine limestone; they are very thick because when they were deposited, this area west of the Utah Hingeline was subsiding steadily. The limestone is often made of an infinite number of tiny animal shells that were crushed and recrystallized and for the most part are unrecognizable. Larger, thicker shells from invertebrates such as brachiopods and crinoids have mostly retained their identities.

Ordovician limestone appears near milepost 26; it contains fossil trilobites, cephalopods, and reef-forming corals and stromatoporoids (a variety of sponge with a calcareous skeleton).

To the south and east, a great deal of the exposed rock for as far as you can see is volcanic. It is part of a vast Tertiary outpouring that covered most of

The Burbank Hills are symmetrical: rock layers dip toward the center in a great syncline that extends from Utah 21 south for 30 miles (48 km) and north for more than 50 miles (80 km). —Lucy Chronic photo

southwestern Utah and extended well into Nevada between 33 and 19 million years ago. It is one of the most voluminous volcanic fields known on Earth. The explosive eruptions produced pale-colored tuffs and breccias, which often contain phenocrysts, crystals that formed in the magma before it erupted. The volcanoes did not fare well through the Basin and Range faulting that began around 17 million years ago, nor through the erosion that continues today. Their forms have mostly vanished from the landscape.

Utah 21 crosses the wide Pine Valley; the playa of this closed basin contains water only in the wettest of years. The road then rises to meet the Wah Wah Mountains. Here, we get a good look at Cambrian rocks. The oldest layer—thick quartzite—comes first: it was nearly pure quartz sand deposited along the beaches and bars of an eastward-creeping shoreline. Above it, forming most of the western face of the range, lie shale and limestone deposited in shallow water close to the shore. Near and east of the saddle, through which the high-way passes, are still younger rocks, mostly limestone, which was deposited in water far from sources of shoreline sediment. Together, these layers record the gradual submergence of the continent's western edge.

These rocks are filled with evidence of rich marine life, especially of trilo-bites. Near the equator, in a quiet ocean basin, trilobites thrived. In places, their remains make up thick coquinas—layers composed almost entirely of fossils.

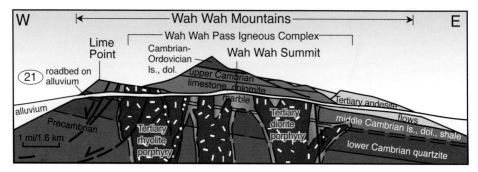

Except for some Tertiary volcanic rocks on their eastern side and intrusions near the summit, the northern Wah Wah Mountains are almost entirely composed of east-tilting Cambrian strata. Hot mineral-rich water from the intrusions interacted with lime-stone to form marble and other minerals.

East of milepost 45, Utah 21 descends rapidly into the Wah Wah Valley, which is continuous northward with the much larger Sevier Desert. During Pleistocene time, these valleys and Pine Valley were part of Lake Bonneville.

In the San Francisco Mountains, east of the valley, Precambrian quartzite and argillite (metamorphosed shale) have been thrust over Cambrian and Ordovician rocks, which were later cut by a Tertiary intrusion. The old mining camp of Frisco is close to the highway. Lead, copper, uranium, beryllium, and fluorite occur here, where hot mineral-rich water escaping the magma reacted with Paleozoic lime-stone and dolomite—a common origin for mineral deposits in Utah.

Very unusual red beryl (emerald) grew in cavities in a 20-to-18-million-year-old rhyolite flow on the east flank of the Wah Wah Mountains. It is mined on privately owned claims. This is the only location in the world where red beryl crystals are large enough and of high enough quality to be cut into gems. Field of view about 1 inch (2.5 cm) across. —Kevin Ward photo, www.themineralgallery.com.

East of the San Francisco Mountains, Utah 21 descends toward Milford in a basin unlike those farther west; it has no playa. Instead, in wet years it drains north around the Cricket Mountains into Sevier Lake.

Geothermal power is produced at Roosevelt Hot Springs, northeast of Milford. Wells between 2,100 and 6,000 feet (640 and 1,830 m) deep tap water that is more than 520°F (271°C). Steam and hot water are run through separate power plants. The spent water is returned to the underground reservoir.

Small ranges north and south of the highway near Milford expose bent and broken Paleozoic rocks and several small Tertiary intrusions. Again, minerals crystallized where the intrusions met limestone.

Milford is an agricultural center. Water for its fields and the hog farms west of Minersville comes from wells and from Minersville Reservoir. As in many areas of the West, the water table is dropping because water is pumped from the aquifers faster than it is replaced, a practice that geologists call "mining the water."

The Black Mountains, south of Minersville, are remnants of Tertiary volcanoes built of sloping layers of lava, tuff, and breccia and are part of the huge Marysvale Volcanic Field that stretches from here to the Tushar Mountains, east of the town of Beaver. Tertiary granite makes up the central mass of the Mineral Mountains north of Minersville; it is the largest exposed area of granite in Utah. A line of little rhyolite volcanoes perches along the range's crest; they erupted between 800,000 and 500,000 years ago.

Disturbed Paleozoic and Mesozoic layers at the southern end of the Mineral Mountains are in the hanging, or upper, wall of the Cave Canyon Fault. They moved west some 30 miles (48 km) as the crust was stretched over the last 17 million years. This fault may be part of the seldom-seen Sevier Desert Detachment Fault, which extends from here west to Nevada below the surface.

Near Beaver, Utah 21 crosses a terrace of Pleistocene alluvium and then several shoreline ridges formed not by Lake Bonneville but by a smaller lake that occupied this basin in Pliocene time.

Utah 30
Snowville—Nevada State Line
93 miles (150 km)

Utah 30 heads west out of Snowville across the Curlew Valley. Small groupings of dark-colored late Tertiary lava flows and cinder cones rise from a floor of Pleistocene-age Lake Bonneville sediment. Many of the mountain ranges Utah 30 passes would have been islands at the highest lake level 18,000 years ago; terraces showing the different shoreline levels look like bathtub rings on their slopes.

North of the valley, an open-pit gold mine is visible on the southern end of the Black Pine Mountains. The gold was deposited by hot water in veins and is scattered through Paleozoic shale. Mining has continued off and on since the 1920s, with most of the production during the early 1950s.

Due west and ahead are the Raft River Mountains; we are looking down the axis of an anticline. These mountains, the Grouse Creek Mountains, and Idaho's Albion Mountains are together a large and well-studied metamorphic core complex.

During the Sevier Orogeny, the rocks of the Raft River and Grouse Creek Mountains were buried deeply, compressed and stretched, and heated and metamorphosed by the movement of thrust sheets above them. During Late Cretaceous time, the overlying unmetamorphosed sheets slid eastward, away from this area, and the metamorphic rocks rose up into a high dome seeking isostatic equilibrium.

The Elba Quartzite is the prominent light-colored rock that, along with darker schist that underlies it, surfaces the slopes of the Raft River Mountains. The overlying Paleozoic beds that slid off the range above the Raft River Detachment Fault make the low hills here and there along the base of the mountains. The Elba Quartzite was deposited as sand on the slowly subsiding western edge

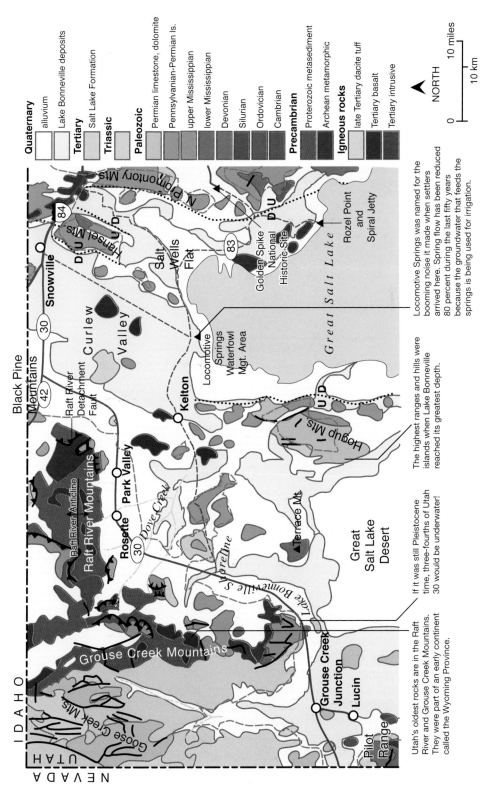

Quaternary
- alluvium
- Lake Bonneville deposits

Tertiary
- Salt Lake Formation

Triassic

Paleozoic
- Permian limestone, dolomite
- Pennsylvanian-Permian ls.
- upper Mississippian
- lower Mississippian
- Devonian
- Silurian
- Ordovician
- Cambrian

Precambrian
- Proterozoic metasediment
- Archean metamorphic

Igneous rocks
- late Tertiary dacite tuff
- Tertiary basalt
- Tertiary intrusive

NORTH

0   10 km
0   10 miles

Locomotive Springs was named for the booming noise it made when settlers arrived here. Spring flow has been reduced 80 percent during the last fifty years because the groundwater that feeds the springs is being used for irrigation.

The highest ranges and hills were islands when Lake Bonneville reached its greatest depth.

If it was still Pleistocene time, three-fourths of Utah 30 would be underwater!

Utah's oldest rocks are in the Raft River and Grouse Creek Mountains. They were part of an early continent called the Wyoming Province.

Geology along Utah 30 between Snowville and the Nevada state line.

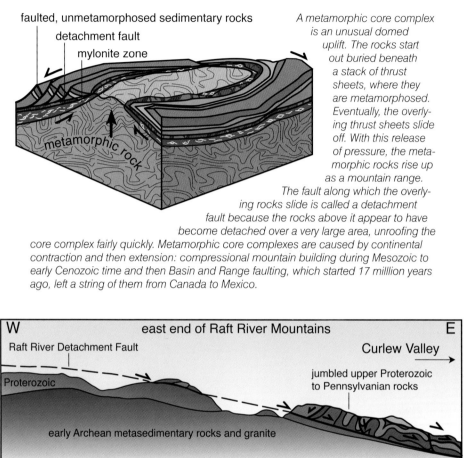

faulted, unmetamorphosed sedimentary rocks

detachment fault

mylonite zone

metamorphic rock

A metamorphic core complex is an unusual domed uplift. The rocks start out buried beneath a stack of thrust sheets, where they are metamorphosed. Eventually, the overlying thrust sheets slide off. With this release of pressure, the metamorphic rocks rise up as a mountain range. The fault along which the overlying rocks slide is called a detachment fault because the rocks above it appear to have become detached over a very large area, unroofing the core complex fairly quickly. Metamorphic core complexes are caused by continental contraction and then extension: compressional mountain building during Mesozoic to early Cenozoic time and then Basin and Range faulting, which started 17 milllion years ago, left a string of them from Canada to Mexico.

W   east end of Raft River Mountains   E

Raft River Detachment Fault

Curlew Valley

Proterozoic

jumbled upper Proterozoic to Pennsylvanian rocks

early Archean metasedimentary rocks and granite

modified from Wells, 2001

The eastern end of the Raft River Mountains is a jumble of sedimentary rock that slid off the range's rising metamorphic core. The fault surface is a layer of mylonite, a carapace of fine-grained minerals welded together by friction.

of the young North American continent, which remained tectonically quiet from late Proterozoic time through most of Paleozoic time.

After Rosette, Utah 30 turns south to skirt the Grouse Creek Mountains. The low mountains to the east are mostly Mississippian through Permian sandstone and limestone deposited in the Oquirrh Basin. The basin was shallow, with abundant marine life. About 2 to 3 vertical miles (3 to 5 km) of sediment accumulated as tectonic pressure from the west warped the basin downward. At the time, the ancient Antler Range was rising in what became Nevada, and some sediment from it reached the basin. Sedimentation kept up with subsidence for the most part, though occasionally the basin dropped more rapidly and ocean circulation slowed, leading to oxygen-poor environments, which preserved fossils very well. The sequence of Mississippian through Permian rocks in this part

The hard, resistant Proterozoic Elba Quartzite, shown here in the Dove Creek Quarries yard near Rosette, is pale green with a mica sheen. Heat and pressure changed the clay between its quartz sand grains into shiny layers of chlorite, a form of mica colored by iron and manganese. The rock splits easily along its layering and is a desirable building stone shipped all over the country. The quartzite forms the slopes of the Raft River Mountains, visible in the distance.
—Lucy Chronic photos

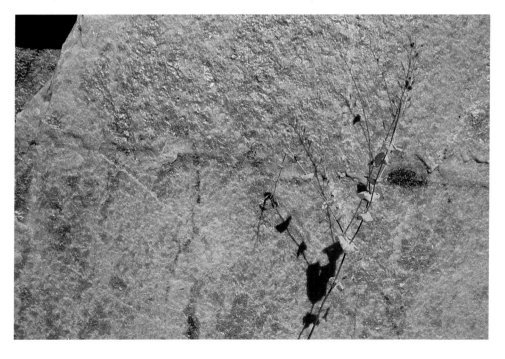

Pliocene volcanic rocks, mostly basalt, mark the landscape along Utah 30 in places—part of a wave of volcanism that crossed Utah from northwest to southeast. The unmistakable terraces are shorelines etched by Lake Bonneville into their slopes. —Lucy Chronic photo

of Utah is among the most time-continuous sections found anywhere in the world.

These Oquirrh rocks were deformed by thrust faulting during the Sevier Orogeny, though they were not metamorphosed. Around 17 million years ago, Basin and Range faulting broke and stretched the land into north-south-trending mountain ranges separated by basins. The basins have filled with sediment, and here, leveled by Lake Bonneville, they form huge, flat, often salty deserts.

The last leg of Utah 30 passes between two mountain ranges. The Pilot Range, to the south, is composed mostly of Precambrian and Paleozoic rocks that were folded into a long, north-south-trending syncline during the Sevier Orogeny. About 26 million years ago a large body of granite intruded these rocks. To the north, the Goose Creek Mountains are mostly Permian rocks, but their southern end, near the highway, is blanketed by dacite tuff that is only 8.5 million years old.

Utah 36
Tintic Junction—I-80
66 miles (106 km)

From Tintic Junction, Utah 36 climbs up and over a small divide, riding continuously on the light-colored Salt Lake Formation. The Salt Lake Formation is ash-rich sedimentary rock deposited rapidly during Miocene and Pliocene time; it is younger than local volcanic rocks.

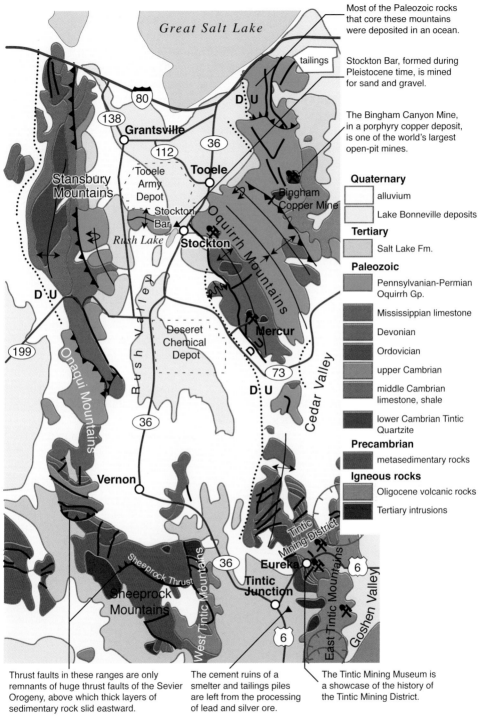

Most of the Paleozoic rocks that core these mountains were deposited in an ocean.

tailings

Stockton Bar, formed during Pleistocene time, is mined for sand and gravel.

The Bingham Canyon Mine, in a porphyry copper deposit, is one of the world's largest open-pit mines.

Great Salt Lake

80

138

Grantsville

112

36

Tooele Army Depot

Tooele

Stansbury Mountains

Stockton Bar

Bingham Copper Mine

Rush Lake

Stockton

Oquirrh Mountains

D U

199

Onaqui Mountains

Rush Valley

Deseret Chemical Depot

Mercur

D U

73

Cedar Valley

36

Vernon

Sheeprock Thrust

Sheeprock Mountains

West Tintic Mountains

36

Tintic Mining District

Eureka

6

Tintic Junction

East Tintic Mountains

Goshen Valley

6

Quaternary

alluvium

Lake Bonneville deposits

Tertiary

Salt Lake Fm.

Paleozoic

Pennsylvanian-Permian Oquirrh Gp.

Mississippian limestone

Devonian

Ordovician

upper Cambrian

middle Cambrian limestone, shale

lower Cambrian Tintic Quartzite

Precambrian

metasedimentary rocks

Igneous rocks

Oligocene volcanic rocks

Tertiary intrusions

Thrust faults in these ranges are only remnants of huge thrust faults of the Sevier Orogeny, above which thick layers of sedimentary rock slid eastward.

The cement ruins of a smelter and tailings piles are left from the processing of lead and silver ore.

The Tintic Mining Museum is a showcase of the history of the Tintic Mining District.

0   10 miles

10 km

NORTH

Geology along Utah 36 between Tintic Junction and I-80.

The road runs between the East and West Tintic Mountains. In both ranges, Paleozoic rocks were thrust east as part of the Sevier Thrust Belt. Thrust faults surface in a few places but are mostly covered by Tertiary volcanic rock. Extensional faulting after the volcanic episode dropped neighboring basins down and fractured the Paleozoic rocks.

At least four volcanoes developed above the Paleozoic rocks in or near the East Tintic Mountains between 33 and 31 million years ago, during Oligicene time. Their tuff, breccia, and ash no doubt covered a far greater area than they do today. In the roots of the volcanoes, hot, acidic metal-rich water interacted with limestone and dolomite to deposit lead, silver, gold, and zinc minerals. Many of them are unusual, and some are unique to the Tintic Mining District.

As we drop onto the flats, notice the stunted vegetation. The Quaternary alluvium here contains salt and other elements left by evaporating water and groundwater, neither of which are conducive to the growth of plants. Deep gullies in Quaternary alluvium edge the road after milepost 8.

The road rises briefly over low hills of Devonian and Mississippian rocks (mostly limestone), an extension of the Sheeprock Mountains to the south, and then drops into Vernon. The Sheeprock Thrust Fault, well exposed high in these mountains and in the West Tintic Mountains, was one of the earliest thrust faults to develop in this part of the Sevier Thrust Belt. Here, Cambrian rocks have been thrust over Precambrian rocks.

Near milepost 38, Utah 36 passes the first of two Tooele Army Depots, which were used to store and then later incinerate US chemical weapons. The second depot is to the north, near Tooele.

Copper, silver, gold, and mercury have been mined from many places in the Oquirrh Mountains, which are east of the highway. The minerals always occur

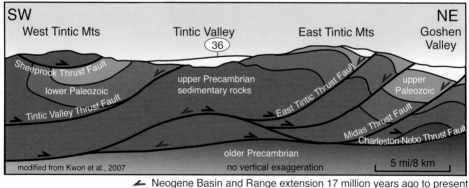

Neogene Basin and Range extension 17 million years ago to present
Cretaceous Sevier Orogeny compression 145–65 million years ago

Great pressure from the west forced immense sheets of rock eastward during the Sevier Orogeny. Once movement along one thrust fault stalled, another fault would form beneath it. The highest thrust sheet here—above the Sheetrock Thrust Fault—slid an incredible 100 miles (160 km) east. More thrust faults below it carried it another 35 miles (56 km). The thrust faults later provided an avenue for movement during Basin and Range extension, when rocks moved back to the west. For simplicity, igneous rocks are not included here.

Historic drawing of the Stockton Bar. Its layers tell a wonderfully accurate story of environmental conditions, including wave energy, lake salinity, and current directions during Utah's last ice age. Drawn by G. K. Gilbert in 1890, in his monograph on Lake Bonneville.

Lake Bonneville's highest shoreline terrace, the distinct line about one-third of the way up the mountain, is nearly 1,000 feet (300 m) above the level of these houses. When the lake cut the terrace 15,500 years ago, it covered more than 20,000 square miles (52,000 square km). It was about the size Lake Michigan is now. —Lucy Chronic photo

where Tertiary intrusions and their hot acidic water met Paleozoic limestone. Mines in the Mercur area produced mercury, which forms a brilliant red sulfide mineral called cinnabar. The mines contained an unusual amount of mercury and extremely fine gold, which could not be recovered economically until cyanide processing was invented in the early 1890s. The town boomed, and then burned. Mining has not been steady since.

To the west, faulted and folded rocks—from Cambrian on the far side through Pennsylvanian on this side—form the Stansbury Mountains. This uplift formed during Devonian time, before the Sevier Orogeny affected the region. Its development may be related to mountain building that occurred in Nevada.

Around milepost 49 the road climbs onto Stockton Bar, a huge bar deposited in this narrow area by the shoreline currents of Lake Bonneville. The bar built gradually over several thousand years. Much of the drive from north of Vernon to I-80 is on fine sediment deposited in Lake Bonneville.

Utah 56
Uvada—Cedar City
61 miles (98 km)

The Indian Peak Caldera Complex north of Uvada includes the eroded remains of many calderas. Their circular faults lie where magma chambers collapsed after feeding volcanic eruptions. Between 33 and 19 million years ago, the eruptions spread ash and volcanic debris over most of southwestern Utah and large areas of surrounding states. This was part of a wave of igneous activity that swept across Nevada and Utah after the eastward pressure of the Sevier Orogeny had waned. It is likely that the Farallon Plate, completely subducted and very close below the North American Plate, began to roll downward, allowing a wedge of hot asthenosphere in against the water-rich base of the lithosphere. The water favored melting—and the formation of the mineral deposits that accompanied this episode of volcanism.

Erupting lava acts in different ways depending on its temperature and composition. Dark-colored mafic lava is very runny and hot, up to 2,372°F (1,300°C), and it usually forms widespread flows of basalt. Light-colored felsic lava, like the magma that forms granite, is thicker, more viscous, and not as hot—about 1,832°F (1,000°C). If felsic magma contains much gas, the gas can't escape quickly, so the magma often erupts explosively in great clouds of ash. Without the gas, the magma forms thick flows that don't go far. The volcanics that spread across Utah between 33 and 19 million years ago were mostly felsic and intermediate (between felsic and mafic) in composition, and their eruptions were mostly violent. The resulting rocks are mostly pink to grayish-tan rhyolite, dacite, and andesite welded tuffs and volcanic breccias.

Rock from each eruption has a slightly different chemical makeup and can be dated using radioactive elements. Each ashfall can thus be traced to its source and measured, and the full area and volume of material erupted from each volcano can be calculated.

South of Modena there is a hill capped by much younger 12-million-year-old lava flows. About 17 million years ago, the intermediate tuffs and breccias disappeared to be replaced by basalt with only minor amounts of rhyolite. The change coincided with the start of Basin and Range normal faulting. The crust began to stretch westward, extending, and basaltic magma rose up from the mantle along faults. Along the way, it melted lighter crustal rock, which has a lower melting point but flows more slowly. At different times, the same vent at the surface disgorged either basalt or rhyolite; the two magmas apparently didn't mix much.

In Utah, the only place to see evidence of the earlier, violent volcanism is in the Basin and Range mountain ranges. In the down-dropped basins the volcanic rock is buried by alluvial fans and lake beds. Basin and Range faulting has dropped the floors of these valleys hundreds and sometimes thousands of feet (up to thousands of meters) relative to the mountains. These faults are all considered active.

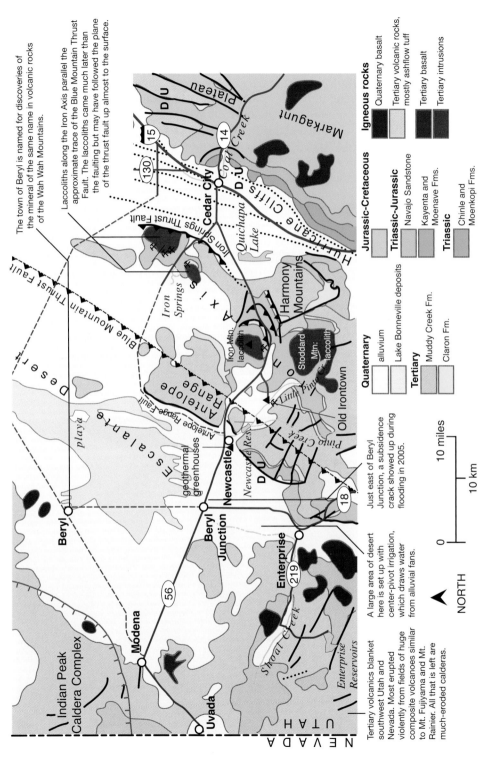

The town of Beryl is named for discoveries of the mineral of the same name in volcanic rocks of the Wah Wah Mountains.

Laccoliths along the Iron Axis parallel the approximate trace of the Blue Mountain Thrust Fault. The laccoliths came much later than the faulting but may have followed the plane of the thrust fault up almost to the surface.

A large area of desert here is set up with center-pivot irrigation, which draws water from alluvial fans.

Just east of Beryl Junction, a subsidence crack showed up during flooding in 2005.

Tertiary volcanics blanket southwest Utah and Nevada. Most erupted violently from fields of huge composite volcanoes similar to Mt. Fujiyama and Mt. Rainier. All that is left are much-eroded calderas.

NORTH

10 miles

10 km

0

Igneous rocks
Quaternary basalt
Tertiary volcanic rocks, mostly ashflow tuff
Tertiary basalt
Tertiary intrusions

Jurassic-Cretaceous
Navajo Sandstone
Triassic-Jurassic
Kayenta and Moenave Fms.
Triassic
Chinle and Moenkopi Fms.

Quaternary
alluvium
Lake Bonneville deposits

Tertiary
Muddy Creek Fm.
Claron Fm.

Geology along Utah 56 between Uvada and Cedar City.

A large area of the desert near Beryl Junction is set up with center-pivot irrigation: a rolling sprinkler arm rotates around a central well, which draws water from alluvial fans. About 0.5 mile (0.8 km) east of Beryl Junction, a subsidence crack showed up clearly during flooding in 2005. Water flowed into the fissure, enlarging and defining it. It turns out the land is subsiding and cracking in many places as irrigation water is pumped out. Well levels have decreased here by 100 feet (30 m) or more, and the land has subsided about 4 feet (1.2 m).

A small geothermal system near Newcastle heats several greenhouses with hot well water. The water may get its heat from still-cooling volcanic rocks along the Antelope Range Fault.

The Iron Axis forms the core of the range east of Newcastle. The area is rich in mining history. The Iron Axis is so named because it comprises a line of intrusions, called laccoliths, that are associated with iron-rich mineralization.

Utah 56 swings south around Iron Mountain near milepost 37. Historic mines with dumps and tailings line the roadside. These mines have been active fairly steadily since 1850, with short-term shutdowns for economic reasons. Old mines are dangerous, so these have been fenced off.

Gray and tan sedimentary rock of the Cretaceous Iron Springs Formation and pink rocks of the Tertiary Claron Formation surround the Iron Mountain and Stoddard Mountain laccoliths. After milepost 42, many of the beds are in a jumbled mess, sometimes overturned, and sometimes steeply dipping. These are the masses of rock that slid off of the Iron Mountain laccolith. As Utah 56

From the Iron Axis, this boulder of magnetite is extremely rich ore. Magnetite and hematite from the mining district average 45 percent iron. —Lucy Chronic photo

At Old Irontown, signs give great information on early ore processing. This charcoal kiln was built when charcoal was the energy of choice. —Lucy Chronic photo

leaves the hills and heads towards Cedar City, two other laccoliths are visible to the north. If you look behind you and to the north, you can see one of the open-pit mines on Iron Mountain.

East of Cedar City is the prominent upsweep of the Hurricane Cliffs defining the edge of the Markagunt Plateau. Red beds of the Chinle and Moenkopi Formations mark part of the slope above town.

Utah 257
Milford—Hinckley
70 miles (113 km)

The Mineral Mountains, east of Milford, are cored by the largest expanse of exposed granitic rock in Utah. The granitic rock rose into buried Paleozoic rocks as a series of intrusions between 15 and 9 million years ago. Granitic magma is light colored; it tends to be very thick and gassy and causes explosive eruptions if it reaches the surface, but often it never reaches the surface and instead forms intrusions like these. The intrusions are the same age as the explosive volcanics of Mt. Belknap, to the east in the Tushar Mountains. The granite and some surrounding rocks were exposed by erosion that followed Basin and Range faulting. Between 800,000 and 500,000 years ago, during Pleistocene time, a line of rhyolitic volcanoes erupted on the very top of the Mineral Mountains.

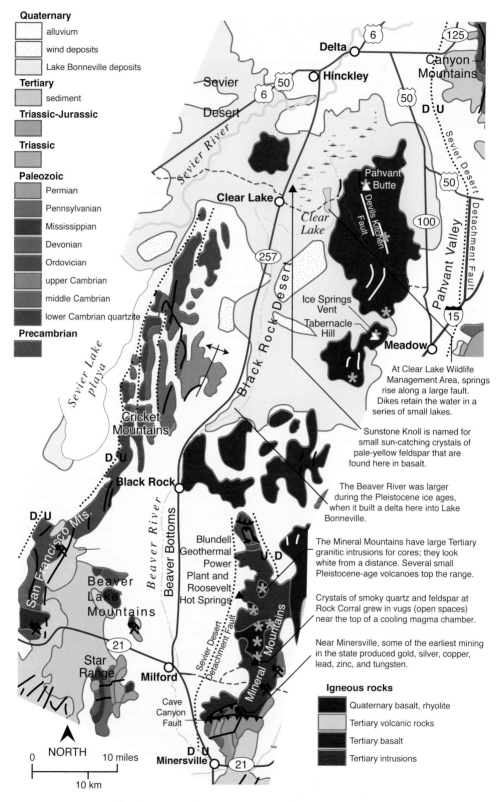

Quaternary
- alluvium
- wind deposits
- Lake Bonneville deposits

Tertiary
- sediment

Triassic-Jurassic

Triassic

Paleozoic
- Permian
- Pennsylvanian
- Mississippian
- Devonian
- Ordovician
- upper Cambrian
- middle Cambrian
- lower Cambrian quartzite

Precambrian

At Clear Lake Wildlife Management Area, springs rise along a large fault. Dikes retain the water in a series of small lakes.

Sunstone Knoll is named for small sun-catching crystals of pale-yellow feldspar that are found here in basalt.

The Beaver River was larger during the Pleistocene ice ages, when it built a delta here into Lake Bonneville.

The Mineral Mountains have large Tertiary granitic intrusions for cores; they look white from a distance. Several small Pleistocene-age volcanoes top the range.

Crystals of smoky quartz and feldspar at Rock Corral grew in vugs (open spaces) near the top of a cooling magma chamber.

Near Minersville, some of the earliest mining in the state produced gold, silver, copper, lead, zinc, and tungsten.

Igneous rocks
- Quaternary basalt, rhyolite
- Tertiary volcanic rocks
- Tertiary basalt
- Tertiary intrusions

Geology along Utah 257 between Milford and Hinckley.

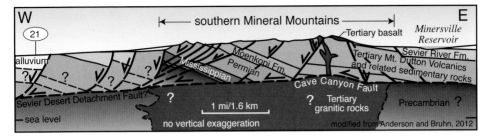

At the southern end of the Mineral Mountains, the Cave Canyon Fault separates Tertiary granite from older sedimentary rocks that lie over it. This low-angle normal fault, exposed in a canyon in the mountains north of the region depicted in this section, may be a splay of the Sevier Desert Detachment Fault.

Discernable in seismic reflection data and well cores taken from an area of 2,700 square miles (7,000 square km), the Sevier Desert Detachment Fault extends far to the west from the west side of the Mineral Mountains, dipping about 10 degrees. Rocks originally above it may have moved as much as 29 miles (47 km) to the west in the last 17 million years. The detachment is classified as active, and like the Wasatch Fault it poses the threat of generating earthquakes of up to magnitude 7. Recently, however, geologists have proposed that rocks above it may move very slowly, by what's called aseismic creep, which cannot be detected by seismometers, so the movement is unlikely to cause an earthquake.

The Beaver Lake and San Francisco Mountains, northwest of Milford, were productive mining areas. Deposits of silver, lead, copper, zinc, and gold formed when mineral-rich fluids from a Tertiary intrusion came into contact with limestone. About 4 miles (6 km) north of Milford, copper mines in the Beaver Lake Mountains are visible to the west.

About 12 miles (19 km) northeast of Milford, the Blundell Power Plant uses hot water—not steam, because it's under high pressure—from wells up to 1,700 feet (520 m) deep to generate electricity. The water is an amazing 464°F to 514°F (240°C to 268°C).

Between Milford and Black Rock, Utah 257 travels along a typical Basin and Range valley: a long north-south-trending graben. The Mineral Mountains, to the east, are an up-faulted block, or horst. The valley fill is almost 1 mile (1.6 km) deep here.

Utah 257 passes a Tertiary basalt flow, to the east, and then curves around the Cricket Mountains, to the west. Pumice and perlite are mined from a nearby basalt flow to the east. When lava reaches the surface, it depressurizes quickly, and the gas in it expands rapidly. Frozen into the cooled rock, the gas bubbles leave holes that are called vesicles. Pumice has so many vesicles that it floats on water. Perlite is water-rich volcanic glass with distinctive concentric cracks. When heated it expands, like popcorn, into lightweight, absorbent grains that are used in building materials and garden soils.

The Cricket Mountains are among several ranges in western Utah with superb Cambrian sedimentary rocks. Here, the Cambrian sequence is about

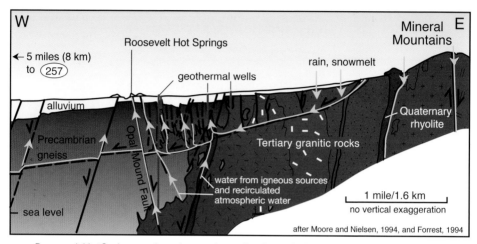

Roosevelt Hot Springs and nearby geothermal wells are fed by water that moves through a complex of faults. The water is heated by still-hot Pleistocene intrusive rocks.

15,000 feet (4,570 m) thick. When the Cambrian layers were deposited, Utah was near the equator. A warm, shallow embayment over the continental shelf teemed with life. The land was virtually barren; no fossils of terrestrial species have been found, though bacteria probably had colonized the continent's surface. Because there were no plants to hold sediment together and absorb water, it is easy to imagine erosion being quite rapid, keeping the continent beveled down to sea level.

How different life was then! Trilobites, small arthropods with three lengthwise body sections, swam in great abundance. Sponges on the seafloor had ornate frameworks of silica spicules. Small brachiopods littered the seafloor with shells. Many species had no skeletons to leave behind, but they left trace fossils: wormlike creatures left burrows in the mud; others were responsible for bite marks left on trilobites. But there were no fishes, no dinosaurs, no mammals, no amphibians, and probably no corals. The completeness of the Cambrian section, and the amazing abundance of trilobites, have allowed geoscientists to study the area's paleobiogeography, which is the interrelationships between animals and the geography in which they lived.

During the earliest Cambrian, 540 million years ago, Earth was undergoing profound changes, one of which was the seemingly rapid diversification of life. Called the Cambrian Explosion for the incredible jump in the variety and number of species, it is now known to be related to the breakup of the supercontinent Rodinia into separate isolated landmasses, and to the melting of the ice that had gripped the globe in late Precambrian time. Relatively quickly, a huge variety of habitats and niches became available.

The highway bends north again, following the west side of the Black Rock Desert. In the desert, the state's most recent volcanic rocks form rough black terrain. Lava flows, cinder cones, lava domes, and volcanic vents range from 2.7

million to only 600 years old. The volcanic features lie along the deep Basin and Range normal faults that their magma followed upward. The highway rides on Lake Bonneville deposits almost to Hinckley.

The Cricket Mountains are long north-south ridges of faulted and folded Cambrian rocks. The faults and a large syncline on the east side cause repeated outcrops of the same beds. —Lucy Chronic photo

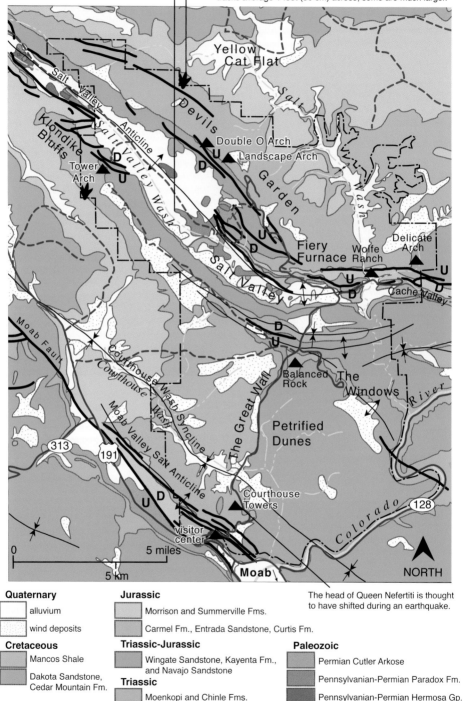

In Salt Valley, the unconsolidated sediments are often hummocky because evaporites flowed and overlying rock collapsed into solution cavities where the evaporites dissolved.

All around the Salt Valley Anticline, the top of the Moab Tongue of the Curtis Formation is known to dinosaur footprint scientists as a megatracksite. Large three-toed dinosaurs left literally billions of tracks on this layer. The tracks average 1 foot (30 cm) across; some are much larger.

Yellow Cat Flat

Salt Valley

Salt Valley Anticline

Klondike Bluffs

Tower Arch

Devils Garden

Double O Arch
Landscape Arch

Salt Valley Wash

Salt Valley

Fiery Furnace

Wolfe Ranch

Delicate Arch

Cache Valley

Moab Fault

Courthouse Wash Syncline

Courthouse Wash

Moab Valley Salt Anticline

313

191

The Great Wall

Balanced Rock

The Windows

Petrified Dunes

Courthouse Towers

Colorado River

128

visitor center

0

5 miles

5 km

Moab

NORTH

The head of Queen Nefertiti is thought to have shifted during an earthquake.

Quaternary
- alluvium
- wind deposits

Cretaceous
- Mancos Shale
- Dakota Sandstone, Cedar Mountain Fm.

Jurassic
- Morrison and Summerville Fms.
- Carmel Fm., Entrada Sandstone, Curtis Fm.

Triassic-Jurassic
- Wingate Sandstone, Kayenta Fm., and Navajo Sandstone

Triassic
- Moenkopi and Chinle Fms.

Paleozoic
- Permian Cutler Arkose
- Pennsylvanian-Permian Paradox Fm.
- Pennsylvanian-Permian Hermosa Gp.

Geology of Arches National Park.

SOMETHING SPECIAL: PARKS, MONUMENTS, AND A RECREATION AREA

Utah is lucky to have so many parks with superb geology, and to have a dry climate, praised by geologists for the way it leaves rocks exposed and plain to see. Of course, it is also a boon that good weather can almost always be counted on for those vacationing in Utah.

The parks are meant to get you outside: their roads, trails, and backcountry are fascinating to explore. Some parks have museums, guide leaflets, introductory movies, and talks and walks led by park personnel. Most of Utah's parks are on the Colorado Plateau, its scenery the draw, and its land big and remote from the busy world, but those outside the Plateau are very much worth a side trip. We describe the largest and most geologically interesting parks in this chapter; if you don't find one here, its geology may be covered in earlier chapters with its closest highway.

Rocks and fossils are protected in Utah's parks, so please do not collect or damage them. Instead, take photographs and enjoy the stories that the rocks tell. The ancient people may have drawn on them, but those drawings are now a part of the region's tales. Be careful to preserve everything for others to enjoy, for these are the unique corners of the Earth.

ARCHES NATIONAL PARK

More than two thousand stone arches, formed in narrow parallel fins of coral-colored sandstone, are the main attractions of Arches National Park. Many are visible from the roads that wind through sculptured miles of folded and faulted Jurassic rocks, which are spectacular in their own right.

From the visitor center, the road climbs up an incline of Navajo Sandstone, the fallen-in side of the Moab Valley Salt Anticline. In Early Jurassic time the Navajo Sandstone was a vast dune field in a low-latitude desert farther south than Utah is today. Its rock has eroded into rolling hills called Petrified Dunes even though their shapes are only small imitations of the Jurassic dunes.

Later, a sea invaded the area. Sand and mud accumulated in tidal flats along the shore. These sediments became the dark red-brown, thinly bedded, often wavy or crumpled Dewey Bridge Formation, which is equivalent to the Carmel Formation elsewhere in Utah. This formation sits on top of the Navajo in the park.

Sand dunes again marched across the floodplain to become the coral-colored Entrada Sandstone, known locally as slickrock because it weathers into

Across from the Arches visitor center a spectacular roadcut exposes the Moab Fault in Permian Cutler Arkose. Most large faults are really zones of closely spaced small faults (dashed line). These are normal faults: the hanging wall (upper side) moved down relative to the footwall (lower side). These faults were formed by the collapse of Moab Valley Salt Anticline. —Felicie Williams photo

The Dewey Bridge layers commonly display evidence of soft-sediment deformation, crenellations that formed before the rock was cemented. Above them, the lower third of the Entrada Sandstone is also deformed; in places it seems to funnel down into the Dewey Bridge. Sometimes, as at the Windows Section of Arches, the Entrada's waves and funnels appear to have guided the development of arches by creating stronger and weaker zones in the rock. —Felicie Williams photo

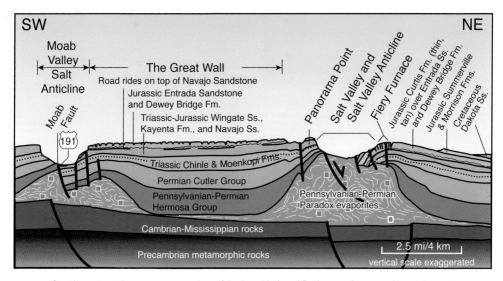

Southwest-northeast cross section of Arches National Park near the paved road. Incredibly, both Moab Valley and Salt Valley are underlain by long evaporite walls that are 10,000 feet (3,050 m) deep.

smooth, rounded, barren surfaces good for hiking and mountain biking. The Courthouse Towers are carved from it, sitting on pedestals of Dewey Bridge. Many of the Entrada's vertical surfaces are darkened with desert varnish.

The story of the park's arches begins with the Paradox Basin. It was a wide basin, warped downward in Pennsylvanian-Permian time (318 to 251 million years ago) ahead of a mountain range that was pushing westward near the Colorado border. The basin filled with seawater, but because sea level fluctuated repeatedly over tens of millions of years, the basin became restricted, almost landlocked, perhaps up to forty times. When the basin was restricted, evaporation made its water so saturated with minerals that salt, potash, and gypsum precipitated onto the basin's floor. The rock composed of these evaporate minerals, and accompanying silty and clayey layers, is called evaporite, though "salt" is also used loosely.

Salt can flow, though it does so more slowly than glacial ice. Less dense than overlying rocks, salt pushes upward, flowing to places with less confining pressure. In the Paradox Basin, the low-pressure areas were along the northwesterly fault scarps; as a result, a series of long, northwest-southeast-trending anticlines (called salt anticlines) formed in beds overlying the salt. Tectonic compression related to the Laramide Orogeny probably made these anticlines narrower and taller. As the anticlines were pushed up, vertical joints formed in the rocks along their flanks. Fields of these jointed rocks extend along both outer flanks of the Salt Valley Anticline. Today, the anticline crests are long valleys because the salt dissolved once it was close to the surface. The overlying rocks caved in, their broken pieces also weathering away. You will not find halite, or common salt, on the surface here—only crusty, crumbly, contorted masses of clay and

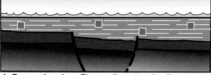

1. Pennsylvanian: Fine sediment and salt are deposited in the Paradox Basin. The salt precipitates from ocean water saturated with salt.

2. Pennsylvanian through Early Triassic: The buoyant salt is pushed westward by alluvial fans that build off the mountain range Uncompahgria. The salt rises over a long graben.

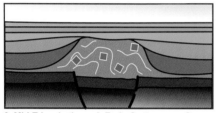

3. Mid-Triassic through Early Cretaceous: Once the salt from the sides is depleted, the salt wall stops rising. New sediment lies uninterrupted over the anticline.

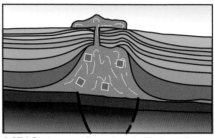

4. Mid-Cretaceous through early Tertiary: The Laramide Orogeny pushes in from the west. The anticline becomes narrower and higher. Salt may have flowed out on the surface.

5. Mid-Tertiary to present: Extension across the West widens the anticline, and its roof is breached. Weathering carries off some of the salt, and the anticline's former roof collapses and rolls in.

The development of a salt anticline in the Paradox Basin.

less-soluble gypsum. Some of these masses are crossed by the dirt road that passes down the Salt Valley.

The Paradox Basin was named for Colorado's Paradox Valley, where the Dolores River flows across, not along, a valley that formed in a salt anticline. This happens a lot in the basin, including at Arches, where Salt Wash cuts across the Cache Valley, and at Moab, where the Colorado River crosses Moab Valley. The rivers cut across the valleys in this manner because their channels were established in younger, higher layers before they cut down into the anticlines.

As salt in the center of the anticlines dissolved, salt flowed toward the center of the anticlines from underneath the rock on their flanks. The rock on the flanks gradually lost its support, and some of it caved in, or rolled over toward the center of the anticline. This process fanned out the jointed rock into fins. These fins, in the Fiery Furnace, Devils Garden, and Klondike Bluffs, are where most of the arches occur.

As it turns out, the combination of Entrada Sandstone underlain by Dewey Bridge mudstone is ideal for the process of arch formation. Water moves along the horizon between the two formations, and when an area of the rock is weaker, a hollow area or tunnel beneath the slabs of more-resistant Entrada Sandstone can be excavated in the mudstone. As erosion excavates the tunnel more and more, the overlying sandstone is undermined, and slabs of increasing size fall off, gradually enlarging the tunnel into a soaring arch.

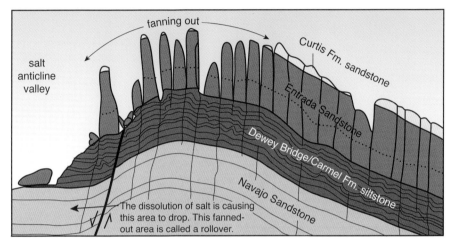

The development of fins.

Rockfalls are common amid these towering cliffs and soaring arches. Part of Landscape Arch fell in 1991, its fall frozen in midair in a lucky tourist's snapshot. Nobody saw the fall of Wall Arch, along the trail west of Landscape Arch, on the night of August 4, 2008. Some day Balanced Rock will lose its balance. Other rockfalls will form new arches and new sculptures.

Arches, windows, and natural bridges are different features. A window, like a window in a house, is a small opening, usually well above ground level. An arch is larger and spans a rock surface; it forms with no help from streams. A natural bridge is formed by stream erosion and spans a watercourse (even a dry one).

In the park, younger rocks are only preserved where they rolled over into the curved valley near Wolfe Ranch. After the Jurassic sand dunes were deposited, the area subsided, and the Curtis Formation and then the red tidal-flat mud of the Summerville Formation were laid down. Dinosaur footprints are common on the white layer at the very top of the Curtis Formation; the very best ones are where the overlying red Summerville beds just lap up on the Curtis. The surface is often rough because there are so many of them.

Then, as the continent drifted northwest, the sea receded. Subduction far to the west pushed up mountains and caused volcanoes to develop. Their ash became part of the Morrison Formation, laid down on a lake-dotted plain with rivers flowing northeastward from the faraway mountains. By Cretaceous time, mountain building had moved closer, bowing down the continent's interior so the sea again covered it: deposits of river-washed sediment and ocean

Delicate Arch is in the upper part of the Entrada Sandstone and lowermost Moab Tongue of the Curtis Formation. Its trail crosses a barren slope marked with furrows and potholes that hold rainwater briefly, harboring tiny animals and plants that are able to live out their entire life cycles in the few days or weeks after a rain. Acids they excrete help enlarge the furrows and potholes by dissolving the rock's calcium carbonate cement. —Felicie Williams photo

mud—the Cedar Mountain Formation—are overlain by Dakota Sandstone shorelines and about 5,000 feet (1,500 m) of marine Mancos Shale.

In Late Cretaceous and early Tertiary time, as the Rocky Mountains rose in Colorado, thick blankets of younger sediment probably covered this area. Their closest outcrops are the Book Cliffs and Roan Cliffs to the north, so their original extent here is unknown.

Bryce Canyon National Park and Cedar Breaks National Monument

Soft pink and white siltstone, mudstone, sandstone, and limestone layers that wall the amphitheaters in Bryce Canyon and Cedar Breaks are eroded in the Tertiary Claron Formation. This brilliantly colored rock was deposited between about 60 and 45 million years ago on a virtually level plain filled with sediment eroded from the Sevier Orogeny mountains, looming high to the west. This was a relatively quiet period between the mountain building of Cretaceous time and the volcanism of Tertiary time.

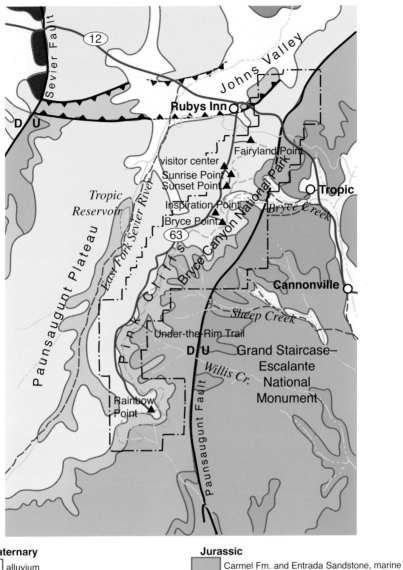

0   5 miles

5 km

NORTH

Geology of Bryce Canyon National Park.

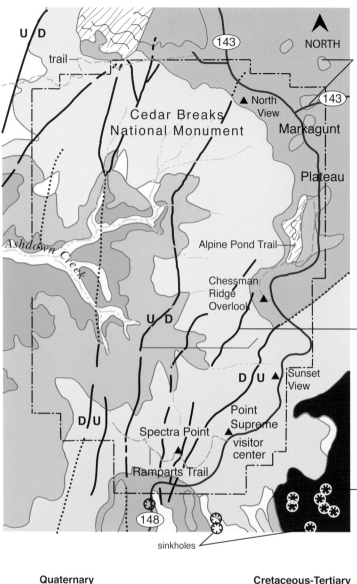

Tertiary volcanic rocks younger than the Brian Head Formation are remnants of a large Miocene-age landslide, the Markagunt Megabreccia.

Many minor north-south-trending normal faults parallel the Hurricane Fault Zone. They are evidence that Basin and Range faulting is occurring here.

Sinkholes form where water sinks through the surface rock, dissolves the soft beds below (here, the Claron Fm.), and leaves the surface with no support.

Quaternary

☐ alluvium

▨ landslide deposits

Tertiary

▨ Markagunt Megabreccia and other Tertiary sediments

▨ Brian Head Fm., soft gray sediment with abundant volcanic ash

☐ Claron Fm., white and salmon-colored lake and river deposits that have been changed by soil-forming processes

Cretaceous-Tertiary

▨ Wahweap and Grand Castle Fms. sandstone and siltstone

Cretaceous

▨ Straight Cliffs Fm. mudstone and sandstone

Igneous rocks

▨ Quaternary basalt

0 ————— 1 mile

0 ————— 1 km

Geology of Cedar Breaks National Monument.

At Bryce Canyon, and at Cedar Breaks, breathtaking natural amphitheaters are carved into soft sedimentary rocks and decorated with towers, turrets, and crenellated ridges so intricate that they must be seen to be believed. —Lucy Chronic photo

The color of the Claron's pink, consistently oxidized beds and their lack of internal layering are evidence that the fine deposits were heavily mixed, or bioturbated, and aerated by the plants and animals that lived on the plain; these layers were soils. Abundant burrows of crayfish, ants, wasps, and other animals have been found. The upper layers of the Claron Formation, exposed best in the cliffs of Cedar Breaks, contain white limestone, laid down in shallow lakes. The Claron also contains sandy and pebbly stream deposits that weren't affected by the soil-building processes.

At Cedar Breaks, the Claron Formation is thicker and quite a bit more colorful than at Bryce (originally it was called the Cedar Breaks Formation). A bonus to Cedar Breaks visitors comes in the form of much younger lava flows along Utah 14, east of the national monument; some look as if they erupted only yesterday.

The areas of Claron Formation in Bryce and Cedar Breaks are high and so have been exposed to erosion. Even though the Paunsaugunt Plateau is on the side of the Paunsaugunt Fault that dropped downward, it is higher than the land to the east due to erosion by headwaters of the Paria River. Though it may not seem possible, the land west of the fault has dropped down 2,000 feet

(610 m). To get a sense of the magnitude of the vertical drop, while in Bryce Canyon look to the northeast, where the pink Claron Formation is exposed along the edge of Table Cliff Plateau; it is much higher than the Claron in the park. Bryce Canyon was carved by the many-branched headwaters of the Paria River into the east side of the Paunsaugunt Plateau.

Cedar Breaks graces the west side of the Markagunt Plateau, west of which the land has dropped along the Hurricane Fault, which, with several thousand feet of movement, separates Utah's Colorado Plateau country from the Great Basin to the west. To the east, the Sevier Fault edges the plateau. The entire block of the plateau is tilted gently to the east because of the way the fault curves, leveling out as it gets deeper.

Viewpoints along the rim drives at Cedar Breaks and Bryce Canyon overlook the deeply eroded Claron Formation, revealing the way it erodes. Hikes along the trails in both parks let you become better acquainted with their ornate ridges, turrets, and cool, dark canyons.

As you can see from the walls of Bryce Canyon and Cedar Breaks, horizontal sandstone, limestone, and siltstone layers of the Claron Formation vary in their ability to resist erosion. Siltstone layers are the most easily eroded. Sandstone and limestone, slightly more resistant, form the canyon rims and prominent ledges seen in both areas. An extensive series of vertical joints has helped guide erosion.

At Cedar Breaks, the upper white member of the Claron Formation rims the amphitheater. It is mostly lake-deposited limestone and contains only occasional fossils of freshwater snails and clams, suggesting that the lake was not all that hospitable to life. This part of the formation has mostly eroded away at Bryce Canyon. —Felicie Williams photo

Overall, the layers erode rapidly. Studies of the growth rings of trees whose roots now hang out over the canyon rims show that the rims are receding between 9 and 48 inches (23 and 122 cm) per century, which is quite a rapid rate. Streams, freeze-and-thaw weathering, the pelting force of rain and hail, landslides, and wind all contribute to the carving of the ornate landscape.

The Claron Formation, though, represents only part of the history of these parks. The oldest rocks exposed in either park are Cretaceous. The Cretaceous Dakota Sandstone, seen at Bryce, was a shoreline deposit of the great Cretaceous Interior Seaway that flooded the downwarped basin that developed east of the Sevier Thrust Belt.

The seaway was deepest as it deposited the overlying Tropic Shale. Frequently called "blue" in this region because of its beautiful shade of bluish gray, it erodes into slopes and badlands. It is very fossiliferous, with the remains of oysters, ammonites, sharks, and many varieties of fishes, crocodiles, turtles, plesiosaurs, pliosaurs, and dinosaurs.

As the seaway retreated, the Straight Cliffs Formation was deposited along its shifting shoreline. Its many layers of coal are remnants of coastal swamps. They alternate with sandstone from coastal bars and shale from lagoons. Above the Straight Cliffs Formation and below the Claron, the Wahweap and Kaiparowits Formations are lake and stream deposits with terrestrial vertebrate fossils.

Rocks that overly the Claron differ in the two parks: In Bryce, only a few layers of conglomerate are exposed; overlying sediment has been eroded. But in Cedar Breaks, two additional deposits provide a glimpse of post-Claron events. The oldest of these, the Brian Head Formation, is soft sedimentary rock containing lots of volcanic ash. By Oligocene and Miocene time, when it was deposited, massive volcanoes of the Marysvale Volcanic Field were erupting. This volcanic field, north and west of both parks, is one of the largest areas of volcanic rocks in the United States.

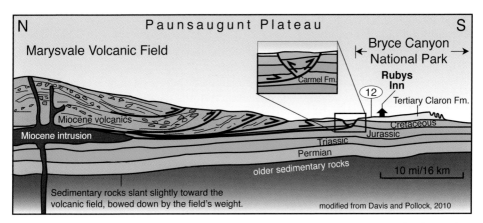

A pair of thrust faults just north of Bryce probably formed when some of the huge pile of volcanic debris to the north slid southward along a fault, partly along weak shale beds of the Jurassic Carmel Formation. Where the main fault curved up to the surface, it splayed into the two thrust faults.

The younger deposit is the Miocene Markagunt Megabreccia, of which a few outcrops remain at Cedar Breaks. This landslide deposit contains a mix of pieces, some bigger than city blocks, and one that is 5 miles (8 km) across. It slid along the Claron Formation and younger clayey beds. Some of the Brian Head Formation was caught up in the slide, and its clayey layers were also glide planes for the slide. The megabreccia probably slid off the south flank of the Marysvale Volcanic Field, though the region's deeply faulted valleys and plateaus formed afterward, so it is hard to tell which way the land sloped at the time of the slide. Because of the soft, wet clayey layers beneath it, only a few degrees of incline were needed for the slide to move.

Canyonlands National Park

Canyonlands National Park offers an escape to a place of heights and depths, an alien world of fins, spires, baking sun, and shadowy silence. The Green and Colorado Rivers, and their confluence in the park, divide the park into three districts: Island in the Sky, a high plateau between the rivers in the north; the Maze and Horseshoe Canyon, west of both rivers; and the Needles, southeast of the rivers. Roads lead into these districts, but there are no bridges between them, just a long drive around the park. Four-wheel-drive roads lead to the remote areas, as do the rivers, which require experience to raft. Be sure to get the appropriate permits before exploring the backcountry. Bring plenty of water and supplies, and know what kind of country you are heading into.

The layer-cake geology of the Colorado Plateau shows up beautifully in Canyonlands National Park. The Colorado and Green Rivers have exposed mile upon mile of sedimentary rocks: Pennsylvanian, Permian, Triassic, and Jurassic, the oldest deep in the canyons. The layered rocks are part of the north end of the Monument Upwarp, so they cross Canyonlands in a broad, gentle arch.

Like its neighbor Arches National Park, Canyonlands is underlain by salty evaporite beds of the Paradox Formation. In Pennsylvanian time, the continental collision that formed the supercontinent Pangaea also pushed up the Appalachians, the Ancestral Rockies, and a range called Uncompahgria in western Colorado. Uncompahgria was thrust westward over other rock. It weighed down land farther west, forming the Paradox Basin, a gulf with restricted circulation. In Paradox Basin, a balance between evaporation and circulation resulted in ideal conditions for the formation of evaporites: there was enough evaporation for minerals to be concentrated in the seawater, and yet the gulf was regularly replenished with fresh supplies of seawater. The basin filled with the Paradox Formation: up to 4,000 feet (1,220 m) of salt, gypsum, potash, and anhydrite along with thin layers of mud and silt, and then with limestone and shale of the Honaker Trail Formation.

Toward the end of Pennsylvanian time, the basin had filled and alluvial fans from the mountains—the Cutler Arkose—began to reach Canyonlands. As the mountains wore down, wind winnowed the alluvium and built dunes, which became the Permian Cedar Mesa Sandstone. Through the rest of Permian and Triassic and into Jurassic time, dune sand alternated with the red sandy mud of tidal flats and floodplains as the climate and sea level changed. Many of the

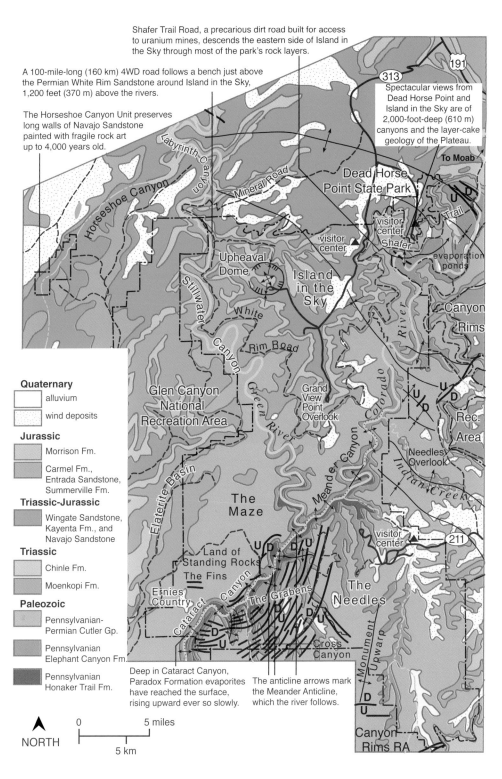

Shafer Trail Road, a precarious dirt road built for access to uranium mines, descends the eastern side of Island in the Sky through most of the park's rock layers.

A 100-mile-long (160 km) 4WD road follows a bench just above the Permian White Rim Sandstone around Island in the Sky, 1,200 feet (370 m) above the rivers.

The Horseshoe Canyon Unit preserves long walls of Navajo Sandstone painted with fragile rock art up to 4,000 years old.

Spectacular views from Dead Horse Point and Island in the Sky are of 2,000-foot-deep (610 m) canyons and the layer-cake geology of the Plateau.

191

313

To Moab

Labyrinth Canyon

Horseshoe Canyon

Mineral Road

Dead Horse Point State Park

visitor center

visitor center

Shafer

Trail

U D

evaporation ponds

Upheaval Dome

Stillwater Canyon

White

Island in the Sky

Canyon Rims

Rim Road

Green River

Colorado River

Quaternary

alluvium

wind deposits

Jurassic

Morrison Fm.

Carmel Fm.,
Entrada Sandstone,
Summerville Fm.

Triassic-Jurassic

Wingate Sandstone,
Kayenta Fm., and
Navajo Sandstone

Triassic

Chinle Fm.

Moenkopi Fm.

Paleozoic

Pennsylvanian-
Permian Cutler Gp.

Pennsylvanian
Elephant Canyon Fm.

Pennsylvanian
Honaker Trail Fm.

Glen Canyon National Recreation Area

Grand View Point Overlook

U D

U D

Rec. Area

Needles Overlook

Elaterite Basin

The Maze

Meander Canyon

Indian Creek

visitor center

211

Land of Standing Rocks

The Fins

Ernies Country

Cataract Canyon

The Grabens

D U

D U

U D

The Needles

Cross Canyon

Monument Upwarp

D U

0   5 miles

NORTH

5 km

Deep in Cataract Canyon, Paradox Formation evaporites have reached the surface, rising upward ever so slowly.

The anticline arrows mark the Meander Anticline, which the river follows.

Canyon Rims RA

Geology of Canyonlands National Park.

formations are interfingered in the area. Occasionally, a period of increased erosion left a conglomerate layer, such as the Shinarump at the base of the Chinle Formation.

The rivers did most of the sculpting in Canyonlands, but their work was aided by the way deep vertical joints split the rocks and by the influence of salt under part of the national park. Erosion was, of course, aided by the great elevation of the high desert. Once the Colorado River had cut headward into this land, it gave the other rivers and streams of the region the energy to cut very deeply.

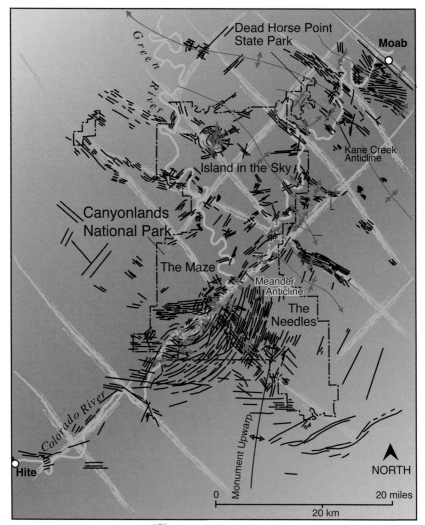

Precambrian faults ≈≈≈≈≈ *controlled the location of Pennsylvanian and later folding* ―↕―, *related to the movement of salt, and upwarping* ―↕―, *related to the Laramide Orogeny. The folding and upwarping caused jointing* ↘//, *and the salt created instability that allowed the joints to extend downward, fragmenting the rock layers so individual pieces could move.*

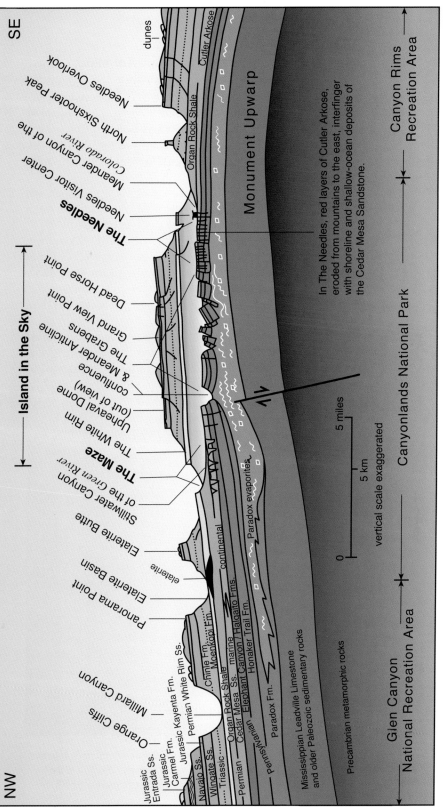

NW

SE

Island in the Sky

The Needles

The Maze

Orange Cliffs
Millard Canyon
Panorama Point
Elaterite Basin
Elaterite Butte
Stillwater Canyon of the Green River
The White Rim
Upheaval Dome (out of view)
confluence of the Green River
& Meander Anticline
The Grabens
Grand View Point
Dead Horse Point
Needles Visitor Center
Meander Canyon of the Colorado River
North Sixshooter Peak
Needles Overlook
dunes

Jurassic Entrada Ss.
Jurassic Carmel Fm.
Jurassic Kayenta Fm.
Navajo Ss.
Wingate Ss.
Triassic
Chinle Fm.
Moenkopi Fm.
Permian White Rim Ss.
Organ Rock Shale
Cedar Mesa Ss.
Elephant Canyon / Halgaito Fms.
Honaker Trail Fm.
marine
continental
elaterite
Paradox evaporites
Pennsylvanian
Paradox Fm.
Mississippian Leadville Limestone
and older Paleozoic sedimentary rocks

Precambrian metamorphic rocks

Cutler Arkose
Organ Rock Shale

Monument Upwarp

In The Needles, red layers of Cutler Arkose,
eroded from mountains to the east, interfinger
with shoreline and shallow-ocean deposits of
the Cedar Mesa Sandstone.

0   5 km
        5 miles
vertical scale exaggerated

Glen Canyon
National Recreation Area

Canyonlands National Park

Canyon Rims
Recreation Area

Northwest-southeast cross section of Canyonlands National Park.

Southeast of the Rivers: Meander Anticline, the Grabens, and the Needles

Salt and gypsum, which become plastic under pressure, are also less dense and thus more buoyant than other rocks; they constantly push up against whatever rocks overly them. This means trouble, and the scenery here is the proof. The paved park roads do not lead all the way to Meander Anticline, the Grabens, or the Needles, but they are accessible on foot, by 4WD roads, or by rafting the rivers.

Along the Colorado, where the river has removed a great deal of rock, salt and gypsum have flowed up, gradually arching the overlying rocks. The result is Meander Anticline, which curves and follows the river's course.

Because near the Grabens the river is west of the crest of the Monument Upwarp, bedding dips westward, toward the river. Huge cracked-off crescents of rock are moving down the slope of the upwarp toward the river above subsurface salt and gypsum, creating the Grabens, an area of parallel-walled valleys. A graben is a slice of rock that has dropped down between two steep normal faults, creating a valley above it in the process.

At the Needles, farther up the slope of the upwarp, the movement hasn't been as extreme, but the underlying salt has still allowed joints in the rock to open up. Adding to the uniqueness of the landscape here, there are two crossing sets of joints along which weathering also occurs. The resulting spires, rounded by weathering, are fun to explore.

The Needles formed in an area of Permian rocks where red alluvial fans from the east interfinger with white dunes from the west, giving the spires red and white layers. The edge of the Needles is near Utah 211; the scenery from the trails is even better. —Felicie Williams photo

Island in the Sky: Between the Green and the Colorado

Utah 313 climbs west from US 191 onto a high cliff-rimmed plateau of river-laid Kayenta Formation with scattered rounded hills of Navajo Sandstone and then divides.

Utah 313 heads east to Dead Horse Point State Park; Island in the Sky Road leads to Canyonlands National Park, winding between brush-covered sand dunes. Trails explore both recent dunes and weathered dunes of Navajo Sandstone. This world of rock is composed of unending ledges and slopes cut by the rivers that meet 2,000 feet (610 m) below. From the overlooks on Island in the Sky, you look over the world as if your sandstone pedestal were a fortress.

The Colorado River flows southwesterly, following a Precambrian line of weakness. The many bends in the channels of both the Colorado and the Green are considered entrenched meanders inherited from Miocene time, when the rivers looped across a gentle plain. Sediment washed in from side canyons also deflects the rivers, altering their courses.

It took somewhere between 2 and 6 million years for the rivers to cut these canyons. Certainly they cut more quickly when they carried meltwater and gravel during the Pleistocene ice ages: stranded bars of gravel eroded from the Rockies during the ice ages lie here and there above today's channels.

On the shelf of Permian White Rim Sandstone that surrounds the Island in the Sky, the White Rim Road is a favorite hiking, 4WD, and mountain biking trail. Dune sand, blown in on windy days, covers the land away from the rim. —Felicie Williams photo

Upheaval Dome

Upheaval Dome, at the north end of the park at the end of Upheaval Dome Road, is surprising and unexpected: it's a circular ring of steeply tilted Triassic and Jurassic layers 3.4 miles (5.5 km) across. Inside the ring there are deformed and altered Permian, Triassic, and lower Jurassic rocks. Erosion has revealed this feature extremely well. It was called a dome because it was long credited to rising salt and gypsum, like the nearby salt anticlines. Now it is generally believed to be a meteorite impact crater. Although no pieces of the meteorite have been found, in 2008 geologists found strained, annealed quartz in the crater, which only forms under extreme explosive pressure. Bands of rock that are very deformed also support the impact hypothesis.

Meteorites strike Earth at speeds of 6 to 12 miles (10 to 20 km) per second. Red-hot from friction with the atmosphere, they shatter and often vaporize. The immense amount of energy they release pushes the ground outward into a

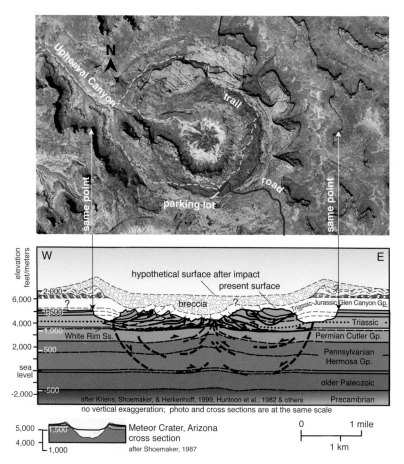

Upheaval Dome is obviously younger than the deformed lower Jurassic rock of the Glen Canyon Group. The earliest the impact may have occurred is 165 million years ago. This cross section depicts a Jurassic-age impact crater, though the crater of Upheaval Dome could be a remnant of a much larger, younger impact. —USDA-FSA-APFO photo

On Utah 313, the road to Island in the Sky, the upper part of the Carmel Formation and the lower part of the overlying Entrada Sandstone were deformed while still unlithified, around 165 million years ago. It may be that the heavy dunes of the Entrada settled unevenly into the saturated sandy mud. Or, both formations could have been shaken by an earthquake. It would make a great story if the features were caused by the Upheaval Dome meteorite impact, but geologists have not yet found strong enough evidence for a connection. —Felicie Williams photo

crater much larger than the meteorite. Meteor Crater, Arizona, is thirty-seven times wider than the meteorite that made it. After the instant of impact, the Earth rebounds. Rock that was blasted outward slides back into the crater, surging up at the center and leaving a ring-shaped valley. Dust falls from the sky and blankets the ground near the impact.

Horseshoe Canyon: A Far-Flung Outlier

Horseshoe Canyon was included in the park to protect its prehistoric pictographs. The people who painted them are known as the Barrier Canyon people, after the canyon's original name. The paintings, made between 1900 BC and AD 300, are on high, smooth, overhanging walls of Navajo Sandstone. The people used iron oxides for reds, oranges, and yellows, clay for white, probably charcoal for black, and on occasion ground copper minerals for greens and blues. All of the pictographs occur in close proximity.

Capitol Reef National Park

Capitol Reef National Park is centered over the 100-mile-long (160 km) Waterpocket Fold, the steep eastern limb of the Circle Cliffs Uplift. The fold formed in Late Cretaceous time, during the Laramide Orogeny. Pressure caused by the subduction of the Farallon Plate beneath the North American Plate along the west coast caused several huge folds like this to develop in southeast Utah. All of them drape over deep faults in hard, crystalline Precambrian rock. In the case of the Waterpocket Fold, rocks were lifted 7,000 feet (2,130 m) higher on the west side of the fold; however, erosion has partly leveled the land by removing Jurassic and younger rocks from that side.

Steeply tilted Triassic and Jurassic rocks form the hogbacks of the Waterpocket Fold, including Capitol Reef itself, which is built of dark-red dune-formed Wingate Sandstone; thinly bedded river deposits of the Kayenta Formation; and the massive, white, dune-formed Navajo Sandstone, its crossbedded, rounded shapes forming the crest of the reef. Contrasting dark-red Carmel Formation shale fronts the east side of the reef.

Both the Wingate and Navajo Sandstones hold many clues to their dune origin. Both are crossbedded, with long, sloping layers. Both contain fine, rounded, even-sized, frosted sand grains—frosted because they hit each other as they bounce along. Both have ripple marks that are higher and less sharp than those of water-deposited dunes.

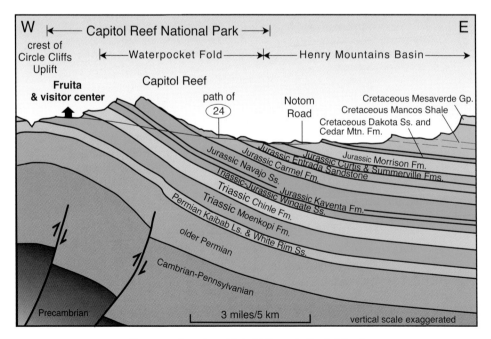

Cross section along Utah 24 through Capitol Reef.

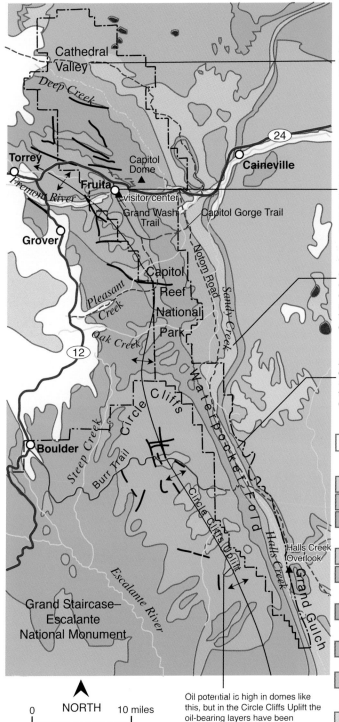

Glass Mountain is a mass of selenite crystals that extruded from a layer beneath the surface.

Black lava boulders on the hill near the visitor center came from high volcanic-capped plateaus not far to to the west; debris flows and streams draining Pleistocene glaciers brought them here.

The Morrison Formation contains pockets of uranium ore in carbon-rich river channel deposits; small prospect pits dating back to the 1950s and '60s dot its outcrops east of Capitol Reef.

At Oyster Shell Reef, oysters in the Dakota Sandstone collected near the shore of a rising Cretaceous sea.

Tertiary-Quaternary
alluvium, landslide, and glacial deposits

Cretaceous
Mesaverde Gp.

Mancos Shale

Dakota Sandstone, Cedar Mountain Fm.

Jurassic
Morrison Fm.

Carmel Fm., Entrada Ss., Curtis Fm., Summerville Fm.

Triassic-Jurassic
Wingate Ss., Kayenta Fm., and Navajo Ss.

Triassic
Moenkopi and Chinle Fms.

Paleozoic
Permian White Rim Ss., Kaibab Limestone

Igneous rocks
Tertiary volcanic rocks

Tertiary intrusions

0   NORTH   10 miles

10 km

Oil potential is high in domes like this, but in the Circle Cliffs Uplift the oil-bearing layers have been breached at the surface, and the little oil that is left is unprofitable.

Cathedral Valley

Deep Creek

Torrey

Fremont River

Capitol Dome

Fruita

visitor center

Grand Wash Trail

Grover

24

Caineville

Capitol Gorge Trail

Noton Road

Sandy Creek

Capitol Reef National Park

Pleasant Creek

Oak Creek

12

Circle Cliffs

Waterpocket Fold

Boulder

Steep Creek

Burr Trail

Circle Cliffs Uplift

Halls Creek

Halls Creek Overlook

Grand Gulch

Escalante River

Grand Staircase–Escalante National Monument

Geology of Capitol Reef National Park.

Capitol Reef National Park is named for the almost impassable "reef" of rocks that rises up along the Waterpocket Fold. Since sandstone, siltstone, mudstone, and limestone layers each erode differently, the Waterpocket Fold appears as a series of ridges—some quite high, some relatively insignificant—with intervening racetrack valleys. —Lucy Chronic photo

Wingate Sandstone cliffs west of Fruita break along vertical joints; the rock maintains it sharp profile while contributing to talus slopes below. The Wingate rests on the sloping, many-hued Triassic Chinle Formation, its former soil layers rich in clay derived from volcanic ash. The bottom of the slope is the dark brick-red mudstone and siltstone of the Moenkopi Formation, which surfaces most of the park west of the reef. —Lucy Chronic photo

Here, in this hot and barren desert, the waterpockets for which the immense fold is named are often the only source of water between storms. They are found where intermittent streams that cut across the reef have worn softer layers more deeply, grinding out bowls with swirling stones below pour-offs. Water also collects after rainstorms in shallow potholes in sandstone surfaces. Standing water, no matter how ephemeral, weakens the rock just a little; wind or more running water removes loosened sand grains, slightly deepening the pools. With more depth, tiny plants and animals come to inhabit them, secreting acids that further weaken the rock. Grain by grain, these pools become large enough to hold water for a considerable time.

Honeycomb weathering is common in the sandstones of Capitol Reef. Rainwater soaks into the rock, dissolving the cement holding the sand grains together and redepositing it near the surface as it evaporates, making the surface of the rock harder. Tiny cavities form in the surface of the rock and eventually reach the inner soft layer, and moisture and wind enlarge them. Eventually, the rock may become honeycombed with closely spaced holes rimmed by the more-resistant outer layer.

East of Capitol Reef there are younger Jurassic rocks. The deep-red Entrada Sandstone erodes into pinnacles and spires in Cathedral Valley, in the northern end of the park. The large north-south-trending valley east of the reef is in the valley-forming Morrison Formation. This shale has a central sandier section, the Salt Wash Member, which forms the next continuous hogback east of the national park. Both the Morrison Formation and the Triassic Chinle Formation (exposed west of the reef) contain abundant volcanic ash. The ash decomposes into bentonite, a clay that swells when wet, discouraging the establishment of plants. Both also contain less-oxidized iron minerals that give them their purplish and greenish hues.

East of the Morrison Formation valley, with its Salt Wash Member hogback, and about 2 miles (3 km) east of the park's eastern boundary, there is a stockade-like ridge of sandstone composed of the Cedar Mountain Formation and Dakota Sandstone. They are the oldest Cretaceous units in the Capitol Reef area. Younger Cretaceous rocks surface farther east; the dark-gray, valley-forming Mancos Shale first, and then the cliff-forming Mesaverde Group sandstones interlayered with shale and coal.

A maze of narrow canyons follows and cuts through the reef in the park. Trails lead along some canyons, and to a number of arches, natural bridges, and the summits of domes. Capitol Reef's rocks are well exposed via trails at Grand Wash and Capitol Gorge.

DINOSAUR NATIONAL MONUMENT

President Woodrow Wilson created Dinosaur National Monument in 1915 to preserve its incredible fossils. Over time, people learned what amazing country—for both geology and scenery—lay in the nearby canyons, and in 1938 the monument was expanded to include the canyons.

The monument is on the northern edge of the Colorado Plateau. Here, the flat-lying rocks of the Plateau country tilt up along the south flank of the Uinta

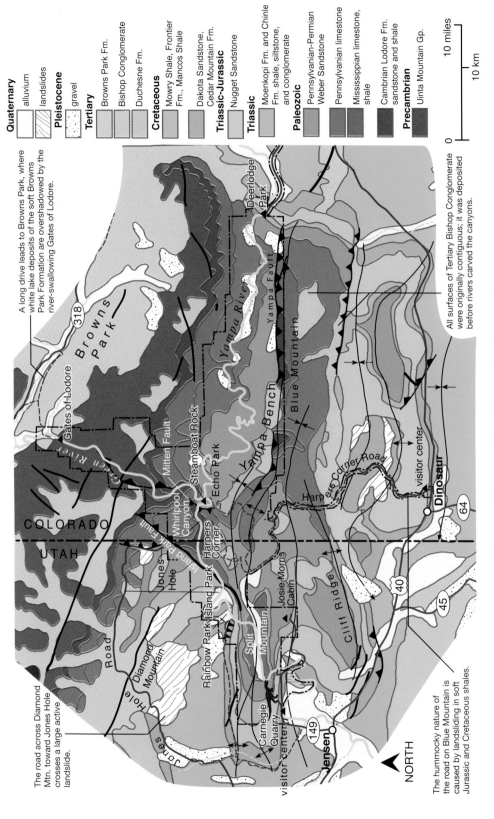

Quaternary
☐ alluvium
▨ landslides

Pleistocene
∴ gravel

Tertiary
☐ Browns Park Fm.
☐ Bishop Conglomerate
☐ Duchesne Fm.

Cretaceous
☐ Mowry Shale, Frontier Fm., Mancos Shale
☐ Dakota Sandstone, Cedar Mountain Fm.

Triassic-Jurassic
☐ Nugget Sandstone

Triassic
☐ Moenkopi Fm. and Chinle Fm. shale, siltstone, and conglomerate

Paleozoic
☐ Pennsylvanian-Permian Weber Sandstone
☐ Pennsylvanian limestone
☐ Mississippian limestone, shale

Precambrian
☐ Cambrian Lodore Fm. sandstone and shale
☐ Uinta Mountain Gp.

0   10 miles

0   10 km

A long drive leads to Browns Park, where white lake deposits of the soft Browns Park Formation are overshadowed by the river-swallowing Gates of Lodore.

The road across Diamond Mtn. toward Jones Hole crosses a large active landslide.

All surfaces of Tertiary Bishop Conglomerate were originally contiguous; it was deposited before rivers carved the canyons.

The hummocky nature of the road on Blue Mountain is caused by landsliding in soft Jurassic and Cretaceous shales.

NORTH

COLORADO
UTAH

Browns Park
318
Gates of Lodore
Green River
Mitten Fault
Steamboat Rock
Echo Park
Whirlpool Canyon
Harpers Corner
Island Park Fault
Jones Hole
Diamond Mountain
Jones Hole Road
Rainbow Park
Split Mountain
Carnegie Quarry
Josie Morris Cabin
Jensen
149
visitor center
40
45
Yampa River
Yampa Fault
Yampa Bench
Blue Mountain
Deerlodge Park
Harpers Corner Road
Cliff Ridge
Dinosaur
visitor center
64

Geology of Dinosaur National Monument and surrounding area.

Mountains, a huge east-west-trending anticline that is faulted and folded along its northern and southern margins. Precambrian rocks form the range's central mass. Two major rivers come together in the southern center of these great mountains: the Yampa, with headwaters in Colorado, flows into the Green, with headwaters in Wyoming. The Green then twists through folds in the range's south flank at Split Mountain, emerging near the famous dinosaur quarry. The rocks laid bare in the deep canyons of Dinosaur National Monument tell its intriguing story.

Between 770 and 740 million years ago, an east-west-trending extensional basin filled with up to 28,000 feet (8,500 m) of sediment carried by rivers flowing west to where the ocean lapped up into the basin. During the next 250 million years, this sediment was buried, cemented, tilted a few degrees, and eventually exposed at the surface and eroded into an irregular surface. This red sandstone, red siltstone, gray shale, and quartzite compose the Uinta Mountain Group, which is exposed at the Gates of Lodore. Between 525 and 505 million years ago the Lodore Formation, 230 to 600 feet (70 to 180 m) of light-brown to green sandstone and shale, was deposited on the Uinta Mountain Group near the shoreline of the Cambrian ocean.

Between 350 and 300 million years ago, about 2,000 feet (610 m) of interbedded limestone, sandstone, and shale was deposited in shallow seas. Fossils are abundant in these Mississippian and Pennsylvanian layers. The sea retreated between 300 and 250 million years ago and sand dunes developed near the Pennsylvanian-Permian shoreline. The dunes became the pale, barren-surfaced Weber Sandstone, which is up to 2,000 feet (610 m) thick. Limestone beds and

Split Mountain, part of an anticline, is 2,600 feet (790 m) high, with slopes of Weber Sandstone. The Green River most likely cut through it in the last 5 million years. —Felicie Williams photo

a cap of shale interrupt the dunes, reflecting a return of the sea. The Weber forms many of the park's wonders, including Steamboat Rock, the cliffs along the canyon of the Yampa River, and the cliffs of Split Mountain.

Between 220 and 95 million years ago, during Triassic through Early Cretaceous time, continental compression caused by subduction along the west coast pushed up mountain ranges to the west and raised the area in and around Dinosaur National Monument above sea level. This area became a low plain; it was much closer to the equator then but was drifting slowly northward. Dark-red estuary and river muds (now shale) gradually gave way to desert dunes (now sandstone) and finally, during the heyday of the dinosaurs, to a savannah with shifting river channels and shallow lakes—now the Morrison and Cedar Mountain Formations.

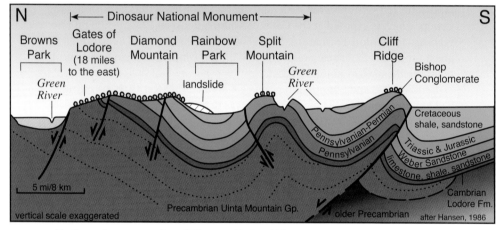

North-south cross section of Dinosaur National Monument and the surrounding area.

The Permian Weber Sandstone dips steeply south off the Split Mountain Anticline. Flatirons of younger rocks lean up against the Weber. Marine sediments containing phosphate make up the Permian Park City Formation. —Felicie Williams photo

Between 110 and 80 million years ago, during Late Cretaceous time, the Sevier Orogeny was approaching. Its mountains weighed down the crust enough to let in the sea. Continental and nearshore deposits—the Cedar Mountain and Dakota sandstones, Mowry Shale, and Frontier Formation—were followed by 5,000 feet (1,500 m) of gray marine Mancos Shale.

Between 70 and 40 million years ago, during the Laramide Orogeny, the Uinta Mountain Group rocks proved a stumbling block to the tectonic forces compressing the continent. They were summarily pushed upward along the old faults that edged the extensional basin, forming the Uinta Mountains. The younger layers above them were folded and faulted and tilted.

The Uintas, much higher than they are now, eroded down to a gentle surface between 40 and 29 million years ago. The well-rounded grit, pebbles, cobbles, and boulders of the Oligocene Bishop Conglomerate mark the end of the orogeny. This surprising conglomerate, up to 400 feet (120 m) thick, rests on the highest places in Dinosaur National Monument. It was deposited in braided, sediment-choked channels of the Yampa River before the modern canyons were cut. At the time, the Continental Divide ran along the Uintas, so the Green River flowed northeastward, far from this area. Layers of volcanic ash run through the conglomerate, probably from volcanoes near the southwest corner of Utah.

A time of great tension began around 17 million years ago when the subduction ended along the west coast. The tension caused extensional faulting (normal faulting) along the Uinta Mountains. A central block dropped down to form Browns Park. The erosion surface of the mountains tilted gently, guiding the Green River into the national monument. It formed a lake, the sediment of which became the pale, ash-rich Browns Park Formation. The lake finally overflowed southeastward into the Yampa. A steep gradient gave the Yampa and Green Rivers great down-cutting power. They easily cut through the Bishop Conglomerate and then, trapped in their channels, continued cutting

downward. Now, the Green River leaves Browns Park by flowing straight into the 2,000-foot-deep (610 m) canyon it cut through the center of a mountain—starting at the Gates of Lodore, named by John Wesley Powell.

In 1909 Dr. Earl Douglass, a geologist from Pittsburgh's Carnegie Museum who was searching the Jurassic Morrison Formation for dinosaur skeletons, found a real bonanza at what became the Carnegie Quarry: the fossilized remains of a water hole and nearby sandbar where drought-stricken dinosaurs had died by the hundreds, along with lizards, turtles, crocodiles, frogs, small mammals, freshwater mollusks, and a variety of plants. The tracks of beetles were also preserved. These dinosaur skeletons are fairly complete, which is extremely unusual. Skulls are fragile and easily fall apart, but here thirteen intact specimens have been found. Many skeletons have been removed for museum displays, but many more remain unearthed. They were roofed over in their original position in 1957.

Dinosaur National Monument has much more to offer than dinosaur fossils. Several roads lead into the park's valleys and overlooks. There are no through roads because canyons bar the way. The only way to cross the park's entire surface is by boat on the Green or Yampa River. This is not only fun but a great way to see the rocks.

Dinosaur bones in the Carnegie Quarry are mostly of large, long-necked, plant-eating sauropods. Note a partial spinal column in the shady area near the center. —Felicie Williams photo

Rounded forms of Jurassic Entrada Sandstone stand north of Rainbow Park. Water seeps into the rock and dissolves the cement between sand grains. Wind blows the sand away, or when nights are below freezing, ice pries them out. Corners of the sandstone, attacked from several directions at once, round relatively quickly. —Felicie Williams photo

The paved road that heads north from the Canyon Visitor Center in Colorado to Harpers Corner passes through Cretaceous, Jurassic, and Triassic rocks that were upturned along the front of Blue Mountain. It then climbs the flank of the anticline that cores Blue Mountain, which is surfaced with the Weber Sandstone's crossbedded dune deposits.

GLEN CANYON NATIONAL RECREATION AREA

Glen Canyon National Recreation Area has scenery so expansive it can be hard to grasp. The recreation area includes not only Lake Powell, but generous reaches of dry, colorful desert. The lake is a highway of water 186 miles (299 km) long with 1,900 miles (3,058 km) of shoreline when full, and floating mile markers and rest areas. Lake Powell fills the Colorado River's Glen Canyon, which was hauntingly beautiful, narrow, and deep before its inundation.

Glen Canyon Dam, which holds back Lake Powell, was built between 1956 and 1963 to store water so an agreed-upon flow could be transmitted to the Lower Colorado River Basin, and to make hydroelectric power. It took

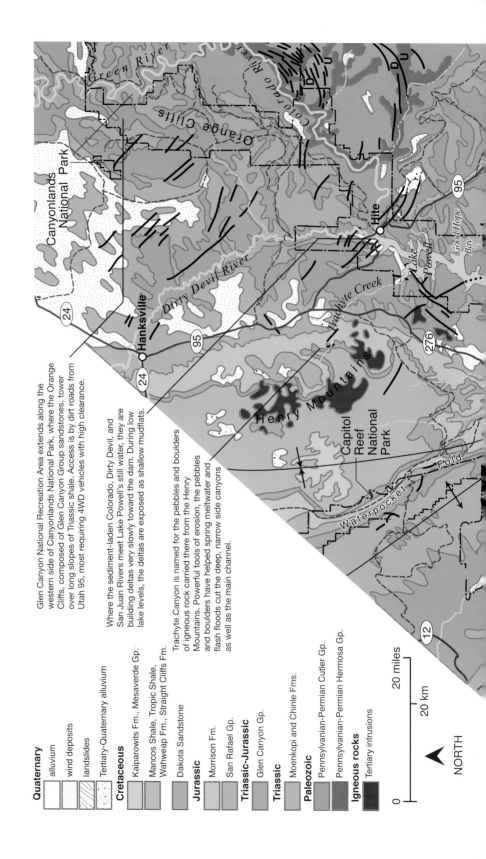

Glen Canyon National Recreation Area extends along the western side of Canyonlands National Park, where the Orange Cliffs, composed of Glen Canyon Group sandstones, tower over long slopes of Triassic shale. Access is by dirt roads from Utah 95, most requiring 4WD vehicles with high clearance.

Where the sediment-laden Colorado, Dirty Devil, and San Juan Rivers meet Lake Powell's still water, they are building deltas very slowly toward the dam. During low lake levels, the deltas are exposed as shallow mudflats.

Trachyte Canyon is named for the pebbles and boulders of igneous rock carried there from the Henry Mountains. Powerful tools of erosion, the pebbles and boulders have helped spring meltwater and flash floods cut the deep, narrow side canyons as well as the main channel.

Quaternary

alluvium

wind deposits

landslides

Tertiary-Quaternary alluvium

Cretaceous

Kaiparowits Fm., Mesaverde Gp.

Mancos Shale, Tropic Shale, Wahweap Fm., Straight Cliffs Fm.

Dakota Sandstone

Jurassic

Morrison Fm.

San Rafael Gp.

Triassic-Jurassic

Glen Canyon Gp.

Triassic

Moenkopi and Chinle Fms.

Paleozoic

Pennsylvanian-Permian Cutler Gp.

Pennsylvanian-Permian Hermosa Gp.

Igneous rocks

Tertiary intrusions

0   20 km
     20 miles

NORTH

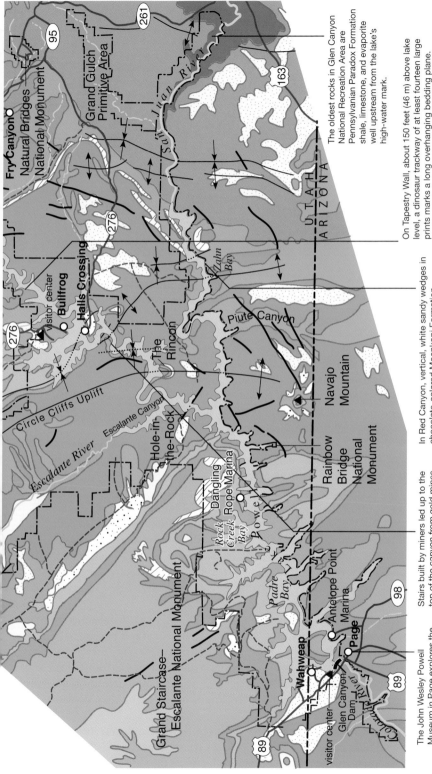

The oldest rocks in Glen Canyon National Recreation Area are Pennsylvanian Paradox Formation shale, limestone, and evaporite well upstream from the lake's high-water mark.

On Tapestry Wall, about 150 feet (46 m) above lake level, a dinosaur trackway of at least fourteen large prints marks a long overhanging bedding plane.

In Red Canyon, vertical, white sandy wedges in chocolate-colored Moenkopi Formation mudstone are the profiles of mud cracks that formed in drying lake beds 240 million years ago.

Stairs built by miners led up to the top of the canyon from gold mines in gravel bars in the river; the mines are now deep under the lake.

The John Wesley Powell Museum in Page explores the area's geology and history.

Geology of Glen Canyon National Recreation Area.

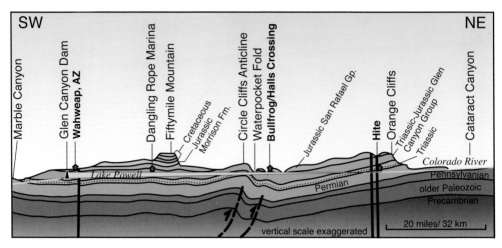

Lake Powell is edged with gently folded rock layers that generally slope slightly north-westward, so as you travel upstream you pass from younger to older layers. Starting at the dam and traveling northeastward, you pass Jurassic, Triassic, and Permian layers.

seventeen years, until 1980, for the lake to fill to capacity because water was allowed to flow downstream to lower-basin users.

Glen Canyon Dam is seated in Navajo Sandstone. Most of Lake Powell's shoreline is in the Glen Canyon Group: the Wingate Sandstone, Kayenta Formation, and Navajo Sandstone. Near the lake's western end, slightly younger rocks are present: Antelope Point Marina is on pale Page Sandstone; Wahweap and Stateline Marina are on the overlying deep-red Carmel Formation, a shoreline deposit of silt and sand; and the massive cliffs and buttes that edge the west end of the lake are Entrada Sandstone, a dune deposit.

East of Rock Creek Bay, around mile buoy 39, the canyon narrows because the Navajo extends below the lake floor; there is no soft shale exposed beneath it that would erode and undermine the Navajo's cliffs, causing the canyon walls to recede. Like the Entrada, the Page and Navajo were deposited in large areas of sand dunes. We know this from the sandstones' long, sweeping crossbeds and their fine, even-sized, well-rounded sand grains, like those that compose modern dunes.

North of Dangling Rope Marina, light-red cliffs of Entrada Sandstone are followed by shorter buff-colored cliffs of Romana Sandstone and pinkish beds of the Morrison Formation, here sandstone and siltstone. The tan rocks above are Cretaceous: a thin bench of Cedar Mountain and Dakota sandstones followed by a slope of Tropic Shale. The Straight Cliffs Formation extends to the skyline. Eight miles (13 km) east of Dangling Rope Marina, Forbidding Canyon leads south to Rainbow Bridge National Monument.

The San Juan River arm of Lake Powell starts in Navajo Sandstone. It winds through several tight meanders, and then older (Triassic) dark-red beds of the Chinle Formation rise above the lake briefly as it crosses the Circle Cliffs Uplift,

Iron oxide has formed concretions known as Moqui marbles (a) that weather out of the Navajo Sandstone. Dinosaur footprint casts (b) on the underside of a sandstone layer formed when fresh tracks were filled by sand. A slab from an oasis in the Navajo-age dune field is surfaced with mud cracks (c). A pothole (d) sits ready for a flash flood to make it larger by swirling around the rocks it holds. —Felicie Williams photos

an anticline. At Piute Canyon, the Chinle returns, and then the Moenkopi Formation appears. Both form slopes below the Navajo Sandstone cliffs. At Zahn Bay, another anticline reveals older Permian rocks: a thin layer of White Rim Sandstone cliffs hang over dark-red Organ Rock Shale, which is underlain by a bench of Cedar Mesa Sandstone.

As these layers sink beneath the surface and then rise again to the east, the lake ends and the San Juan River flows through several miles of muddy delta. The delta was built up when the lake was higher. Now, the river is cutting a new channel through it, not always in the same place it flowed in the past. There are waterfalls over sandstone shelves where before the channel could be rafted. The recreation area continues along the north side of the river until close to Goosenecks State Park and US 163.

Along the Colorado River arm, or main channel of Lake Powell, Hole-in-the-Rock, a steep, narrow defile on the west bank near mile buoy 66, and Cottonwood Canyon southeast of it, follow a northwesterly zone of roughly parallel faults, within which the rock is deformed or broken up; this structure dates back to Precambrian time. In 1880 settlers lowered their wagons here to cross the river. Near mile buoy 69, Escalante Canyon, walled with Navajo Sandstone, leads far to the north. It has its own side canyons and several natural arches.

Above Escalante Canyon, the main channel of Lake Powell continues between towering cliffs of Navajo Sandstone. Older rocks appear briefly where the lake crosses the Circle Cliffs Uplift at the Rincon. Below the Navajo lies the Kayenta Formation, with its horizontal shelves of river-deposited sand and mud, followed by cliffs of red Wingate Sandstone, another dune deposit; its weathered surface is coated with black desert varnish. The slopes beneath the Wingate are varicolored shales of the Chinle Formation and dark-red shale of the Moenkopi Formation. Both the Chinle and Moenkopi accumulated on broad plains. The contact between the two is an unconformity, a period of erosion marked by the ledge-forming Shinarump Conglomerate. The Chinle's weathered volcanic ash, or bentonite, swells when it is wet. Then when it dries, it crumbles away,

A couple is dwarfed by La Gorce Arch, partway up Davis Gulch off Escalante Canyon, and by the large white ring, the top of which is 65 feet (20 m) above the water in this photo. The faint, even higher rings are from the 1983 and 1984 floods. The rings are clay and silt the lake deposited, combined with calcite that is leached from the rock by lake water. La Gorce Arch formed when rock spalled from both sides of a meander. —Felicie Williams photo

undermines the Wingate, and causes rock slides. Ongoing slides, which make the hummocky slopes around the Rincon, are also numerous around Good Hope Bay, where there is more of the Chinle Formation.

The Halls Crossing area is at the axis of a gentle syncline. It makes a wide northwest-southeast-trending valley because the top of the resistant Navajo Sandstone dips below lake level, and the softer rocks above it have eroded back from the channel. Beds dip sharply into the syncline along the Waterpocket Fold.

High terraces help geologists deduce Glen Canyon's history. Gravel from the canyon rim, which contains Precambrian rock from Colorado, indicates the Colorado River began cutting this canyon 6 to 5 million years ago. —Felicie Williams photo

The Rincon, south of mile buoy 77, is an abandoned river channel on the crest of the Circle Cliffs Uplift. Rock layers slope to east and west away from it, exposing older beds in the middle. The rising sun highlights sand (hazy sky) carried into the Rincon by prevailing southwesterly winds. Behind the Rincon is the distant dome of Navajo Mountain. —Felicie Williams photo

Upstream from Halls Crossing and Bullfrog the Navajo Sandstone rises up and the lake narrows again. An occasional bench is covered with cobbles, many composed of dacite porphyry washed here from the intrusions of the Henry Mountains. Layers of gray limestone up to 10 feet (3 m) thick record oases that developed between sand dunes, when the water table was above the surface for a time.

Around mile buoy 110A the Kayenta and Wingate sandstones are quickly followed by older rocks: soft slopes of red and light-gray Chinle Formation, chocolate-brown Moenkopi Formation, Permian White Rim Sandstone (at mile marker 136), red-brown Organ Rock Shale, and wind-deposited Cedar Mesa Sandstone.

In recent years, Hite Marina, at the northern end of the lake, has closed due to low lake levels brought on by drought.

GRAND STAIRCASE–ESCALANTE NATIONAL MONUMENT

The Escalante River cuts a southeasterly course across Grand Staircase–Escalante National Monument, ending in Lake Powell. In 1873, John Wesley Powell named this river in honor of Father Escalante, the Spanish explorer who, with Father Dominguez, set out in 1776 to find a route from Santa Fe, New Mexico, to Monterey, California. They crossed western Colorado, entered Utah near Dinosaur, Colorado, described the fertile land near Utah Lake, and then traveled south along the Wasatch Front before winter storms forced them to return to Santa Fe. They found a crossing of the Colorado at Padre Bay and brought back an account and map of their route through this country.

Grand Staircase–Escalante National Monument spreads across 2,938 square miles (7,609 square km). It lies in a broad area west of the Waterpocket Fold. The Grand Staircase itself is a succession of cliffs that drops down to Lake Powell and the Colorado River, to the south. The northwest corner of the national monument adjoins Bryce Canyon National Park. Capitol Reef National Park lies along its eastern edge, and its southern boundary is shared with Glen Canyon National Recreation Area. Paved highways access the southwestern and northern edges of this immense monument, and there are numerous dirt routes as well. (See the road guides for US 89: Arizona State Line—Kanab and Utah 12: Torrey—US 89 near Panguitch for more information.)

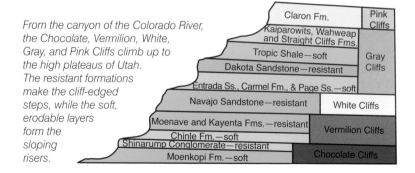

From the canyon of the Colorado River, the Chocolate, Vermilion, White, Gray, and Pink Cliffs climb up to the high plateaus of Utah. The resistant formations make the cliff-edged steps, while the soft, erodable layers form the sloping risers.

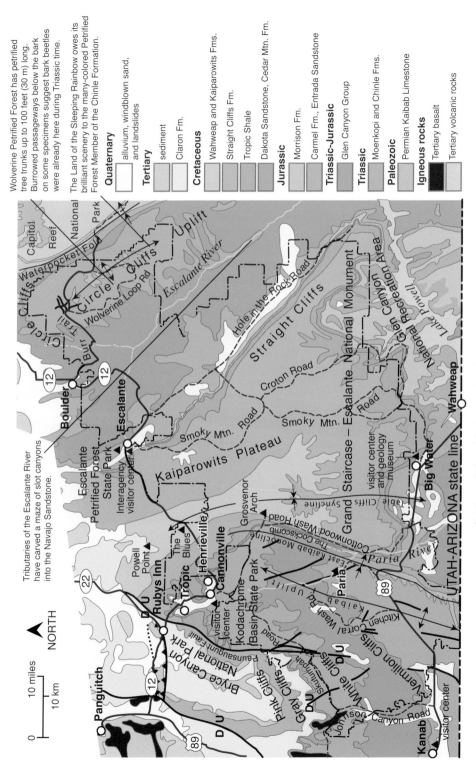

Geology of Grand Staircase–Escalante National Monument.

Wolverine Petrified Forest has petrified tree trunks up to 100 feet (30 m) long. Burrowed passageways below the bark on some specimens suggest bark beetles were already here during Triassic time.

The Land of the Sleeping Rainbow owes its brilliant scenery to the many-colored Petrified Forest Member of the Chinle Formation.

Quaternary

alluvium, windblown sand, and landslides

Tertiary

sediment

Claron Fm.

Cretaceous

Wahweap and Kaiparowits Fms.

Straight Cliffs Fm.

Tropic Shale

Dakota Sandstone, Cedar Mtn. Fm.

Jurassic

Morrison Fm.

Carmel Fm., Entrada Sandstone

Triassic-Jurassic

Glen Canyon Group

Triassic

Moenkopi and Chinle Fms.

Paleozoic

Permian Kaibab Limestone

Igneous rocks

Tertiary basalt

Tertiary volcanic rocks

Tributaries of the Escalante River have carved a maze of slot canyons into the Navajo Sandstone.

NORTH

0   10 miles

10 km

The western side of the Kaibab Uplift, a gentle upsweep of rock east of Kanab, is where the Chocolate and Vermilion Cliffs of the Grand Staircase are best exposed. The Chocolate Cliffs are in the Triassic Moenkopi Formation and the Shinarump Member of the Chinle Formation, composed of tidal flat, river, and floodplain sediments with the resistant ledge of Shinarump Member conglomerate at the top. The overlying Chinle Formation shale and Moenave and Kayenta Formations, sandstones deposited in dunes along a shoreline during Triassic and Jurassic time, form the Vermilion Cliffs. The White Cliffs above them are built of Jurassic Navajo Sandstone, which was laid down in a desert dune field that spread across most of Utah and into neighboring states.

The east edge of the Kaibab Uplift ends precipitously in a fold named the Cockscomb, where the Triassic and Jurassic layers, stretched thin, dive down in a narrow strip of wild cliffs. Here, the steps are instead vertical ridges, and the soft layers form valleys. East of the Kaibab Uplift, on the Kaiparowits Plateau, the layers form a very gentle syncline—a basin. Cretaceous rocks that make the Gray Cliffs of the Grand Staircase surface the plateau.

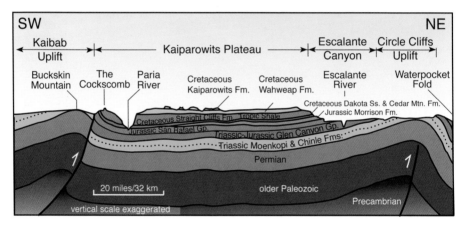

Southwest-northeast cross section across Grand Staircase–Escalante National Monument. The formations here span 200 million years. They were bent into two uplifts over deep faults that are separated by a gentle basin and have been incised by tributaries of the Colorado River.

The rock layers rise slowly to the east, almost imperceptibly at first, toward the Circle Cliffs Uplift. Cretaceous rocks have been eroded away east of the Straight Cliffs, except for a small patch of the Cretaceous Dakota Sandstone. East of the Cretaceous rocks, older Jurassic rocks surface: the Wingate, Kayenta, and Navajo sandstones. The Escalante River flows through these Jurassic rocks.

The Straight Cliffs' straightness is due to a series of deposits left along a straight shoreline of the Cretaceous Interior Seaway. The sandstone layers, deposited as sandbars or barrier islands, form a straight line of beds resistant to erosion. These alternate with shales deposited in lagoons and floodplains behind the barrier islands. The cliffs appear striped, with alternating light-colored sandstone cliffs and dark, tree-covered shale slopes.

Grosvenor Arch, 17 miles (27 km) southeast of Cannonville, is cut in the Cretaceous Cedar Mountain Formation and Dakota Sandstone. —Lucy Chronic photo

Utah 12 snakes across a sea of Navajo Sandstone, straddling sometimes-narrow ridges; this photo barely gives a feel for the amazing labyrinth of canyons below that feed the Escalante River. Hikers take ten days to hike 70 miles (113 km) from the north end of the river to its south end, where it joins Lake Powell. Most of the way, the Escalante is walled in by the Navajo sandstone's sweeping white crossbeds, and below them the red beds of the Wingate and Kayenta sandstones. —Lucy Chronic photo

The Burr Trail travels in a canyon of Wingate Sandstone. The trail is an alternate route to Capitol Reef National Park and the Waterpocket Fold; it is paved until near its precipitous drop across the Waterpocket Fold. —Lucy Chronic photo

The Navajo owes its white color to groundwater, which flowed through the rock some time after it had turned to stone, most likely during the Laramide Orogeny or an episode of Tertiary volcanism. The water must have contained small amounts of hydrocarbons or hydrogen sulfide from organic material or from igneous vapors. The water removed hematite from the sandstone and redeposited it wherever the water became oxidized, probably when it mixed with oxygen-rich water. Spherical concretions (Moqui marbles) often formed. They are very resistant to weathering, so they often end up loose on the surface.

Cliffs of the Triassic-Jurassic Wingate Sandstone ring the crest of the Circle Cliffs Uplift. The Wingate is the same age as the Moenave Formation, but it formed farther inland, in a dune field rather than along a shoreline. The Circle Cliffs Uplift is gently domed.

East of the crest of the Circle Cliffs Uplift the layers drop steeply into the Waterpocket Fold. Capitol Reef National Park is dedicated to preserving the fold's incredible hogback ridges, or "reefs."

The Kaibab and Circle Cliffs Uplifts rose during the Laramide Orogeny at the same time as the Rocky Mountains, even though they are far from the mountain chain. These gentle uplifts are characteristic of the Colorado Plateau. Faults far beneath them determined their geometry. The uplifts were pushed up by pressure from the west, from the subduction of the Farallon Plate along the western edge of the continent. All of the uplifts have gentle western slopes that lead up to steep drop-offs on the east. Long after the Laramide Orogeny, probably between 6 and 5 million years ago, rivers started the canyon cutting that has exposed the beautiful geology we enjoy today.

NATURAL BRIDGES NATIONAL MONUMENT

In Natural Bridges National Monument, meandering White and Armstrong Creeks have cut deeply into the Permian Cedar Mesa Sandstone. Weak horizontal layers of silt and clay, distant fingers of the Cutler Arkose washed eastward from mountains that existed during Permian and Pennsylvanian time, interrupt the sandstone beds. It's the erosion of these weaker layers that helped undermine the necks between the meanders, leading to the development of the bridges.

Though often dry, streams in the national monument occasionally rush through the narrow canyons, pounding against their outside curves with sand, pebbles, and boulders and whittling the curves toward one another until finally a passage is broken through. Repeated freezing and thawing helps, loosening sand grains, scabby flakes, and even large blocks of rock. Studies done elsewhere have found that rivers can cut down through sandstone 3 to 7 inches (8 to 18 cm) per 1,000 years if the channel has a good gradient, or slope. With so little water, the down-cutting at Natural Bridges may be slower. Side-cutting is faster—perhaps ten times as fast. Weathering is the slowest force of erosion, removing only 1.5 to 3 inches (4 to 8 cm) of rock per 1,000 years. Of course, these rates vary with the amount of water, the strength of the rock, and even the orientation of the rock. Sandstone facing the afternoon sun weathers twice as fast as sandstone that faces away from the sun. Rockfalls, the largest contributor to erosion in terms of volume, add even more unpredictability to how fast the sandstone erodes in Natural Bridges National Monument.

Other interesting geologic features can be found at Natural Bridges. After rain, small potholes hold pools of water. Tiny plants and animals live in them, enlarging the holes by secreting acid that dissolves the rock's limy cement. Lichens, algae, and bacteria form dark tapered streaks below seepage lines and also help dissolve the rock's cement. Desert varnish is absent here; the pale sandstone does not contain enough manganese and iron for it to form.

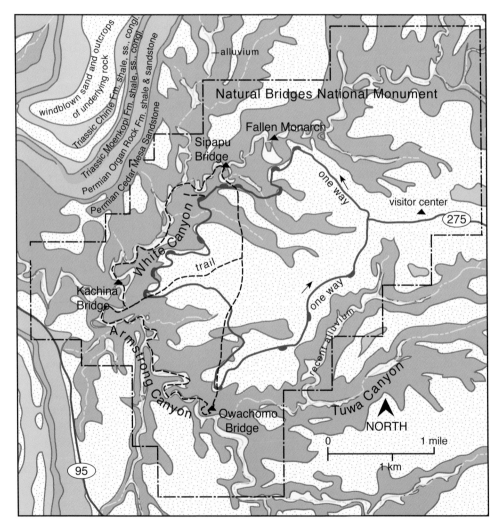

Geological map of Natural Bridges National Monument. The orientation of the canyons was guided by two sets of joints, one northeasterly and one northwesterly. Both also influenced the winding of the stream channels, setting the stage for the rivers to cut natural bridges. Fallen Monarch would have been the largest natural bridge in North America—in the neighborhood of 330 to 360 feet (100 to 110 m) across—had it not fallen.

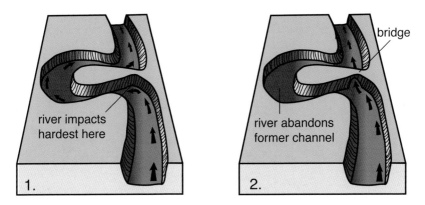

Streams in the national monument occasionally pound against the outside curves of the narrow canyons with sand, pebbles, and boulders, whittling the curves toward one another until finally a passage is broken through.

Kachina Bridge represents an early stage of bridge development; it is thick and massive, with a relatively small passage below it. The pile of rock beneath it crashed down from the bridge in 1992.

Sipapu Bridge is thinner, and the stream below it is no longer wearing away its abutments. Rain, snow, freeze-and-thaw weathering, and rockfalls will cause it to grow thinner.

Owachomo Bridge is in a late state of development; it is slender and increasingly fragile. The creek has cut a new, deeper channel since this old bridge formed. Now the bridge is high and dry. As these bridges continue to erode and ultimately weaken and fall, others will form at sites where streams still pound the curves of the winding canyons.

Natural bridges over time.

The fine textures in the Cedar Mesa Sandstone are fascinating. Thin bands of fine and coarser grains formed when individual ripples, with fine sand on their crests and coarse sand in their troughs, migrated up the dunes. Crumpled bedding sometimes interrupts the smooth pattern of the crossbeds, revealing where sand slumped down the steep leeward slopes of the dunes. Fossil mud cracks in mudstone layers look like white carrot-like wedges from the side.

In 2006, Natural Bridges National Monument was declared the world's first International Dark Sky Park. Here, with low humidity and no light pollution, the desert night sky reveals far more of the universe than most people will ever see.

RAINBOW BRIDGE NATIONAL MONUMENT

Access to Rainbow Bridge is easy from a dock along Lake Powell. Land surrounding the national monument is in the Navajo Nation, so you must apply for hiking permits with the nation.

Spanning Bridge Canyon, a narrow gorge that descends from the great dome of Navajo Mountain to the south, the graceful arch of Rainbow Bridge is composed of Navajo Sandstone. This sandstone was made strong enough to sustain the bridge by the thickness and cleanness of its sand layers, their uniform fine-grained texture, and the strength of the calcium carbonate that cemented the sand grains together.

The National Park Service measures the bridge at 275 feet (84 m) wide and 290 feet (88 m) high. The Natural Arch and Bridge Society, using their more recently established standards, lists it as the world's eighth largest known natural bridge: 234 feet (71 m) wide and 245 feet (75 m) high. The bridge is almost as tall as the Statue of Liberty. Navajo Mountain is the catchment for Bridge Creek's waters, which run beneath the arch.

Beyond the east end of Rainbow Bridge is the deeply carved looping path of an abandoned channel. Rushing around its curves, Bridge Creek would have thrown a full arsenal of sand, pebbles, and boulders against the rock that jutted into its path, eroding and thinning it from both sides until finally breaking through and choosing the shorter route downhill. Judging by the many abandoned curves leading down to Rainbow Bridge, it seems possible that other huge arches have stood and fallen here in times long gone.

The Kayenta Formation, the slabby, ledgy layer on which Rainbow Bridge sits, did its part in the formation of the bridge. It is a stream-deposited sandstone with thin clayey layers between its sand beds. A line of springs and seeps along the top of this formation shows that water percolating down through the Navajo Sandstone stops at this less-permeable rock and flows out along bedding planes. This means the lowest portion of the Navajo remains wet. The water dissolves and carries off the Navajo's cement. It also freezes and thaws many times each winter, loosening grains and sheets of sand. Given enough years, even frost can undermine strong cliffs and lead to rockfalls. These processes weakened rock below the arch, helping Bridge Creek to punch its way through.

Look for the 14-inch-wide (36 cm), three-toed footprint of a dinosaur in the Kayenta Formation at one of the viewing areas. 200 million years ago a fearsome carnivore walked here!

Rainbow Bridge. —Felicie Williams photo

From the air, Rainbow Bridge, center, is dwarfed by Bridge Canyon. Though Bridge Creek generally flows northwest, toward the top in this photo, its channel winds because the creek must find its way across a series of northeasterly joints. The creek formed the gravel terrace, green at the upper left, between 780,000 and 500,000 years ago; it is 1,200 feet (370 m) higher than the creek. This means the channel was cut at a geologically fast rate of 0.018 to 0.028 inches (0.045 to 0.071 cm) per year.
—Falk Weihmann photo

At first, Rainbow Bridge must have been massive, a smallish opening allowing the stream to pass beneath a heavy span. As weathering undermined and loosened slabs of sandstone, thin curving sheets and probably some large blocks fell away, widening the aperture. Gradually, the stream, still cutting downward, carried off the pieces. Weathering continues to perfect the span and will inexorably make it thinner until one day it falls.

Part of the scenery at Rainbow Bridge is Navajo Mountain, to the south. The mountain's core is a lone laccolith, an intrusion that pushed up the sedimentary layers, possibly during the Laramide Orogeny. The layers still arch over the mountain.

TIMPANOGOS CAVE NATIONAL MONUMENT

Joined by excavated passages, three natural limestone caverns—Hansen, Middle, and Timpanogos—make up Timpanogos Cave National Monument, north of Orem. All three caverns are finely adorned with stalactites, stalagmites, cave popcorn, helictites, and other ornaments.

The Mississippian Deseret Limestone, in which the caves developed, was deposited far to the west in the Oquirrh Basin, which developed west of the Utah Hingeline on the gradually subsiding edge of the continent. The limestone was part of a sequence of carbonate mud and sand that is some 30,000 feet (9,100 m) thick. During the Sevier Orogeny of Late Cretaceous time, 100 to 65 million years ago, the rocks deposited in the Oquirrh Basin were thrust 30 to 60 miles (48 to 100 km) eastward over layers of similar age. The Oquirrh Basin rocks, in the hanging, or upper, wall of the great Charleston-Nebo Thrust Fault, were extensively faulted.

The trail to the caverns, which are high on the wall of American Fork Canyon, ascends from 650-million-year-old Precambrian rocks to 330-million-year-old Mississippian rocks. Ordovician and Silurian rock layers are missing. These sedimentary rocks have a complex geologic history, having ended up steeply tilted and faulted and far from where they formed. It is surprising that other than their slant and a few faults, the beds here are relatively undisturbed.

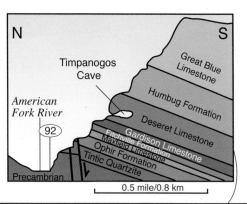

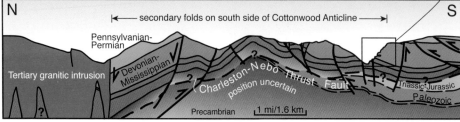

The Laramide Orogeny, coming on the heels of the Sevier Orogeny 65 to 45 million years ago, caused the upward fold of the Uinta-Cottonwood Anticline to the north, tilting the Oquirrh Basin layers southward on the south limb of a secondary fold, and, of course, faulting them more. Even more faulting followed the end of the orogeny as the rocks settled back along the old faults.

Starting about 17 million years ago, extension across the West caused Basin and Range faulting, which broke the rocks along north-south-trending normal faults and exposed the Wasatch Range with its beds of Deseret Limestone. This extensional episode continues today.

The stretching forces that have held sway over the last 17 million years have caused gaps to open up along some of the faults. Fault surfaces smoothed by the movement of rock against rock remain visible in the caves today. Water has enlarged these open spaces wherever they are. Rain and snow absorb carbon dioxide from both the atmosphere and soil; the carbon dioxide reacts with the water to form weak carbonic acid. The acid dissolves the limestone, enlarging joints, faults, and natural pore spaces.

The Timpanogos caverns were enlarged before most of the cave ornaments formed. The water that dissolved the limestone would not have contained enough calcium carbonate for ornaments to form. But as the water flow lessened, and the water became more saturated with calcium carbonate, cave popcorn crystallized on some of the cave walls.

Calcite crusts and draperies developed where water ran across cavern surfaces in sheets. The red coloring is from iron oxide.
—Lucy Chronic photo

Not until quite recently, geologically speaking, possibly only 335,000 years ago, did the ancestor of the American Fork River cut down far enough to lower the water table below the level of the caves. Only then could the calcium carbonate decorations that are dependent on evaporation form. After the caves were exposed to air, seeping water laden with dissolved calcium carbonate dripped from the cavern ceilings and walls, depositing the mineral calcite or aragonite as it evaporated. Stalactites developed where water dripped from the ceilings, and stalagmites built up where drips splashed on the cavern floors. Other cave ornaments have also formed, like frostwork, draperies, and helictites.

Timpanogos is a wet cave, so its ornamentation is ongoing. The growth is infinitesimally slow—0.5 inch (1.3 cm) in 150 years. Each drop of water adds a miniscule bit of calcite to an ornament. Little bumps, beginning stalactites, have developed on the ceilings of the manmade tunnels completed in the 1930s. Aragonite frostwork covers older ornaments. Calcite lily pads grow outward into silent pools. It can be all too easily destroyed, so heavy entrance doors maintain the cave's humidity, and visitors are guided through the caverns on walkways designed to preserve the fragile ornaments.

This cave is particularly well known for its curly helictites, very thin calcite tubes that grow as water passes through tiny pores or breaks in a cave wall because of hydrostatic pressure. At this very small scale, capillary action and hydrostatic pressure, not gravity, draws water along the developing helictite tubes, so they grow in spaghetti-like squiggles in all directions. —Lucy Chronic photo

ZION NATIONAL PARK

Zion National Park lies at the edge of the Colorado Plateau, in the transition zone between the Plateau and the Basin and Range. Erosion is very fast here, sped up by the difference in elevation between the two regions. Erosion has created much of Zion's wondrous scenery by carving into two Jurassic formations: the Kayenta Formation and the Navajo Sandstone.

The Navajo, with sweeping crossbeds and fine, rounded sand grains, forms the massive walls and towers of Zion Canyon. It was deposited in a huge field of sand dunes. The horizontal layers that interrupt its slanting crossbeds were silt and clay blown or washed onto flat surfaces between the dunes.

The Kayenta Formation, below the Navajo, has more easily eroded beds of sandstone, siltstone, and mudstone, with a short sandstone cliff partway down. As the Kayenta beds wear away, the Navajo Sandstone is undermined until it breaks loose and falls, coming to rest as steep slopes of rubble below the cliffs.

The Navajo and Kayenta are only the middle part of the 7,000-foot (2,130 m) thickness of sedimentary rocks in the park. Written in thick layers here is a record of 270 million years of continental and worldwide change.

Zion lies on a fault block between two active faults. The Hurricane Fault, just west of the Kolob Canyons, has down-faulted the area to the west 3,600 to

At the south end of the Markagunt Plateau, the North Fork of the Virgin River has carved through 2,000 feet (610 m) of white and pink Navajo Sandstone to create the walls of magnificent Zion Canyon. —Chris Hinze photo

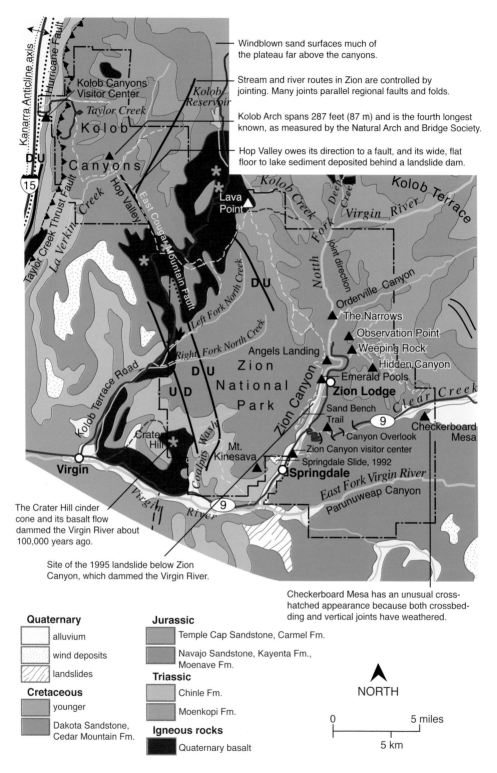

Windblown sand surfaces much of
the plateau far above the canyons.

Stream and river routes in Zion are controlled by
jointing. Many joints parallel regional faults and folds.

Kolob Arch spans 287 feet (87 m) and is the fourth longest
known, as measured by the Natural Arch and Bridge Society.

Hop Valley owes its direction to a fault, and its wide, flat
floor to lake sediment deposited behind a landslide dam.

Kanarra Anticline axis
Hurricane Fault

Kolob Canyons
Visitor Center

Kolob
Reservoir

Taylor Creek

K o l o b

15

D U

C a n y o n s

Taylor Creek Thrust Fault

La Verkin Creek

Hop Valley

East Cougar Mountain Fault

Lava
Point

Kolob Creek

Deep Creek

Kolob Terrace

Virgin River

North Fork

joint direction

Left Fork North Creek

D U

Orderville Canyon

The Narrows

Right Fork North Creek

Observation Point

Weeping Rock

Angels Landing

Hidden Canyon

Z i o n

Emerald Pools

D U

N a t i o n a l

Zion Canyon

Zion Lodge

Clear Creek

U D

P a r k

Sand Bench
Trail

9

Checkerboard
Mesa

Kolob Terrace Road

Crater
Hill

Coalpits Wash

Mt.
Kinesava

Canyon Overlook

Zion Canyon visitor center

Springdale Slide, 1992

Virgin

Springdale

East Fork Virgin River

9

Virgin River

Parunuweap Canyon

The Crater Hill cinder
cone and its basalt flow
dammed the Virgin River about
100,000 years ago.

Site of the 1995 landslide below Zion
Canyon, which dammed the Virgin River.

Checkerboard Mesa has an unusual cross-
hatched appearance because both crossbedding and vertical joints have weathered.

Quaternary

alluvium

wind deposits

landslides

Cretaceous

younger

Dakota Sandstone,
Cedar Mountain Fm.

Jurassic

Temple Cap Sandstone, Carmel Fm.

Navajo Sandstone, Kayenta Fm.,
Moenave Fm.

Triassic

Chinle Fm.

Moenkopi Fm.

Igneous rocks

Quaternary basalt

NORTH

0   5 miles

5 km

Geology of Zion National Park.

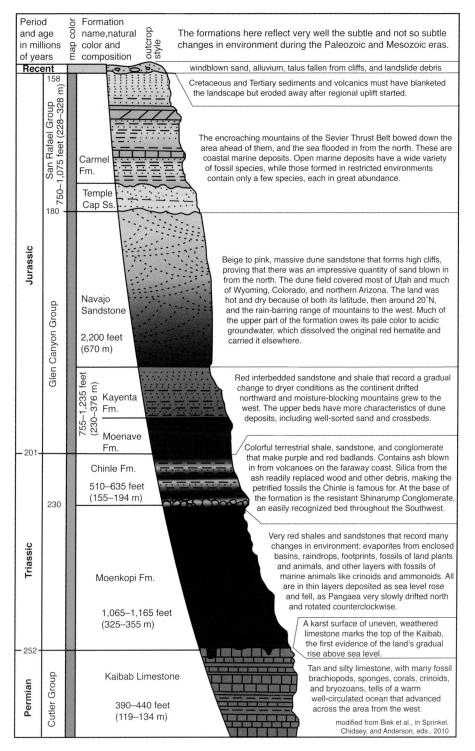

Period and age in millions of years	map color	Formation name, natural color and composition	outcrop style	The formations here reflect very well the subtle and not so subtle changes in environment during the Paleozoic and Mesozoic eras.

Recent — windblown sand, alluvium, talus fallen from cliffs, and landslide debris

158 — Cretaceous and Tertiary sediments and volcanics must have blanketed the landscape but eroded away after regional uplift started.

San Rafael Group 750–1,075 feet (228–328 m)

Carmel Fm. — The encroaching mountains of the Sevier Thrust Belt bowed down the area ahead of them, and the sea flooded in from the north. These are coastal marine deposits. Open marine deposits have a wide variety of fossil species, while those formed in restricted environments contain only a few species, each in great abundance.

Temple Cap Ss.

180

Jurassic

Glen Canyon Group

Navajo Sandstone 2,200 feet (670 m) — Beige to pink, massive dune sandstone that forms high cliffs, proving that there was an impressive quantity of sand blown in from the north. The dune field covered most of Utah and much of Wyoming, Colorado, and northern Arizona. The land was hot and dry because of both its latitude, then around 20°N, and the rain-barring range of mountains to the west. Much of the upper part of the formation owes its pale color to acidic groundwater, which dissolved the original red hematite and carried it elsewhere.

755–1,235 feet (230–376 m)

Kayenta Fm. — Red interbedded sandstone and shale that record a gradual change to dryer conditions as the continent drifted northward and moisture-blocking mountains grew to the west. The upper beds have more characteristics of dune deposits, including well-sorted sand and crossbeds.

Moenave Fm.

201

Chinle Fm. 510–635 feet (155–194 m) — Colorful terrestrial shale, sandstone, and conglomerate that make purple and red badlands. Contains ash blown in from volcanoes on the faraway coast. Silica from the ash readily replaced wood and other debris, making the petrified fossils the Chinle is famous for. At the base of the formation is the resistant Shinarump Conglomerate, an easily recognized bed throughout the Southwest.

230

Triassic

Moenkopi Fm. 1,065–1,165 feet (325–355 m) — Very red shales and sandstones that record many changes in environment: evaporites from enclosed basins, raindrops, footprints, fossils of land plants and animals, and other layers with fossils of marine animals like crinoids and ammonoids. All are in thin layers deposited as sea level rose and fell, as Pangaea very slowly drifted north and rotated counterclockwise.

A karst surface of uneven, weathered limestone marks the top of the Kaibab, the first evidence of the land's gradual rise above sea level.

252

Permian

Cutler Group

Kaibab Limestone 390–440 feet (119–134 m) — Tan and silty limestone, with many fossil brachiopods, sponges, corals, crinoids, and bryozoans, tells of a warm well-circulated ocean that advanced across the area from the west.

modified from Biek et al., in Sprinkel, Chidsey, and Anderson, eds., 2010

The stratigraphy of Zion National Park encompasses 270 million years of geologic history.

Shifting winds in Jurassic time formed the intricate patterns of dune crossbedding revealed here in weathered Navajo Sandstone; the sand was deposited on leeward dune faces. —Lucy Chronic photo

4,900 feet (1,100 to 1,490 m). The Sevier Fault, east of Zion in the Sevier Valley, has dropped the Zion side down. Both developed as a result of Basin and Range extension, and both are active faults.

The erosion that has resulted in Zion's amazing scenery is directly related to the down-cutting of the Colorado River. The Colorado we know today began flowing to the Gulf of California only about 5.5 million years ago. Energized by the huge difference in elevation between the Colorado Plateau and the Gulf of California, the river rapidly cut headward into tributary drainage systems, often following weak zones left by faults, until its basin extended all the way through Utah into Wyoming and Colorado.

Erosion had exposed most of Zion by about 2 million years ago, about when rivers and creeks began to cut its canyons. Locally, the down-cutting ability of the Virgin River is directly controlled by offset on the Hurricane Fault.

The two main sections of Zion National Park are Zion Canyon and Kolob Canyons, both accessible on paved roads. Several of the park's many remarkable features lie outside of the canyons, including Checkerboard Mesa, with bare-rock slopes marked by crossbedding and deeply weathered vertical joints; a petrified forest accessible by trail in the southern portion of the park; and several cinder cones in the region separating Zion Canyon from Kolob Canyons.

Zion Canyon

Though usually quiet, the North Fork of the Virgin River takes on fearsome proportions during heavy summer thundershowers. Picking up sand, pebbles, and even large boulders, it pounds its channel and the flanking cliffs. A basalt flow remnant that caps a mesa 1,300 feet (400 m) above the Virgin River is 1 million years old. Since the lava originally flowed down the river channel, the river has cut downward at an average rate of 1,300 feet (400 m) per million years, or about 1.6 inches (4 cm) per century. Not bad, for canyon cutting. In Pleistocene time, when rain and snow were more abundant, the river must have been torrential much of the time. Erosion doubtless was much more rapid then.

Landslides, falling rock, and basalt flows have blocked Zion's canyons many times, making temporary lakes. Earthquakes trigger many of the landslides. In 1995 a landslide below Zion Lodge dammed the Virgin River. Its lake overflowed, washed out the dam, and, with a rush of suddenly released water, carried away a section of road and stranded people at the lodge. Landslides have occurred at the same place several times during the last 7,000 years. At times, their lakes lasted longer, and you can see fine lake sediment upstream of the landslides.

The Kayenta Formation plays a role in the width of Zion Canyon. Just at the end of the park road at river level, a recent rockslide of broken blocks of Navajo Sandstone occurred where erosion of the underlying Kayenta Formation undermined the Navajo. But just beyond the end of the paved trail, where the Kayenta Formation is below the surface and not exposed to river erosion, vertical Navajo Sandstone cliffs plunge right down to the river, framing a narrow gorge up which neither road nor permanent trail can be built. This is the Narrows of Zion Canyon.

In most of Zion Canyon, a line of springs and seeps marks the contact between the Navajo Sandstone and the Kayenta Formation. Rainfall and snowmelt soak easily down through the porous Navajo but are forced sideways by the much-less-permeable Kayenta, forming springs and seeps where the Kayenta is exposed in the canyon walls. Similar but smaller springs and seeps also develop in the Navajo Sandstone above some of the siltstones that were deposited between the dunes. These wet spots create hanging gardens high on the canyon walls. Springwater speeds weathering, paticularly of the freeze-and-thaw type.

A number of trails zigzag up the walls of Zion Canyon or thread its intricate maze of tributary canyons. Trails to the east and west rims and Observation Point provide especially good opportunities to look at the Navajo Sandstone— its garland-like crossbedding and the various shapes and forms into which it erodes. Hidden Canyon Trail is as promising as its name sounds; it leads to a little canyon formed along one of the many joints that cut the Navajo Sandstone. The "trail" through the Narrows is both adventure and geology lesson (except for sandbars, you wade up the stream). At its tightest point, the Narrows of Zion Canyon is 16 feet (5 m) wide and 1,000 feet (300 m) deep. Don't go in threatening weather, though. Angels Landing Trail is fun unless heights bother you.

Kolob Canyons

The road into Kolob Canyons crosses the east limb of the Kanarra Anticline and the Hurricane Fault, and then climbs through the Moenkopi, Chinle, Moenave, and Kayenta Formations before reaching the Navajo Sandstone. Views from the road and the short scenic trail show all of the west side of Zion and the huge expanse of land to the south.

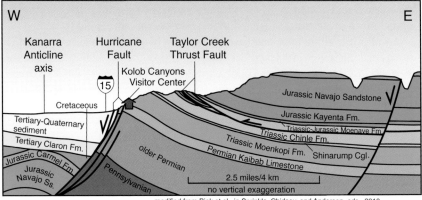

modified from Biek et al., in Sprinkle, Chidsey, and Anderson, eds., 2010

West-east cross section just north of the road into Kolob Canyons shows how the layers were bowed up in the Kanarra Anticline. The Hurricane Fault cuts the anticline, dropping the anticline's western limb far beneath the surface.

The eastern limb of the Kanarra Anticline is visible in the layering of the Navajo Sandstone of the cliff and the underlying Kayenta Formation. —Lucy Chronic photo

Listen!
Hear the tender trickling call
That mellow drops from canyon wall!
If I could dwell forever in the
beauty of this place,
I, too, would sing a never-ending song.

—Halka Chronic

—Halka Chronic painting

Glossary

alkali. A mixture of calcium, potassium, and sodium carbonates that is common in playa deposits.

alluvial apron. An apron-like slope of gravel and sand formed of merged alluvial fans. Also known as a bajada.

alluvial fan. A sloping, fan-shaped mass of gravel and sand deposited by a stream as it issues from a mountain canyon.

alluvium. Sediment deposited by rivers or streams.

alunite. A soft, gray sulfate mineral that forms when volcanic vapors alter volcanic rock, usually inside a cooling caldera. A source of potassium and aluminum.

ammonite. An extinct group of cephalopods with external shells that often had ornate chambers. Similar to a chambered nautilus.

andesite. An extrusive igneous rock that is made mostly of plagioclase with lesser amounts of dark minerals.

angular unconformity. See **unconformity, angular**

anhydrite. A calcium sulfate mineral similar to gypsum except it contains no water.

anticline. An upward fold. When eroded, the oldest rocks are in its center.

aquifer. A porous, permeable rock or layer of sediment from which water may be obtained.

aragonite. A form of calcium carbonate common in cave minerals and mollusk shells.

arch. A fairly large opening formed by the erosion of a rock fin; does not bridge a watercourse.

arkose. A sedimentary rock, often gritty, deposited close to an eroding mountain range and containing at least 25 percent feldspar.

artesian water. Water that rises above a water-bearing layer because of hydrostatic pressure.

ash, volcanic. See **volcanic ash**

ashfall tuff. See **tuff, ashfall**

ashflow tuff. See **tuff, ashflow**

asthenosphere. The ductile part of the upper mantle below the lithosphere. It extends about 430 miles (700 km) below the lithosphere.

badlands. A dry geological terrain, often barren of vegetation, that develops when wind and water have extensively eroded soft sedimentary rock.

bajada. See **alluvial apron**

bar. An elongated sedimentary deposit sculpted by the flow of water and wind.

barchan dune. A crescent-shaped sand dune with curving arms pointed downwind.

basalt. A dark-gray to black volcanic rock poor in silica and rich in iron and manganese.

basin. A broad, enclosed depression, commonly with no drainage to the outside.

Basin and Range. A region characterized by rank upon rank of long, narrow mountain ranges that alternate with basins; developed as a result of tectonic extension. Includes the deserts of western Utah, much of Nevada, and areas in other surrounding states.

batholith. A mass of intrusive rock with over 40 square miles (104 square km) of surface area.

bed. A single layer of sedimentary rock distinguishable from beds above and below.

bedrock. The solid rock that lies below loose surface material.

bentonite. Soft, porous, light-colored clay formed by the decomposition of volcanic ash.

beryl. A pale-green pegmatite mineral.

biotite. Black iron- and magnesium-rich mica.

bioturbation. The disturbing of sedimentary bedding by animals and/or plants.

boulder. A rounded rock fragment with a diameter greater than 10 inches (25 cm).

brachiopod. A marine invertebrate with two bilaterally symmetrical shells. Common fossils.

braided stream. A stream that divides into an interlacing network of small channels.

breakaway zone. The zone where rocks of the hanging (upper) wall break away and move along a fault.

breccia. Rock consisting of broken, angular rock fragments embedded in finer material.

bryozoan. A group of tiny marine shellfish that live in coral-like colonies. Common fossils.

butte. An isolated hill or small mountain, usually with a horizontal top and steep sides.

calcite. The most common form of calcium carbonate. The principal mineral in limestone and travertine and a common cement holding sediment together in sedimentary rocks.

caldera. A basin-shaped, cliff-edged depression formed by volcanic eruption and the subsequent collapse of rock into the emptied magma chamber.

caliche. A crusty, whitish rock or cement of calcite and other minerals that makes a hard layer near the surface in porous sediment. Common in deserts.

carbonaceous. Containing carbon or coal derived from organic material.

carbonate. The calcium carbonate minerals calcite and aragonite, and dolomite (magnesium-calcium carbonate), and rocks composed of them.

chert. A hard, compact variety of quartz with microscopic or submicroscopic grains.

cinder. A chunk of porous basalt ejected from a vent.

cinder cone. A small conical volcano composed of bits of basalt cinder ejected from a vent.

cirque. A steep-walled semicircular valley scooped from a mountainside by the head of a glacier.

clay. The smallest sediment particle, those less than 0.00016 inch (0.004 mm) in diameter.

cobble. A rounded rock fragment with a diameter between 2.5 and 10 inches (6.4 and 25 cm).

Colorado Plateau. A region of relatively undisturbed, flat-lying sedimentary rocks that includes the high plateaus and deep canyons of southeastern Utah and parts of Arizona, New Mexico, and Colorado.

columnar jointing. A joint pattern that forms in volcanic flows as the rock shrinks with cooling. Cracks form first on the flow surface and then extend into the flow as it cools, forming columns that are often hexagonal.

composite volcano. See **volcano, composite**

concretion. A spherical or nodular concentration of minerals deposited within preexisting sediment. Usually harder than the surrounding rock.

conglomerate. Rock composed of rounded, water-worn gravel and finer sediment.

continental crust. The crust that makes up the continents. Lighter in color and density than oceanic crust.

coral. A hard calcium carbonate skeleton of colonial marine organisms. Often forms masses, or reefs.

correlate. To be of the same age.

crater. The circular hollow at the top of a volcano from which volcanic material is ejected. Also the circular depression made by a meteorite impact.

creep. The gradual downhill movement of soil or rock due to gravity.

Cretaceous Interior Seaway. A sea that extended north to south across North America from mid-Cretaceous to Paleocene time, filling the downwarped region east of the growing mountains of the Sevier Orogeny.

crinoid. A group of marine animals with jointed stems and arms. Often called sea lilies.

crossbedding. Fine diagonal layering within larger, originally horizontal layers of sedimentary rock. Formed as surfaces of ripple marks, dune slopes, and sandbar slopes.

crust. The outermost, cooled and hardened part of Earth. It is 3 to 25 miles (5 to 40 km) thick.

cuesta. A ridge that, in profile, has a long, gentle slope of resistant caprock and a short, steep slope that is the eroded edge of the underlying rock layer.

dacite. A volcanic rock made of quartz, feldspar, and more dark minerals than granite.

debris flow. A dense mudflow containing abundant coarse material.

delta. A deposit that develops as a river carrying sediment enters a larger body of water, causing the current to slow and the sediment to settle out of it.

desert pavement. A veneer of tightly packed pebbles that is left when sand and silt are blown away.

desert varnish. A dark, shiny surface of iron and manganese oxides found on many exposed rock surfaces in desert regions.

detachment fault. See **fault, detachment**

diatreme. A volcanic pipe filled with breccia, formed by subterranean explosions caused when hot magma hits groundwater. The origin of the magma is often the upper mantle, so it can contain diamonds.

differential erosion, or **differential weathering**. The variation in rates of erosion or weathering that results from differences in strength between different kinds of rock.

dike. A sheetlike igneous intrusion that cuts across bedding or other igneous rocks.

diorite. An intrusive igneous rock that contains more dark minerals and less quartz than granite.

dip. The angle at which a rock layer is inclined, measured from the horizontal.

dolomite. A carbonate of calcium and magnesium; the sedimentary rock made up of that mineral.

dome. An anticline in which sedimentary rocks dip away in all directions.

earthflow. A slow flow of soil and weathered rock lubricated with water.

earthquake. Ground shaking resulting from movement along a fault.

echinoderm. A type of invertebrate marine animal, including starfish, crinoids, and sea urchins, with an external shell made up of many radially arranged calcite plates, often with five-point symmetry.

entrenched meander. See **meander, entrenched**

erosion. The gradual destruction of rock or soil by wind, water, or ice. See also **headward erosion**

erosion surface. The surface left after rock or soil has been eroded from it.

escarpment. A cliff or steep slope that drops off from a higher region.

evaporite. A mineral, notably salt (halite), gypsum, anhydrite, or potash, deposited from water that is saturated with the mineral due to the water body's partial or complete evaporation.

extrusive igneous rock. See **volcanic rock**

facet. A triangular face on the front of a mountain that is the uneroded surface of a fault. Facets show a fault's exact position and indicate quite recent movement along a fault.

fault. A rock fracture along which displacement has occurred.

fault, detachment. A normal fault that curves with depth until it is gently sloping and above which pieces of the upper, solid crust have slid a great distance over ductile lower layers.

fault, normal. A steep fault along which the overhanging fault block moves down. Forms in an extensional tectonic setting.

fault, reverse. A steep fault along which the overhanging fault block moves up. Forms in a compressional tectonic setting.

fault, thrust. A low-angle fault (less than 45°) in which the upper fault block is pushed over the lower one. Forms in a compressional tectonic setting.

fault block. A segment of Earth's crust bound on two or more sides by faults.

fault scarp. A slope or cliff formed by movement along a fault and marking the fault's location.

fault zone. A width of many small, roughly parallel faults that is markedly thicker than a single fault.

feldspar. A group of common light-colored minerals that contain potassium, aluminum, and calcium.

fill, valley. Gravel, sand, and other sediment washed or blown into a valley.

flatiron. A roughly triangular erosional remnant of an inclined, resistant rock layer.

floodplain. A level surface next to a river channel, onto which the river overflows during floods.

fold. A curve or bend in rock strata.

footwall. The lower fault block along a fault.

formation. A named, recognizable, mappable unit of rock.

fossil. Remains or traces of a life-form preserved in rock.

fossil, trace. An impression left by biological activity, such as burrows, borings, footprints, feeding marks, and root cavities.

freeze-and-thaw weathering. The breakdown of rock by repeated freezing and thawing of water held in pores and cracks.

fuller's earth. Very fine-grained clay used for bleaching or the absorption of impurities in oils.

fusulinid. A single-celled marine organism with a shell usually the size of a grain of rice. Lived during Silurian and Permian time and a useful indicator of time because of their rapid evolution and widespread occurrence.

geothermal. Pertaining to Earth's heat.

geyser. A type of spring that intermittently erupts jets of water and vapor.

glaciation. The formation and movement of glaciers or large sheets of ice.

glacier. A large mass of ice formed by the compaction and recrystallization of snow, which, because of its weight, creeps slowly downslope or outward from its center.

gneiss. Coarse-grained metamorphic rock with bands of lighter- and darker-colored minerals.

gooseneck. A stream or river meander where the channel turns back on itself.

graben. A valley that formed where a fault block dropped down along faults on either side of it.

Grand Staircase. A particular sequence of geological formations that form large steps (soft layers) and risers (harder layers). Grand Staircase–Escalante National Monument was named for the sequence that rises north of the Colorado River in southwest Utah.

granite. A coarse-grained intrusive igneous rock composed of quartz and feldspar and peppered with dark biotite and hornblende.

granitic. Of or similar to granite.

graptolite. An extinct microscopic animal that formed floating, twig-like colonies that are often carbonized as fossils that resemble pencil writing.

gravel. Loose, rounded sediment larger than 0.08 inch (2 mm) and smaller than 2.5 inches (64 mm).

groundwater. Subsurface water that fills spaces in rock or sediment.

group. Several adjacent formations that usually occur together and can be mapped as one.

gypsum. A common evaporite mineral; often colorless to white but can be colored due to impurities.

halite. The mineral name for common salt, sodium chloride, which is an evaporite mineral.

hanging garden. A spring-nourished garden on the side of a cliff.

hanging wall. The upper fault block along a fault.

headward erosion. Erosion at the head of a stream that expands the stream's drainage basin.

hematite. A common red iron oxide mineral.

hogback. A ridge with a sharp crest that developed as a result of the differential erosion of steeply tilted or vertical layers.

hoodoo. A fantastic, often grotesque column or pinnacle of rock.

hornblende. A dark mineral that forms needle-shaped crystals in igneous and metamorphic rock.

horst. An uplifted fault block bound on at least two sides by faults.

hot spring. A spring with water hotter than body temperature.

hydrostatic pressure. Pressure created by the weight of overlying water confined in pipes or rock layers.

hydrothermal alteration. The alteration of a rock's original minerals by hot, often acidic groundwater.

hydrothermal water. Water heated by still-warm igneous rock or magma.

ice cap. Glacial ice that spreads in all directions over a relatively flat surface.

igneous. Formed from molten rock.

interbedded. When layers of a particular lithology alternate with layers of a different lithology.

intrusion. A body of igneous rock emplaced, while molten, into older rock.

intrusive rock. Igneous rock created from magma that cooled below the surface.

island arc. A belt of volcanic islands formed above a subducting, melting tectonic plate.

isostatic rebound. The rise of land masses that were depressed by the weight of ice sheets, water (as in Lake Bonneville), or overlying sediment that was subsequently removed.

joint. A rock fracture along which no significant movement has taken place.

karst. A very irregular surface that develops as water dissolves limestone.

kerogen. A solid hydrocarbon found in oil shale.

kimberlite. Rock containing minerals from Earth's mantle, intruded very quickly in an explosive, gaseous eruption through Earth's crust. Sometimes an ore of diamonds.

laccolith. An intrusion that spread between rock layers, doming those above it.

landslide. The downhill sliding of a relatively dry mass of earth or rock.

Laramide Orogeny. A mountain building event that occurred between 70 and 35 million years ago, in which thick blocks of crust were thrust over other rock a distance of a few tens of miles or less due to tectonic compression; formed the Rocky Mountains.

latite. A light-colored volcanic rock that contains a lot of feldspar and very little quartz.

lava. Magma that erupted onto Earth's surface.

lava dome. A dome-shaped body of volcanic rock formed from very thick lava.

leached. Depleted of elements and minerals because slowly moving water dissolved them.

lichen. A composite organism, composed of fungus growing with green algae or cyanobacteria, that encrusts surfaces.

limestone. A sedimentary rock composed of calcium carbonate; often formed from the shells of marine organisms and calcium-secreting plants.

limonite. A yellow-brown, hydrous iron oxide mineral.

liquefaction. The liquid behavior of an uncemented water-rich sediment that is shaken violently.

lithify. To turn to stone, usually by grains being cemented together.

lithology. The physical characteristics visible in a rock.

lithosphere. The solid outer portion of Earth, consisting of the crust and uppermost mantle.

magma. Molten rock below Earth's surface.

magma chamber. An underground pool of molten rock; often the source for volcanic eruptions.

magnetite. A heavy black or dark-gray magnetic iron mineral.

magnitude. See **Richter scale**

mantle. The zone between Earth's core and crust, with a lower portion thought to be solid and an upper part that deforms plastically.

marble. Metamorphic rock consisting of recrystallized limestone.

marlstone. Lime-rich mudstone.

massif. The entire mass of a large mountain or group of mountains that rose as an integral block.

massive. Without internal structure or layers; homogenous.

mass wasting. The process by which rock and sediment move downward by the force of gravity. Includes slides, soil creep, and rockfalls.

meander. A looplike curve in a river.

meander, entrenched. A meander cut so deeply that the river can no longer change course.

megabreccia. Breccia that covers a large area and contains extremely large pieces.

meltwater. Water derived from a melting glacier.

member. A subdivision of a formation based on distinctive sedimentary characteristics.

mesa. A flat-topped hill or mountain, often edged with cliffs or steep slopes.

metamorphic. Changed through a variable combination of great heat, pressure, and chemical fluids.

metamorphic core complex. A dome-shaped uplift that forms when movement on a detachment fault unroofs part of the lower crust, which then rises because of its own buoyancy.

mica. A group of minerals easily separated into thin, shiny, flexible flakes.

mineral. A naturally occurring inorganic substance with a characteristic chemical composition and predictable properties that include color, texture, hardness, mass, and crystal form.

monocline. A steplike fold in which gently dipping layers rise steeply and then level off again.

monzonite. An intrusive igneous rock with abundant feldspar and very little quartz or dark minerals. Quartz monzonite contains more quartz.

moraine. Completely unsorted sediment deposited by ice.

moraine, ground. A broad moraine deposited beneath a melting glacier.

moraine, lateral. A narrow, linear moraine deposited along the side of a valley by a glacier.

moraine, recessional. A narrow, linear moraine deposited at the front of a receding glacier behind its terminal moraine.

moraine, terminal. A narrow, linear moraine deposited at the lowermost limit of a receding glacier.

mud crack. A shrinkage crack in drying mud.

mudflow. A moving mass of soft, wet earth and debris.

mudflow breccia. The deposit left by a mudflow.

mudstone. A sedimentary deposit made up of clay- and silt-sized particles.

muscovite. White mica.

mylonite. A metamorphic rock formed under great shearing stress along a slow-moving fault zone. It has a distinctive fine-grained, flowing texture around the occasional large, rounded mineral grains.

natural bridge. A rock span that crosses a watercourse. Usually formed when a stream cuts through the rock defining the bend in a meander.

obsidian. Black volcanic glass.

oceanic crust. Rocks that floor the oceans. Usually darker and denser than continental rocks.

oil shale. Fine-grained rock containing a waxy hydrocarbon called kerogen, from which oil can be distilled.

oolites. Sand-sized spheres that form when minerals dissolved in water build up around grains as the grains are agitated along a wave-lapped shore.

Oquirrh Basin. A basin that formed in western Utah during Pennsylvanian and Permian time.

ore. Rock containing enough valuable minerals to be economically mineable.

orogeny. An episode of mountain building that involves faulting, folding, uplift, and sometimes intrusive igneous activity, usually a result of tectonic events.

ostracod. A small crustacean, typically about 0.04 inch (1 mm) in size, with a flat body protected by a bivalve-like shell.

outcrop. Bedrock exposed at Earth's surface.

outwash. A deposit of gravel and sand that was carried from a glacier by meltwater.

oxbow lake. A crescent-shaped lake left when a stream or river abandons a meander.

oxide. A compound formed when an element such as iron combines with oxygen.

oxidizing environment. An environment rich in oxygen that causes a chemical change. For example, in the Southwest iron is oxidized into hematite or limonite, which color the rocks red or orange.

paleontology. A branch of geology that focuses on the fossilized remains of past life.

parabolic dune. A horseshoe-shaped sand dune in which long arms, partly stabilized by vegetation, point in a windward direction.

pavement. See **desert pavement**

pebble. A rock fragment, commonly rounded, that is 0.16 to 2.5 inches (4 to 64 mm) in diameter.

pediment. A gently inclined erosion surface carved in bedrock at the base of a mountain range.

pegmatite. Very coarse-grained igneous rock formed in veins and dikes; usually granitic in composition, with additional rare minerals.

petroglyph. A man-made image created by removing part of a rock surface.

pictograph. A man-made image painted on a rock surface.

plate. See **tectonic plate**

plateau. A large, relatively high area bound on one or more sides by cliffs or steep slopes.

playa. A flat-floored, dry lake bottom in a desert basin with no outlet.

playa lake. A shallow temporary lake that forms on a playa after rain.

plesiosaur. A Jurassic-Cretaceous marine reptile with a short tail, a wide body, four flippers, and often a long neck.

porphyry. An igneous rock with two distinct grain sizes: large minerals in a finer groundmass.

potash. Water-soluble potassium salts, usually potassium chloride, used in making fertilizer.

pothole. A rounded hollow excavated by water whirling stones and sand.

pumice. Frothy volcanic rock that is light enough to float on water.

pyrite. A metallic, brass-colored, brittle iron mineral (iron sulfide); often called fool's gold.

quartz. A hard, glassy, rock-forming mineral composed of crystalline silica (silicon dioxide).

quartzite. Metamorphosed sandstone, which breaks across the grains instead of around them.

racetrack valley. A long valley between upturned sedimentary beds; formed by differential erosion.

radiometric date. The approximate absolute age of a rock obtained by measuring one or more of its radioactive elements and their decay products.

red beds. Sedimentary rocks colored red by the mineral hematite.

reducing environment. An environment in which there is little oxygen, so elements like iron stay unaltered or lose oxygen. In the Southwest, sediments deposited in a reducing environment are generally black, gray, or greenish in color.

reef. A mining term synonymous with vein, or a rugged long outcrop that is difficult to cross.

regress. A fall in sea level with a corresponding shift in the shoreline toward lower ground.

reversed topography. Ground surface that starts as a valley and, over time, becomes a ridge. Commonly forms when lava flows down a valley and solidifies, and then the surrounding softer rocks erode away.

rhyolite. A light-colored volcanic rock that is the extrusive equivalent of granite.

Richter scale. A measure of the magnitudes of earthquakes. Each number is ten times greater than the previous one, so the waves of a magnitude 8 earthquake are ten times greater than those of a magnitude 7 quake, and one hundred times greater than the waves of a magnitude 6 quake.

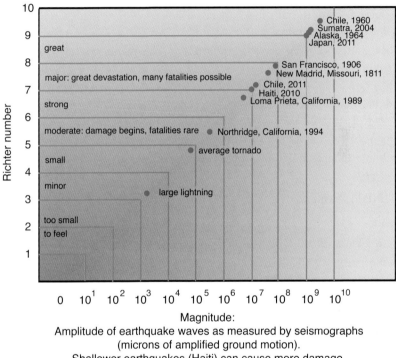

Magnitude:
Amplitude of earthquake waves as measured by seismographs
(microns of amplified ground motion).
Shallower earthquakes (Haiti) can cause more damage.

rift. A deeply faulted valley where two parts of Earth's crust are separating.

river piracy. When a stream cuts headward until it intercepts another stream, causing the second stream's flow to be diverted into the first stream's channel.

salt anticline. An anticline formed by upward-flowing salt, which lifts overlying rocks.

sand. Rock and mineral particles between 0.0025 and 0.08 inch (0.0625 and 2 mm) in diameter. Larger than silt, smaller than pebbles.

sandstone. Sedimentary rock composed of sand.

scarp. A cliff or steep slope formed by either faulting or erosion.

schist. Metamorphic rock with mineral grains oriented parallel to each other.

sedimentary rock. Rock composed of particles of other rock or minerals that have been transported and deposited, usually by water, wind, or ice.

seismic. Vibrations related to earthquakes or human activity.

selenium. An element that is often a by-product of ore refining. Used commercially in glassmaking, as pigment in paints, and in photocells.

Sevier Orogeny. A mountain building event that occurred between 145 and 65 million years ago, in which thin thrust sheets were thrust eastward many tens of miles.

Sevier Thrust Belt. The belt of thrust faults and thrust sheets associated with the Sevier Orogeny.

shale. A fine-grained sedimentary rock composed of clay, silt, or mud.

shear zone. A group of roughly parallel faults, within which the rock is deformed or broken up.

silica. Silicon dioxide; occurs as quartz and as a major part of many other minerals.

sill. A thin body of igneous rock intruded between sedimentary rock layers.

silt. Rock and mineral particles between 0.00016 and 0.0025 inch (0.004 and 0.0625 mm) in diameter. Larger than clay, smaller than sand.

siltstone. Rock composed of particles smaller than sand and larger than clay.

sinkhole. A depression caused by the collapse of the ground into an underlying cavern.

slate. A fine-grained metamorphic rock, formed from shale, that tends to split into flat sheets.

slickenside. A grooved and polished surface that forms as rock moves against rock along a fault.

slump. A small landslide, usually in wet soil, with a relatively small amount of downslope movement.

slurry. A wet, highly mobile mixture of sediment and water that is capable of carrying large rocks.

soil horizon. A soil layer parallel to the surface that differs chemically and physically from layers above and below it.

sorting. The distribution of grain size in sediment or sedimentary rock. *Well sorted* refers to sediment that has been sorted by wind or water, such as beach sand. *Poorly sorted* refers to sediment that has not been sorted by wind or water, such as glacial till.

splay. Secondary faults that form on the sides of a primary fault, usually as it comes to the surface.

stock. An igneous intrusion with a surface exposure of less than 40 square miles (100 square km).

strata. Layers of sedimentary rock.

stratovolcano. See **volcano, composite**

stromatolite. A rounded, layered, calcareous form made by algae continuously trapping and cementing together ocean or lake sediment.

subduction. The downward movement of an oceanic plate as a continental plate overrides it.

supercontinent. A large continent formed by the joining together of multiple continents through tectonic processes.

syncline. A downward fold. When eroded, the youngest rocks are in its center.

tableland. A landscape characterized by having multiple plateaus.

tafoni. Small hollows weathered in rock by moisture, freeze-and-thaw cycles, and wind.

tailings. Fine waste material from an ore-processing mill.

talus. A mass of fallen angular rocks, usually piled below the steep outcrop that was their source.

tar sand. Sand or sandstone saturated with tar or very thick oil.

tectonic. Related to the deformation of Earth's rock due to forces within the Earth.

tectonic plate. A section of Earth's crust that moves as one coherent, relatively rigid piece.

terrace. A bench bordering a valley; part of an earlier river floodplain or lake shoreline that is stranded above the former (or current) water level.

thrust fault. See **fault, thrust**

thrust belt. A region with several large, similarly oriented thrust faults.

thrust sheet. A relatively thin, broad fault block that is thrust over another fault block along a thrust fault.

till. Unsorted sediment deposited by glaciers.

tillite. Rock composed of till.

trace fossil. See **fossil, trace**

trackway. A fossilized series of footprints left by an animal in motion.

transgress. A rise in sea level with a corresponding shift in the shoreline toward higher ground.

transverse dune. A long, often sinuous sand dune that develops at right angles to the prevailing wind.

trap. A structure in rock layers, such as an anticline, that tends to capture oil.

travertine. Calcium carbonate deposited by water in caves or around springs.

trilobite. An extinct marine arthropod with a three-lobed oval body.

tufa. Porous calcium carbonate rock deposited around a spring or along a lakeshore.

tuff. Rock composed of volcanic ash and/or cinders.

tuff, ashfall. Rock composed of ash and/or cinders that fell from an ash cloud.

tuff, ashflow. Rock composed of ash and/or cinders that flowed down the flank of a volcano and welded together as they stopped. Also called welded tuff.

turbidite. Rock deposited by turbidity currents, underwater avalanches on the slopes that edge continents. Typically they grade upward from conglomerate to shale, with distinctive crossbedded and laminar layers.

type locality. The place where a rock unit was originally studied, described, and named.

type section. The section of rock where a formation was originally studied, described, and named.

Uinta Basin. A large depression that formed where Earth's crust sank downward south of the Uinta Mountains during the Laramide Orogeny. It filled with as much as 22,000 feet (6,700 m) of sedimentary rock during Tertiary time.

unconformity. A surface separating rock units of significantly different age. Represents a period of erosion or a period during which no sediment was deposited.

unconformity, angular. An unconformity separating flat-lying rocks from underlying layers that are angled due to previous tectonic activity.

uplift. The vertical elevation of a region of Earth due to tectonic stresses.

Utah Hingeline. A roughly north-south-trending structure in Utah separating the Colorado Plateau from the Basin and Range. Thought to be related to rifting during Precambrian time.

valley fill. See **fill, valley**

varnish. See **desert varnish**

vein. A body of minerals precipitated from hydrothermal water along a joint or fault plane.

vent. An opening at the surface through which lava, cinders, volcanic ash, or volcanic gases escape.

volcanic ash. Finely pulverized rock thrown out by an explosive volcanic eruption.

volcanic rock. Rock solidified from lava on or above Earth's surface; also called extrusive igneous rock.

volcano. A mountain or hill built by a pileup of extrusive igneous rock.

volcano, composite. A steep-sided volcano composed of a combination of lava flows and ashfall. Also called a stratovolcano.

volcano, shield. A wide, dome-shaped volcano composed of moderately fluid lava.

Wasatch Front. The steep western front of the Wasatch Range that developed where the land west of the front dropped down along the Wasatch Fault.

wash. A broad, shallow streambed that is dry much of the year.

water table. The surface below which rocks, sediment, and soil are saturated with groundwater.

weathering. The disintegration of rock due to exposure to the atmosphere.

welded tuff. See **tuff, ashflow**

window. A small opening through a rock fin.

zoning. The grouping of minerals in bands, for example, in a pegmatite or an ore deposit. Determined by variations in the chemical environment in which the minerals crystallized.

Museums, Maps, and General References

Museums

Museums that cover Utah's geology are numerous and excellent. The list of addresses and websites below was current when this volume went to press.

Arches National Park Visitor Center
www.nps.gov/arch/index.htm

Bingham Canyon Mine Visitors Center
12800 S Bacchus Hwy. (Utah 111)
www.kennecott.com/visitors-center

Brigham Young University Earth Science Museum
1683 N Canyon Rd.
Provo

Canyonlands National Park
Island in the Sky and Needles Visitor Centers
www.nps.gov/cany/index.htm

Cleveland-Lloyd Dinosaur Quarry
South of Price
www.blm.gov/wo/st/en/res/Education_in_BLM/Learning_Landscapes/For_
Travelers/go/geology/cleveland-lloyd.html
www.utah.com/playgrounds/cleveland_lloyd.htm

Dead Horse Point State Park Visitor Center and Interpretive Museum
www.stateparks.utah.gov/parks/dead-horse

The Dinosaur Museum
754 S 200 W
Blanding
www.dinosaur-museum.org

Dinosaur National Monument
Quarry Exhibit Hall and Quarry Visitor Center
Utah 149 north of Jensen
www.nps.gov/dino

Fairview Museum of History and Art
85 N 100 E
Fairview

Glen Canyon National Recreation Area
Bullfrog Visitor Center
Utah 276 north of Bullfrog Marina

Carl Hayden Visitor Center at Glen Canyon Dam
US 89 on west side of Glen Canyon Dam

John Wesley Powell River History Museum
1765 E Main St.
Green River
www.jwprhm.com

Museum of Moab
118 E Center St.
Moab
www.moabmuseum.org

Museum of the San Rafael Swell
70 North 100 E
Castle Dale
www.emerycounty.com/sanrafaelmuseum/index.html

Natural History Museum of Utah
301 Wakara Way
Salt Lake City
www.nhmu.utah.edu

North American Museum of Ancient Life
3003 N Thanksgiving Point
Lehi
www.thanksgivingpoint.org/visit/museumofancientlife

Ogden's George S. Eccles Dinosaur Park
1544 E Park Blvd.
Ogden
www.dinosaurpark.org

Oquirrh Mountains Mining Museum
Deseret Peak Recreation Complex
Utah 112 between Tooele and Grantsville
www.co.tooele.ut.us/ht19_oquirrhmtn.html

Park City Museum
528 Main St.
Park City
www.parkcityhistory.org

The Powell Museum
6 N Lake Powell Blvd.
Page, Arizona
www.powellmuseum.org

St. George Dinosaur Discovery Site at Johnson Farm
2180 E Riverside Dr.
St. George
www.dinosite.org

Tintic Mining Museum
241 W Main St.
Eureka

Utah Field House of Natural History State Park Museum
496 E Main St.
Vernal
www.stateparks.utah.gov/parks/field-house

Utah State University Eastern Prehistoric Museum
155 E Main St.
Price
www.usueastern.edu/museum

Weber State University Museum of Natural Science
2705 Harrison Blvd.
Ogden
www.webersci.org

Wells Fargo–Silver Reef Museum
1903 Wells Fargo Rd.
Leeds
www.miningutah.com/id317.html

Western Mining and Railroad Museum
296 S Main St.
Helper
www.wmrrm.com

Maps

Maps often tell the geologic story most clearly. Utah is fortunate: many, many geological maps are available. Rather than list them all here, we recommend that you visit the Utah Geological Survey website (www.geology.utah.gov), where the geological map keys are up-to-date and easy to follow. The maps are at several different scales, from ones that cover the entire state to ones that cover a 7.5′ quadrangle, an area about 6.7 miles wide by 8.5 miles long (10.8 km by 13.7 km). Separate maps have been made for almost every large park. The US Geological Survey has mapped some areas not mapped by the Utah Geological Survey.

Many maps of Utah include cross sections and text. You can buy maps at many of the parks, from the Utah Geological Survey's store (located at 1594 W North Temple in Salt Lake City), and from their website, and most of them can be downloaded from the website.

General References

Entries with an asterisk (*) are those we recommend for further reading. Other sources were used only as references for the figures in this volume. Special mention is due to the Utah Geological Survey website, which contains a wealth of information far too large to list here.

*Anderson, P. B., and D. A. Sprinkel, eds. Continuously updated. *Geologic Road, Trail, and Lake Guides to Utah's Parks and Monuments.* Utah Geological Association Publication 29. Chapters from out-of-print 2000 edition available at www.utahgeology.org/uga29Titles.htm.

Anderson, W., and R. L. Bruhn. 2012. *Implications of Thrust and Detachment Faulting for the Structural Geology of Thermo Hot Springs KGRA, Utah.* Utah Geological Survey Open-File Report 587. Available at www.geology.utah .gov/online/ofr/ofr-587.pdf.

Atkinson, W. W., Jr., and M. T. Einaudi. 1978. "Skarn Formation and Mineralization in the Contact Aureole at Carr Fork, Bingham, Utah." *Economic Geology* 73 (7): 1326–65.

*Averett, W. R., ed. 1983. *Northern Paradox Basin—Uncompahgre Uplift.* Grand Junction Geological Society Field Trip.

*Baars, D. L. 1993. *Canyonlands Country: Geology of Canyonlands and Arches National Parks.* University of Utah Press.

*Baars, D. L. 2002. *A Traveler's Guide to the Geology of the Colorado Plateau.* University of Utah Press.

*Baars, D. L., and C. M. Molenaar. 1971. *Geology of Canyonlands and Cataract Canyon.* Canyonlands Research Bibliography Paper 331.

Baars, D. L., and G. M. Stevenson. 1981. "Tectonic Evolution of the Paradox Basin, Utah and Colorado." In *Geology of the Paradox Basin*, Rocky Mountain Association of Geologists 1981 Field Conference Guidebook, edited by D. L. Wiegand, 23–31, figure p. 27.

*Baldridge, W. S. 2004. *Geology of the American Southwest: A Journey Through Two Billion Years of Plate-Tectonic History.* Cambridge University Press.

*Biek, B., Willis, G., and B. Ehler. 2010. "Utah's Glacial Geology." *Utah Geological Survey Survey Notes* 42 (3). Available at http://geology.utah.gov /surveynotes/articles/pdf/utah_glacial_geology_42-3.pdf.

Biek, R. F., et al. 2003. *Geologic Map of the Center Creek Quadrangle, Wasatch County Utah.* Utah Geological Survey M-192, scale 1:24,000.

Biek, R. F., et al. 2010. "Geology of Zion National Park." In *Geology of Utah's Parks and Monuments*, Utah Geological Association Publication 28, edited by D. A. Sprinkel et al., 53.

Biek, R. F., et al. 2010. *Geologic Map of the St. George 30′ x 60′ Quadrangle and East Part of the Clover Mountains 30′ x 60′ Quadrangle, Washington and Iron Counties, Utah.* Utah Geological Survey Map 242DM, scale 1:100,000.

*Billingsley, G. H., Huntoon, P. W., and W. J. Breed. 1987. *Geologic Map of Capitol Reef National Park and Vicinity, Emery, Garfield, Millard, and Wayne Counties, Utah.* Utah Geological and Mineral Survey Map 87, scale 1:24,000.

*Blakey, R. Colorado Plateau Stratigraphy and Geology and Global and Regional Paleogeography. Last updated March 2011. http://jan.ucc.nau.edu/rcb7/.

*Blakey, R., and W. Ranney. 2008. *Ancient Landscapes of the Colorado Plateau.* Grand Canyon Association.

Boutwell, J. M., and L. H. Woolsey 1912. *Geology and Ore Deposits of the Park City District, Utah.* US Geological Survey Professional Paper 77.

*Bowers, W. E. 1991. *Geologic Map of Bryce Canyon National Park and Vicinity: Southwestern Utah.* US Geological Survey Map I-2108.

Bryant, B. 1990. *Geologic Map of the Salt Lake City 30′ x 60′ Quadrangle, North-Central Utah, and Uinta County, Wyoming.* US Geological Survey Miscellaneous Investigation Series Map I-1944, scale 1:100,000.

Bryant, B. 2003. *Geologic Map of the Salt Lake City 30′ x 60′ Quadrangle, North-Central Utah and Uinta County, Wyoming.* Utah Geological Survey Map 190DM, scale 1:100,000.

*Bureau of Land Management. Hanksville-Burpee Dinosaur Quarry. www.blm.gov/ut/st/en/prog/more/cultural/Paleontology/utah_paleontology /color_country_paleontology/hanksville-burpee.html.

*Bureau of Land Management. "2012 Oil Shale and Tar Sands Programmatic EIS." http://ostseis.anl.gov/guide/oilshale/index.cfm.

*Campbell, R. "The Evaporation Basin." Earthshots: Great Salt Lake, Utah. Last modified August 5, 2013. http://geochange.er.usgs.gov/sw/changes/anthro-pogenic/gsl/.

*Climb-Utah.com. "Copper Ridge Sauropod Dinosaur Trackway, Moab, Utah." www.climb-Utah.com/Moab/moabdino.htm.

*Climb-Utah.com. "Nine Mile Canyon Guide and Information." www.climb-utah.com/Misc/ninemile.htm.

*Climb-Utah.com. "Potash Petroglyphs and Dinosaur Tracks: Moab Area Roadside Attraction." www.climb-utah.com/Moab/potash.htm.

*Currey, D. R., Atwood, G., and D. R. Mabey. 1984. *Major Levels of Great Salt Lake and Lake Bonneville.* Utah Geological and Mineralogical Survey Map 73, scale 1:750,000.

Davis, G. H., and G. L. Pollock. 2010. "Geology of Bryce Canyon National Park, Utah." In *Geology of Utah's Parks and Monuments*, Utah Geological Association Publication 28, edited by D. A. Sprinkel et al., 53.

DeCelles, P. G., and J. C. Coogan. 2006. "Regional Structure and Kinematic History of the Sevier Fold-and-Thrust Belt, Central Utah." *Geological Society of America Bulletin* 118 (7–8): 841–64.

DeCelles, P. G., Lawton, T. F., and G. Mitra. 1995. "Thrust Timing, Growth of Structural Culminations, and Synorogenic Sedimentation in the Type Sevier Orogenic Belt, Western United States." *Geology* 23 (8): 699–702.

*Doelling, H. H. 1985. *Geology of Arches National Park.* Utah Geological and Mineral Survey Map 74, scale 1:50,000.

*Doelling, H. H., et al. 1990. *Geologic Map of Antelope Island, Davis County, Utah.* Utah Geological and Mineral Survey. Utah Geological Survey Map 127, scale 1:24,000.

*Doelling, H. H., and G. C. Willis. 2007. *Geologic Map of the Lower Escalante River Area, Glen Canyon National Recreation Area, Eastern Kane County, Utah.* Utah Geological Survey Miscellaneous Publication 06-3DM, scale 1:100,000.

Dover, J. H. 1995. *Geologic Map of the Logan 30′ x 60′ Quadrangle, Cache and Rich Counties, Utah, and Lincoln and Uinta Counties, Wyoming.* US Geological Survey Miscellaneous Investigations Series Map I-2210, scale 1:100,000.

*Eaton, J. G., et al. 2001. *Cretaceous and Early Tertiary Geology of Cedar and Parowan Canyons, Western Markagunt Plateau, Utah.* Utah Geological Association Field Trip Road Log, Publication 30, p. 337–63.

Forrest, R. J. 1994. "Geothermal Development at Roosevelt Hot Springs Geothermal Area, Beaver County, Utah 1972–1993." In *Cenozoic Geology and Geothermal Systems of Southwestern Utah,* Utah Geological Association Publication 23, edited by R. E. Blackett and J. N. Moore, 37–44.

Geological Society of America. Geologic Time Scale. Last updated November 2012. www.geosociety.org/science/timescale/.

Gilbert, G. K. 1890. *Lake Bonneville.* US Geological Survey Monograph 1, plate II.

*Gould, S. J. 1989. *Wonderful Life: The Burgess Shale and the Nature of History.* W. W. Norton & Co.

Grand Canyon Trust. "History of the Atlas Mine Project." www.grandcanyontrust .org/utah/uranium_history.php.

Gruen, G., Heinrich, C. A., and K. Schroeder. 2010. "The Bingham Canyon Porphyry Cu-Mo-Au Deposit. II. Vein Geometry and Ore Shell Formation by Pressure-Driven Rock Extension." *Economic Geology* 105 (1): 69–90, figure p. 74.

*Gwynn, J. W. 2007. "Taking Another Look at Utah's Tar Sand Resources." *Utah Geological Survey Survey Notes* 39 (1): 8–9. Available online at http://geology .utah.gov/surveynotes/articles/pdf/tarsand_resources_39-1.pdf.

Hacker, D. B., et al. 2007. "Shallow Level Emplacement Mechanisms of the Miocene Iron Axis Laccolith Group, Southwestern Utah." In *Field Guide to Geologic Excursions in Southern Utah,* Geological Society of America Rocky Mountain Section Annual Meeting. Utah Geological Association Publication 35, edited by W. R. Lund, figure p. 23.

*Hamblin, W. K. 2004. *Beyond the Visible Landscape: Aerial Panoramas of Utah's Geology.* Department of Geology, Brigham Young University.

*Hamilton, W. L. 1995. *Geological Map of Zion National Park, Utah.* Zion Natural History Association, Zion National Park.

*Hansen, W. R. 1969. *The Geologic Story of the Uinta Mountains.* US Geological Survey Bulletin 1291.

Hansen, W. R. 1986. *Neogene Tectonics and Geomorphology of the Eastern Uinta Mountains in Utah, Colorado, and Wyoming.* US Geological Survey Professional Paper 1356, p. 42.

*Hansen, W. R., Rowley, P. D., and P. E. Carrara. 1983. *Geologic Map of Dinosaur National Monument and Vicinity, Utah and Colorado.* US Geological Survey Miscellaneous Investigations Series Map I-1407, scale 1:50,000.

Hayden, J. M., Lawton, T. F., and D. L. Clark. 2008. *Provisional Geologic Map of the Champlin Peak Quadrangle, Juab and Millard Counties, Utah.* Utah Geological Survey Miscellaneous Publication 08-1, scale 1:24,000.

Hecker, S. 1993. *Quaternary Tectonics of Utah with Emphasis on Earthquake Hazard Characterization.* Utah Geological Survey Bulletin 127.

*Hintze, L. F. 1975. *Geological Highway Map of Utah.* Department of Geology, Brigham Young University.

*Hintze, L. F., and B. J. Kowallis. 2009. *Geologic History of Utah.* Brigham Young University Geology Studies Special Publication 9.

*Hoffman, J. F. 1985. *Arches National Park: An Illustrated Guide.* Western Recreational Publications.

*Huntoon, P. W., Billingsley, G. H., and W. J. Breed. 1982. *Geologic Map of Canyonlands National Park and Vicinity, Utah.* Canyonlands Natural History Association, scale 1:62,500.

Kriens, B. J., Shoemaker, E. M., and K. K. Herkenhoff. 1999. "Geology of the Upheaval Dome Impact Structure, Southeast Utah." *Journal of Geophysical Research: Planets* 104 (E8): 18867–87.

Kwon, S., et al. 2007. "Effect of Predeformational Basin Geometry in the Kinematic Evolution of a Thin-Skinned Orogenic Wedge: Insights from Three-Dimensional Finite Element Modeling of the Provo Salient, Sevier Fold-Thrust Belt, Utah." *Journal of Geophysical Research: Solid Earth* 112 (B2).

Laabs, B. J. C., and E. C. Carson. 2005. "Glacial Geology of the Southern Uinta Mountains." In *Uinta Mountain Geology*, Utah Geological Association Publication 33, edited by C. M. Dehler et al., 235–53.

Laine, M. D., et al. 2007. *Covenant Oil Field, Central Utah Thrust Belt: Possible Harbinger of Future Discoveries.* Poster presented at AAPG Rocky Mountain Section Meeting, Snowbird, Utah. Available online at http://geology.utah. gov/emp/pump/pdf/pump_snowbird07.pdf.

Lochman-Balk, C. 1972. "Cambrian System." In *Geologic Atlas of the Rocky Mountain Region*, Rocky Mountain Association of Geologists, Denver, Colorado, edited by W. W. Mallory, p. 60–75.

Maldonado, F., and V. S. Williams. 1992. *Geologic Map of the Parowan Gap Quadrangle, Iron County, Utah.* US Geological Survey Geologic Quadrangle 1712, scale 1:24,000.

McNeil, B. R. and R. B. Smith. 1992. *Upper Crustal Structure of the Northern Wasatch Front, Utah, from Seismic Reflection and Gravity Data.* Utah Geological Survey Contract Report 92-7.

*Mitchell, J. R. 2006. *Gem Trails of Utah*, expanded 2nd ed. Gem Guides Book Co.

*Moabtailings.org. "Moab UMPTRA Project." 2011. http://www.moabtailings .org.

Moore, J. N., and D. L. Nielson. 1994. "An Overview of the Geology and Geochemistry of the Roosevelt Hot Springs Geothermal System, Utah." In *Cenozoic Geology and Geothermal Systems of Southwestern Utah*, Utah Geological Association Publication 23, edited by R. E. Blackett and J. N. Moore, 25–36.

Munroe, J. S. 2005. "Glacial Geology of the Northern Uinta Mountains." In *Uinta Mountain Geology*, Utah Geological Association Publication 33, edited by C. M. Dehler et al., 215–34.

*National Park Service. "Geologic Map of Cedar Breaks National Monument." February 2006. http://nature.nps.gov/geology/inventory/publications/map _graphics/cebr_geology-layout.pdf.

*National Park Service. "National Park Service Museum Collections: Dinosaur! National Monument." Dinosaur National Monument. Last modified December 12, 2003. www.nps.gov/history/museum/exhibits/dino/index.html.

*Navajo Nation Parks and Recreation. Monument Valley Navajo Tribal Park. http://navajonationparks.org/htm/monumentvalley.htm.

Nelson, S. T., et al. 2005. "Emerald and Fibrous Calcite Mineralization in the Southwestern Uinta Mountains." In *Uinta Mountains Geology*, Utah Geological Association Publication 33, edited by C. M. Dehler et al.

*Nichols, T. 1999. *Glen Canyon: Images of a Lost World.* Museum of New Mexico Press.

Oviatt, C. G. 2012. Current Data on Great Salt Lake and Lake Bonneville Levels, personal communication.

Peterson, F. 1988. "Pennsylvanian to Jurassic Eolian Transportation Systems in the Western United States." *Sedimentary Geology* 56: 207–60.

*Rankin, H. "Glen Canyon Dam and the 1983 Floods." *Lake Powell Answers* (blog), December 18, 2009. http://lakepowellrealty.net/glen-canyon-1983-floods.

Schelling, D. D., and J. V. Vrona. 2007. "Structural Geology of the Central Utah Thrust Belt: A Geological Field Trip Road Log." In *Central Utah—Diverse Geology of a Dynamic Landscape*, Utah Geological Survey Publication 36, edited by G. C. Willis et al., 483–518.

Shoemaker, E. M. 1987. "Meteor Crater, Arizona." In *Geological Society of America Rocky Mountain Section Centennial Field Guide*, vol. 2, edited by S. S. Beus, 399–404.

Sprinkel, D. A., et al., eds. 2010. *Geology of Utah's Parks and Monuments.* Utah Geological Association Publication 28.

Sprinkel, D. A., Park, B., and M. Stevens. 2010. "Geology of the Sheep Creek Area." In *Geology of Utah's Parks and Monuments,* Utah Geological Association Publication 28, edited by D. A. Sprinkel et al., 542.

Sprinkel, D. A., Yonkee, W. A., and T. C. Chidsey, Jr. 2011. *Sevier Thrust Belt: Northern and Central Utah and Adjacent Areas.* Utah Geological Association Publication 40.

*Stokes, W. L. 1987. *Geology of Utah.* Utah Museum of Natural History Occasional Paper 6.

*Thaden R. E., et al. 2008. *Geologic Map of the White Canyon—Good Hope Bay Area, Glen Canyon National Recreation Area, San Juan and Garfield Counties, Utah.* Utah Geological Survey Miscellaneous Publication 08-3DM, 1:100,000. Available online at www.geology.utah.gov/online/mp/mp08-3.pdf.

*University of Kansas Natural History Museum, Division of Invertebrate Paleontology. "Utah's Cambrian Life: Evolution and Biogeography of Burgess Shale Type Fossils." 2008. http://kumip.ku.edu/cambrianlife/index .html.

*University of Utah. Earthquake Education Services. www.seis.utah.edu /edservices/edservices.html.

*USDA Forest Service, Region 4. "Mammoth Cave and Ice Cave" under "Geologic Points of Interest by Activity—Caves and Sinkholes." http://www .fs.usda.gov/detail/r4/learning/nature-science/?cid=fsbdev3_016205.

*US Geological Survey. "Colorado Plateau Province." Geologic Provinces of the United States. Last updated October 10, 2000. http://www2.nature.nps.gov /geology/usgsnps/province/coloplat.html.

*Utah.com. "Mill Canyon/Copper Ridge." Other Playgrounds Section. http:/Utah .com/playgrounds/mill_canyon.htm

*Utah Geological Survey. Regularly updated. http://geology.utah.gov.

*The Virtual Fossil Museum. "Fossils Across Geological Time and Evolution." 2002–2010. www.fossilmuseum.net.

Von Tish, D. B., Allmendinger, R. W., and J. W. Sharp. 1985. "History of Cenozoic Extension in Central Sevier Desert, West-Central Utah, from COCORP Seismic Reflection Data." *American Association of Petroleum Geologists Bulletin* 69: 1077–87.

*Walker, G. 2003. *Snowball Earth: The Story of the Great Global Catastrophe That Spawned Life as We Know It.* Crown Publishers.

*Weaver, L. UtahGeology.com. Updated regularly. www.utahgeology.com.

Wells, M. L. 2001. "Rheological Control on the Initial Geometry of the Raft River Detachment Fault and Shear Zone, Western United States." *Tectonics* 20 (4): 435–57.

Willis, G. C. 2000. "The Utah Thrust System." In *Geology of Northern Utah and Vicinity*, Utah Geological Association Publication 27, edited by L. E. Spangler.

*Willis, G. C. 2004. *Interim Geologic Map of the Lower San Juan River Area, Eastern Glen Canyon National Recreation Area and Vicinity, San Juan County, Utah.* Utah Geological Survey Open-File Report 433DM, scale 1:50,000. http://geology.utah.gov/maps/geomap/non_quad/pdf/ofr-443.pdf.

WISE Uranium Project. "Decommissioning of Moab, Utah, Uranium Mill Tailings." Last updated January 30, 2014. www.wise-uranium.org/udmoa .html.

Witkind, I. J., and M. P. Weiss. 1991. *Geologic Map of the Nephi 30'x 60' Quadrangle, Carbon, Emery, Juab, Sanpete, Utah, and Wasatch Counties, Utah.* US Geological Survey Miscellaneous Investigations Series Map I-1937, scale 1:100,000.

INDEX

Page numbers in boldface refer to photos or information in captions.

Felicie Williams started learning about geology at a young age when she accompanied her geologist parents during summer field seasons. She earned a BA in geology from the University of Colorado in Boulder and an MS in exploration geology from the University of British Columbia.

Felicie worked for years as a mineral exploration geologist and mapper both for mining companies and for the Colorado Geological Survey. She rewrote *Roadside Geology of Colorado* for the 2002 edition with Halka Chronic, her mother.

Felicie married geologist Mike Williams in 1982. In 1992 they moved to Grand Junction, Colorado, with their children and started visiting Utah and learning about its amazing geology.

Growing up and traveling around the world with two parents who had PhDs in geology, **Lucy Chronic** participated in discussions and adventures focusing on her parents' amazingly varied interests. The family's love of geology won out, and she earned a BA in geology from Carleton College in Minnesota and an MS in paleontology from the University of Wyoming.

Lucy's professional life has been eclectic, with writing and science the threads that tie her various pursuits together. She has worked as an archaeologist, educator, scientific writer, fire lookout, and interpreter in state and national parks. She and Halka coauthored the second edition of *Pages of Stone: Geology of the Grand Canyon and Plateau Country National Parks and Monuments.*

Lucy resides in the mountains of central Idaho with her husband, Chris Hinze, and enjoys her daughters, Betsy and Haley, during their all-too-infrequent visits home.

Halka Chronic grew up in Tucson, Arizona, and the desert Southwest. She earned degrees from the University of Arizona and Stanford and then a PhD in geology from Columbia University in 1949. Her career took her all over the globe—from teaching at Haile Selassie University in Ethiopia to identifying geologic landmarks in the southern Rockies for the US National Park Service. She spent more than thirty years in Boulder, Colorado, working as an editor and raising her daughters.

Halka lived for many years in Sedona, Arizona. Semiretired, she wrote, took children on field trips, and gave talks about the region's geology. Among the ten geology guidebooks she wrote were three other Roadside Geology guides for Mountain Press: Arizona, Colorado, and New Mexico.

Halka passed away in Grand Junction, Colorado, in April 2013.